Mammalian Cell Mutagenesis

Row 1: G. Adair, M. Meuth; J. Hozier, H. Evans
Row 2: L. Chasin; E. Chu; M. Moore
Row 3: F. de Serres; A. Hsie, D. Clive, D. DeMarini; L. Russell
Row 4: K. Tindall, P. O'Neill; F. Hutchinson, K. Dixon

Mammalian Cell Mutagenesis

Edited by

MARTHA M. MOORE
U.S. Environmental Protection Agency

DAVID M. DEMARINI
U.S. Environmental Protection Agency

FREDERICK J. DE SERRES
Research Triangle Institute

KENNETH R. TINDALL
National Institute of Environmental Health Sciences

COLD SPRING HARBOR LABORATORY
1987

Banbury Report 28: Mammalian Cell Mutagenesis

Printed in the United States of America
Cover and book design by Emily Harste

Library of Congress Cataloging-in-Publication Data

Mammalian cell mutagenesis.

(Banbury report, ISSN 0198-0068 ; 28)
Includes index.
1. Mutagenesis. 2. Mutation (Biology) 3. Mammals—
Cytology. I. Moore, Martha M. II. Series.
QH462.A1M36 1987 599'.01592 87-21784
ISBN 0-87969-228-6

All Cold Spring Harbor Laboratory publications may be ordered directly from Cold Spring Harbor Laboratory, Box 100, Cold Spring Harbor, New York 11724. (Phone: 1-800-843-4388) in New York State (516) 367-8325.

Banbury Report Series

Banbury Report 1: Assessing Chemical Mutagens
Banbury Report 2: Mammalian Cell Mutagenesis
Banbury Report 3: A Safe Cigarette?
Banbury Report 4: Cancer Incidence in Defined Populations
Banbury Report 5: Ethylene Dichloride: A Potential Health Risk?
Banbury Report 6: Product Labeling and Health Risks
Banbury Report 7: Gastrointestinal Cancer: Endogenous Factors
Banbury Report 8: Hormones and Breast Cancer
Banbury Report 9: Quantification of Occupational Cancer
Banbury Report 10: Patenting of Life Forms
Banbury Report 11: Environmental Factors in Human Growth and Development
Banbury Report 12: Nitrosamines and Human Cancer
Banbury Report 13: Indicators of Genotoxic Exposure
Banbury Report 14: Recombinant DNA Applications to Human Disease
Banbury Report 15: Biological Aspects of Alzheimer's Disease
Banbury Report 16: Genetic Variability in Responses to Chemical Exposure
Banbury Report 17: Coffee and Health
Banbury Report 18: Biological Mechanisms of Dioxin Action
Banbury Report 19: Risk Quantitation and Regulatory Policy
Banbury Report 20: Genetic Manipulation of the Early Mammalian Embryo
Banbury Report 21: Viral Etiology of Cervical Cancer
Banbury Report 22: Genetically Altered Viruses and the Environment
Banbury Report 23: Mechanisms in Tobacco Carcinogenesis
Banbury Report 24: Antibiotic Resistance Genes: Ecology, Transfer, and Expression
Banbury Report 25: Nongenotoxic Mechanisms in Carcinogenesis
Banbury Report 26: Developmental Toxicology: Mechanisms and Risk
Banbury Report 27: Molecular Neuropathology of Aging
Banbury Report 28: Mammalian Cell Mutagenesis

Corporate Sponsors

Abbott Laboratories
American Cyanamid Company
Amersham International plc
Becton Dickinson and Company
Cetus Corporation
Ciba-Geigy Corporation
CPC International Inc.
E.I. du Pont de Nemours & Company
Eli Lilly and Company
Genentech, Inc.
Genetics Institute
Hoffmann-La Roche Inc.
Monsanto Company
Pall Corporation
Pfizer Inc.
Schering-Plough Corporation
Smith Kline & French Laboratories
Tambrands Inc.
The Upjohn Company
Wyeth Laboratories

Core Supporters

The Bristol-Myers Fund, Inc.
The Dow Chemical Company
Exxon Corporation
International Business Machines Corporation
The Procter & Gamble Company
Rockwell International Corporation Trust
Texas Philanthropic Foundation Inc.

Special Program Support

James S. McDonnell Foundation
Alfred P. Sloan Foundation

This meeting was also funded by the Environmental Protection Agency.

Participants

Gerald M. Adair, University of Texas System Cancer Center, Smithville

Richard J. Albertini, University of Vermont College of Medicine, Burlington

Marcia L. Applegate, Florida State University, Tallahassee

Charles R. Ashman, Department of Radiation Oncology, University of Chicago Medical Center, Illinois

Angela E. Auletta, U.S. Environmental Protection Agency, Washington, D.C.

Herman E. Brockman, Department of Biological Sciences, Illinois State University, Normal

Michele P. Calos, Department of Genetics, Stanford University School of Medicine, California

Lawrence A. Chasin, Department of Biological Sciences, Columbia University, New York, New York

Ernest H.Y. Chu, Department of Human Genetics, University of Michigan Medical School, Ann Arbor

Donald Clive, Burroughs Wellcome Company, Research Triangle Park, North Carolina

Richard L. Davidson, Center for Genetics, University of Illinois College of Medicine, Chicago

Pieter J. de Jong, Biomedical Sciences Division, Lawrence Livermore National Laboratory, California

David M. DeMarini, Genetic Toxicology Division, U.S. Environmental Protection Agency, Research Triangle Institute, Research Triangle Park, North Carolina

Frederick J. de Serres, Center for Life Sciences and Toxicology, Research Triangle Institute, Research Triangle Park, North Carolina

Kathleen Dixon, National Institute of Child Health and Human Development, Bethesda, Maryland

Helen H. Evans, Division of Radiology, Biochemical Oncology, Case Western Reserve University, Cleveland, Ohio

Richard A. Gibbs, Baylor College of Medicine, Texas Medical Center, Houston

Barry W. Glickman, Department of Biology, York University, Downsville, Ontario, Canada

John C. Hozier, Department of Biological Science, Florida State University, Tallahassee

Abraham W. Hsie, Department of Preventive Medicine and Community Health, University of Texas Medical Branch, Galveston

Franklin Hutchinson, Department of Molecular Biophysics and Biochemistry, Yale University, New Haven, Connecticut

Ronald H. Jensen, Biomedical Sciences Division, Lawrence Livermore National Laboratory, California

Howard L. Liber, Laboratory of Radiobiology, Harvard School of Public Health, Boston, Massachusetts

John B. Little, Harvard School of Public Health, Boston, Massachusetts

Veronica M. Maher, Departments of Microbiology and Biochemistry, Michigan State University, East Lansing

Mark Meuth, Imperial Cancer Research Fund, Clare Hall Laboratories, South Mimms, Hertfordshire, England

Martha M. Moore, Genetic Toxicology Division, U.S. Environmental Protection Agency, Research Triangle Park, North Carolina

Janice A. Nicklas, Genetics Laboratory, University of Vermont, Burlington

J. Patrick O'Neill, Genetics Laboratory, University of Vermont, Burlington

Toby G. Rossman, Institute of Environmental Medicine, New York University Medical Center

Lianne B. Russell, Biology Division, Oak Ridge National Laboratory, Tennesee

Alain Sarasin, Laboratory of Molecular Mutagenesis, Institut de Recherches Scientifiques sur le Cancer, Villejuif, France

Leon F. Stankowski, Jr., Pharmakon Research International, Inc., Waverly, Pennsylvania

Larry H. Thompson, Biomedical Sciences Division, Lawrence Livermore National Laboratory, California

Kenneth R. Tindall, Laboratory of Genetics, National Institute of Environmental Health Sciences, Research Triangle Park, North Carolina

Preface

Eight years since the second Banbury conference entitled, "Mammalian Cell Mutagenesis: The Maturation of Test Systems," we have again convened to discuss issues relevant to mutagenesis in mammalian cells. Significant technical advancements in the past few years have allowed more sophisticated analyses of mutations in mammalian cells, both complementing studies in established systems as well as providing a basis for the development of a variety of elegant, new assays. These sessions are intended to update *Banbury Report 2* and to provide current detailed genetic, biochemical, and molecular insights regarding mammalian cell mutational processes. Unfortunately, a comprehensive evaluation of all possible cellular pathways of mutation is impossible due to limitations of time and space. Pathways that share enzymatic and/or mechanistic similarities to gene amplification or immunoglobulin rearrangements, insights regarding the role of chromatin structure on mammalian cell mutagenesis, or implications derived from studies of oncogene activation have been left as enticing topics for future Banbury conferences. Rather, these sessions present current insights derived from some of the better-studied systems and in some of the best-defined areas of mutation research.

This volume begins with an historical review and presents new insights regarding some of the most commonly used loci in mammalian mutagenesis studies. Section 2 presents analyses of the quantitative differences in observed mutant frequencies at different loci. Data are presented suggesting that such quantitative differences may have a fundamental genetic basis common to lower eukaryotes (*Neurospora crassa*), cultured mammalian cells, and the mouse. Section 3 elaborates the advances in the analysis of mutations derived in vivo. Independent mutational events in vivo can now be defined, and studies allowing quantitative measures of in vivo spontaneous and induced mutant frequencies are underway. Sections 4 through 7 have been organized to discuss molecular analyses of genomic rearrangements and DNA base-sequence alterations in chromosomal genes and shuttle vectors. These data provide the first collective presentation of mutational spectra generated in mammalian cells. Finally, Section 8 discusses mammalian DNA repair and recombination. Data are presented discussing the cloning of mammalian DNA repair genes as well as analyses of mutagen-induced recombination and gene conversion. Whereas these processes currently have an obscure relationship to mutational pathways, it is obvious that species and gene specific differences in the repair and processing of potentially mutagenic lesions will be an important aspect of future analyses of mutation in mammalian cells.

Two issues that influenced the organization of this conference are worthy of specific comment. The first has to do with the impact of molecular biology on the analysis of mutations in mammalian cells. The inability to perform Mendelian genetic analyses to prove the genetic basis of a particular phenotype has long been a limitation of studies in mammalian cell culture. This limitation has led to heated debates concerning the correlation of phenotype and genotype in mammalian tissue culture cells. While some selectable markers used in somatic cell genetic studies remain poorly defined, the correlation of phenotype and genotype is no longer the issue that it was at the time of *Banbury Report 2*. Techniques in molecular biology have established a clear genetic basis for most of the mutations discussed at this conference. Indeed, advances in molecular biology have provided the means whereby mutations at mammalian genetic loci can be defined at the DNA base-sequence level. The papers to follow reflect the diversity of molecular approaches being applied to these analyses. We can now derive mechanistic insight regarding mutational pathways from DNA sequence analyses of isolated mutants. These data must be viewed with some caution, however, since any gene may have sequence-specific and/or chromosomal position-specific biases. Only as molecular insights are confirmed between systems will general rules regarding pathways of mutation in mammalian cells be discerned. It is equally important, however, that differences between systems be thoroughly explored. As more molecular data are generated, it is the differences between systems that have the potential to provide insight into the unexpected. We must carefully assess any differences observed in well-defined systems and continue with the development of new systems for these analyses. The inherent limitation of any one system to provide comprehensive insight into mutational mechanisms and the possibility of generating unique observations in different systems are two compelling reasons for the development of multiple systems for such studies. This volume provides both confirming and differing data regarding mutagenesis in mammalian cells. The presentations are intended to provide a stimulus for future research efforts as well as for new system development.

A second issue concerns the use of mammalian mutagenesis systems as assays to detect potential carcinogens. Although some of the systems reported here have been used as short-term bioassays, this volume was not envisioned as a forum to debate the predictive value of any of these assay systems. This debate deserves more thoughtful consideration than was considered possible without focusing on this issue. The following papers are intended to provide mechanistic insights relevant to the induction of mutations in mammalian cells. Such studies serve to augment our understanding of basic cellular processes and need not necessarily be rationalized in terms of their predictive value.

Finally, we acknowledge the countless hours of research represented in the following presentations, and we thank the various funding agencies for their support of these important studies. We thank the conference attendees for their participation as well as those individuals whose dedicated efforts have helped to make these presentations possible. In addition, we thank the U.S. Environmental Protection Agency and the Banbury Center, Cold Spring Harbor, N.Y. for providing funding for this conference. In addition, we extend our special thanks to Ralph Battey, Beatrice Toliver, Steve Prentis, Inez Sialiano, Katya Davey, Shirley Milton, and Joyce Skovronski for their assistance in planning and coordinating this meeting and in the preparation of this volume.

The papers to follow represent studies that will inevitably give rise to new insights and, hopefully, to future Banbury conferences on mammalian cell mutagenesis. In anticipation of future research efforts in the area of mammalian cell mutagenesis, we present the following progress report.

K.R. Tindall
M.M. Moore
D. DeMarini
F.J. de Serres

Contents

Review of Genetic Markers for the Study of Mutation in Mammalian Cells

Analysis of Mutation at the Chinese Hamster *aprt* Locus

GERALD M. ADAIR
Science Park Research Division
The University of Texas System Cancer Center
Smithville, Texas 78957

OVERVIEW

The Chinese hamster ovary (CHO) *aprt* locus represents an attractive system for the molecular analysis of induced mutation in mammalian cells. We provide a brief overview of this locus, discuss the genetic basis for the high-frequency spontaneous generation of $aprt^{+/0}$ hemizygotes in untreated CHO cell populations, and illustrate some of the strengths and limitations of the hemizygous CHO *aprt* locus as a target for mutation. We briefly describe the application of two different systems employing the *aprt* locus. The $aprt^{+/0}$ hemizygote CHO-AT3-2 and its derivative sublines permit molecular analysis and comparison of induced mutation spectra in repair-proficient versus repair-deficient genetic backgrounds. R10A-1 is heterozygous for the *aprt* locus.

INTRODUCTION

Adenine phosphoribosyl transferase (APRT; E.C.2.4.2.7) is a purine salvage pathway enzyme (m.w. 19,415) that catalyzes the conversion of adenine to adenosine 5′-monophosphate. This enzyme is coded by an autosomal, "housekeeping" gene locus (for review, see Taylor et al. 1979, 1985), which appears to be constitutively expressed throughout the cell cycle (Hordern and Henderson 1982). The *aprt* locus has been mapped in both CHO (Adair et al. 1984) and normal Chinese hamster cells (Adair et al. 1983a).

APRT$^-$ mutants can be readily isolated by selection with adenine analogs such as 8-azaadenine (AA) or 2,6-diaminopurine (DAP); APRT$^+$ revertants can be selected by blocking the de novo purine biosynthetic pathway with an inhibitor such as amethopterin, azaserine, or alanosine while supplying exogenous adenine as a purine source (Chasin 1974; Jones and Sargent 1974; Taylor et al. 1985). Since both *aprt* alleles are normally expressed in CHO cells (Meuth and Arrand 1982; Simon et al. 1982, 1983; Adair et al. 1984), direct selection of APRT$^-$ mutants from wild-type CHO cell populations is impractical. However, functionally heterozygous ($aprt^{+/-}$) or hemizygous ($aprt^{+/0}$) cells arise spontaneously at relatively high frequencies (10^{-4} to

10^{-3}) in untreated CHO cell populations (Jones and Sargent 1974; Adair et al. 1980; Thompson et al. 1980; Bradley and Letovanec 1982); these functionally heterozygous or hemizygous cells have reduced levels of APRT activity and can be selected on the basis of their resistance to intermediate concentrations of AA or DAP. Such cell lines permit single-step selection of fully AA- or DAP-resistant (AA^r or DAP^r) $APRT^-$ mutants at frequencies comparable to those for mutants for the X-linked hypoxanthine-guanine phosphoribosyl transferase (*hgprt*) locus (Adair et al. 1980; Carver et al. 1980; Thompson et al. 1980; Adair and Carver 1983). Phenotypic expression kinetics for AA-resistance are more rapid than for thioguanine (TG)-resistance; only 2–4 days is typically required for maximal expression of induced mutations at the *aprt* locus (Carver et al. 1980; Thompson et al. 1980; Adair and Carver 1983).

The CHO *aprt* locus offers several distinct advantages as a system for the molecular analysis of mutation in mammalian cells: (1) The entire gene spans less than 2.5 kb of DNA (Lowy et al. 1980); (2) a detailed restriction map is available (Lowy et al. 1980; Nalbantoglu et al. 1983; Grosovsky et al. 1986; G. Adair, unpubl.); (3) complete sequence data have been published (Nalbantoglu et al. 1986a); (4) one can utilize hemizygous, $aprt^{+/0}$, CHO cell lines that contain only a single copy of the *aprt* gene (Adair et al. 1983b; Nalbantoglu et al. 1983); (5) there are no *aprt* pseudogenes in CHO (Nalbantoglu et al. 1983; Grosovsky et al. 1986; G. Adair, unpubl.); and (6) the entire *aprt* gene can be efficiently reintroduced into $APRT^-$ cells by transfection (Lowy et al. 1980; G. Adair, unpubl.).

RESULTS

High-frequency Deletion of the Z4 *aprt* Allele in CHO Cells

The *aprt* locus has been mapped to the very distal portion of Chinese hamster chromosome 3, in the region 3p2→pter (Adair et al. 1983a). In the CHO cell line, both chromosome-3 homologs have undergone rearrangement; the Z3 and Z7 CHO-marker chromosomes are the products of a reciprocal translocation between Chinese hamster chromosomes 3 and 4, whereas the Z4 chromosome appears to have arisen by pericentric inversion following a break in the proximal region of the short arm of the other chromosome-3 homolog (Adair et al. 1984). Despite these rearrangements, both *aprt* alleles are normally expressed in CHO cells. We have regionally mapped the two CHO *aprt* genes to the distal portion of chromosome Z7p and to a region adjacent to the inversion break-junction of chromosome Z4q, respectively (Adair et al. 1984).

The spontaneous generation of *aprt*$^{+/0}$ hemizygotes in untreated CHO cell populations appears to be due to the high-frequency spontaneous deletion of one *aprt* allele (Bradley and Letovanec 1982; Simon et al. 1982, 1983; Adair et al. 1983b; Simon and Taylor 1983; Taylor et al. 1985). Close proximity of the CHO Z4 *aprt* allele to the inversion break-junction region appears to predispose specifically that gene to high-frequency spontaneous deletion. In CHO-AT3-2, loss of the Z4 *aprt* allele was accompanied by interstitial deletion of a portion of the long arm of chromosome Z4 in the region proximal to and including the inversion break-junction (Adair et al. 1983b). We have observed preferential loss or inactivation of the Z4 *aprt* allele in 19 out of 20 independently-derived CHO *aprt*$^{+/-}$ or *aprt* $^{+/-}$ cell lines examined. In most cases, the entire Z4 *aprt* gene has been deleted; three cell lines retain an apparently intact but inactive Z4 allele.

Comparison of Mutation Frequencies at the *aprt* and *hgprt* Loci in CHO-AT3-2 Cells

Both *aprt* and *hgprt* are constitutively expressed housekeeping genes coding for closely related purine salvage enzymes of similar size and subunit structure that perform similar catalytic functions (Stout and Caskey 1985; Taylor et al. 1985). Since neither gene product is required for cell viability under normal culture conditions, both loci should theoretically permit the detection of a broad spectrum of genetic damage, including intragenic deletions, rearrangements, and frameshifts, as well as base-pair substitution mutations. However, physical hemizygosity for these two drug-resistance-marker loci in CHO-AT3-2 would present certain mutational constraints in the case of larger, multilocus deletions or gross chromosomal rearrangements.

Induced mutation frequencies at the *aprt*, *hgprt*, and ouabain (*oua*) marker loci following treatment of CHO-AT3-2 cell populations with equitoxic doses of eight different direct-acting mutagens are compared in Table 1. The observed differences in AA^r:TG^r mutant frequency ratios for ethylnitrosourea (ENU) (0.93), ethylmethanesulfonate (EMS) (0.67), ICR-191 (1.18), benzo[a]pyrene-diol-epoxide (B[a]PDE) (0.31), and UV (2.00) are particularly interesting, because these mutagens are all thought to produce mainly point mutations. Although these two gene loci differ markedly in size (the Chinese hamster *aprt* gene spans only about 2.5 kb, compared to ~44 kb for the *hgprt* gene), their effective target sizes for point mutations (exon sequences, intron-splice junctions, and promoter sequences) are nearly identical.

Table 1
Differential Mutability of Drug Resistance Marker Loci in CHO-AT3-2 Cells

Mutagen	Induced mutation frequencies ($\times 10^5$, $\bar{S} = 0.37$)		
	AA^r	TG^r	Oua^r
ENU[a]	135	145	15
EMS[b]	119	179	41
ICR-191[b]	118	100	3
B[a]PDE[c]	17	54	19
UV[a]	16	8	3
MMS[b]	12	16	4
MMC[b]	9	9	1
AAF[d]	3	4	<1

[a]Data from (G. Adair and R. Humphrey, in prep.).
[b]Data from (Adair and Carver 1983).
[c]Data from (M. MacLeod et al., in prep.).
[d]Data from (M.-S. Tang et al., in prep.).

Southern Blot Analysis of Mutation at the *aprt* Locus in CHO-AT3-2

The CHO-AT3-2/*aprt* system we have developed offers some unique advantages for the molecular analysis of spontaneous or induced mutations. We have used the $aprt^{+/0}$ hemizygote CHO-AT3-2 to isolate a series of UV-hypersensitive, DNA-repair-deficient mutant sublines. These cell lines, representing several different complementation groups, are hypersensitive to both the cytotoxic and mutagenic effects of 254 nm UV light and are specifically deficient in their ability to remove UV-induced (6–4) photoproducts from their DNA (Mitchell et al. 1987; G. Adair and R. Humphrey, in prep.). They are cross-sensitive to a wide variety of chemical mutagens, including agents that form bulky adducts with DNA, such as B[a]PDE, or DNA cross-linking agents, such as mitomycin C (MMC). Together, CHO-AT3-2 and its hypersensitive derivatives permit molecular analysis and comparison of induced mutational spectra at the endogenous *aprt* locus in repair-proficient versus repair-deficient genetic backgrounds.

We have isolated and characterized over 472 spontaneous and 151 UV-induced AA^r mutants of CHO-AT3-2 and its repair-deficient derivative sublines. For Southern blot analysis of these mutants, DNA samples are digested with a battery of 8–12 different restriction enzymes, subjected to agarose gel electrophoresis, blotted onto nitrocellulose, and hybridized with a ^{32}P-labeled, nick-translated CHO *aprt* probe. As a probe, we routinely use the *pHaprt* 3.9-kb *Bam*HI fragment, which carries the entire CHO *aprt* gene plus some 5′- and 3′-flanking sequences. As shown in Figure 1, we have detected restriction site loss and/or site gain restriction-fragment-length polymorphism (RFLP) mutations at 28 sites along the *aprt* gene: within each of the five *aprt* exons, at several different intron-splice junctions, and at one site in a possible promoter sequence. A summary of the frequencies of the various classes of mutational alterations that were detected is presented in Table 2. We have recovered 19 deletions, ranging in size from <30 bp to >5.5 kb. Curiously, although five of these deletions extend upstream (in the 5′ direction) for considerable distances, none extend beyond the *aprt* gene in the 3′ direction, suggesting that there may be an essential gene just downstream from this locus. Grosovsky et al. (1986) and Breimer et al. (1986) have also noted the same polarity of deletions. The most striking difference we have seen between mutational spectra for repair-proficient and repair-deficient cell lines is

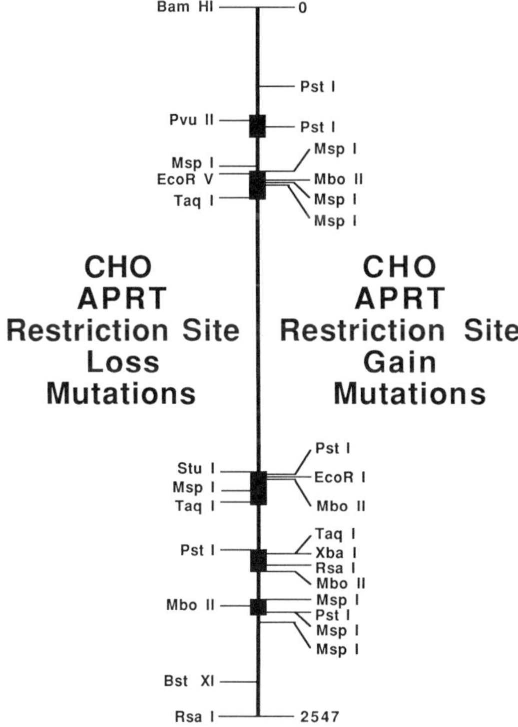

Figure 1

The distribution of restriction-site-loss or restriction-site-gain mutations along the CHO-AT3-2 *aprt* gene.

Table 2
Southern Blot Analysis of Mutations at the *aprt* Locus in Repair-proficient versus Repair-deficient CHO Cells

Cell line mutagenic treatment	Total AA^r clones screened	AA^r clones showing *aprt* RFLPs					
		site loss	site gain	partial deletion	complete loss of gene	insertion	rearrangement
CHO-AT3-2							
Spontaneous	161	15	4	8	0	1	0
254 nm UV	50	2	9	0	0	0	0
Class I repair-deficient							
Spontaneous	167	8	1	5	0	0	9
254 nm UV	51	4	2	0	0	0	0
Class II repair-deficient							
Spontaneous	169	12	6	5	0	0	7
254 nm UV	50	6	8	1	0	0	0

a class of spontaneous rearrangement events that has only been observed in repair-deficient cells. Most of these rearrangements have a single breakpoint within the *aprt* gene and extend in the 5′ direction.

Analysis of Mutation at the Heterozygous *aprt* Locus in V79-R10A-1 Cells

In contrast to CHO cells, in which both hamster chromosomes carrying an *aprt* gene have undergone rearrangement, the V79 Chinese hamster cell line has two normal, unrearranged chromosome-3 homologs. We have isolated a V79 $aprt^{+/-}$ heterozygote, designated R10A-1, which is heterozygous for an *aprt* *Msp*I RFLP reflecting loss of the exon-3 *Msp*I site. This cell line also has a chromosomal polymorphism that permits distinction of the two chromosome-3 homologs. Preliminary experiments suggest that spontaneous mutant frequencies for the *aprt* locus in this cell line are much higher than frequencies obtained in CHO-AT3-2, and there appears to be a bimodal distribution of R10A-1 AA^r colony sizes, with many of the colonies arising on mutation plates being extremely small and slow-growing.

We have isolated a total of 52 independent, spontaneous AA^r mutants of R10A-1 from Luria-Delbrück fluctuation analysis experiments. Southern blot analysis of these mutant DNAs has revealed one partial deletion and either complete loss of the wild-type *aprt* allele or homozygosis of the mutant allele in 25 out of 52 AA^r mutants examined to date (Table 3). None of the AA^r mutants isolated from CHO-AT3-2 or its derivatives or from other CHO *aprt* hemizygotes have shown complete loss of the *aprt* gene (Tables 2 and 3).

DISCUSSION

The CHO *aprt* locus is an ideal system for the molecular analysis of induced mutation in mammalian cells, offering several distinct advantages over alternative systems. The most unique feature of the *aprt* gene is its size; the entire gene spans less than 2.5 kb of DNA (Lowy et al. 1980) compared to ~44 kb for the hamster *hgprt* gene, ~26 kb for the *dhfr* gene, ~9 kb for the thymidine kinase (*tk*) gene, and ~32 kb for the *ada* gene. The relatively small size of the *aprt* locus greatly facilitates the sequencing of mutant genes (Breimer et al. 1986; Nalbantoglu and Meuth 1986; Nalbantoglu et al. 1986a,b; P.J. de Jong et al.; E.A. Drobetsky et al.; both in prep.). The availability of repair-deficient mutant sublines of CHO-AT3-2 that are hypersensitive to a broad range of physical and chemical mutagens further enhances the power of this system for the molecular analysis of induced mutation.

It is curious that at the same time technological advances are rapidly opening new doors for the molecular analysis of intragenic mutations in

Table 3
Southern Blot Analysis of Mutations at the *aprt* Locus in Hemizygous versus Heterozygous Cell Lines

Cell line/ genotype	Total AA^r clones screened	AA^r clones showing *aprt* RFLPs					
		site loss	site gain	partial deletion	complete loss of gene	insertion	rearrangement
CHO-AT3-2 $aprt^{+/0}$	161	15	4	8	0	1	0
CHO-7A1-1 $aprt^{+/0}$	31	3	0	0	0	0	0
CHO-7A10-1 $aprt^{+/-}$	30	0	0	0	0	0	0
V79-R10A-1 $aprt^{+/-}$	52	0	0	1	25[a]	0	0

[a]We have not yet determined whether these clones reflect simple loss of the wild-type *aprt* allele or homozygosis of the mutant allele.

mammalian cells, there is a growing body of evidence that suggests that multilocus or chromosomal events may be even more important than intragenic mutations in the overall picture of mammalian cell mutagenesis (Eves and Farber 1983; Moore et al. 1985; Evans et al. 1986; Yandell et al. 1986). Expression of a recessive mutant allele at a heterozygous autosomal locus can occur as a result of any one of a variety of different chromosomal events (e.g., mitotic recombination, regional inactivation, and chromosomal rearrangements) that would lead to either homozygosity or hemizygosity of the mutant allele. Such events may be quite common at mammalian autosomal loci but would probably be undetectable by conventional assay systems employing physically or functionally hemizygous target loci. Any large deletions or rearrangements that resulted in the functional loss of an essential gene located in the region of hemizygosity flanking the target gene locus would be lethal.

ACKNOWLEDGMENTS

This work was supported by National Institutes of Health grants CA-28711 and CA-04484, and facilitated by the collaborative interactions of Drs. Michael Siciliano, Raymond Stallings, Ronald Humphrey, Rodney Nairn, Christy MacKinnon, and Julia Scheerer, with technical support from Ann Brotherman and Patricia Kimmitt.

REFERENCES

Adair, G.M. and J.H. Carver. 1983. Induction and expression of mutations at multiple drug-resistance marker loci in Chinese hamster ovary cells. *Environ. Mutagen.* **5:** 161.

Adair, G.M., J.H. Carver, and D.L. Wandres. 1980. Mutagenicity testing in mammalian cells. Derivation of a Chinese hamster ovary cell line heterozygous for the adenine phosphoribosyltransferase and thymidine kinase loci. *Mutat. Res.* **72:** 187.

Adair, G.M., R.L. Stallings, and M.J. Siciliano. 1984. Chromosomal rearrangements and gene expression in CHO cells: Mapping of the alleles for eight enzyme loci on CHO chromosomes Z3, Z4, Z5, and Z7. *Somatic Cell Mol. Genet.* **10:** 283.

Adair, G.M., R.L. Stallings, K.K. Friend, and M.J. Siciliano. 1983a. Gene mapping and linkage analysis in Chinese hamster: Assignment of the genes for APRT, LDHA, IDH2 and GAA to chromosome 3. *Somatic Cell Genet.* **9:** 477.

Adair, G.M., R.L. Stallings, R.S. Nairn, and M.J. Siciliano. 1983b. High frequency structural deletion as the basis for functional hemizygosity of the aprt locus in CHO cells. *Proc. Natl. Acad. Sci.* **80:** 5961.

Bradley, W.E.C. and D. Letovanec. 1982. High-frequency nonrandom mutational event at the adenine phosphoribosyltransferase (*aprt*) locus of sib-selected CHO variants heterozygous for *aprt*. *Somatic Cell Genet.* **8:** 51.

Breimer, L.H., J. Nalbantoglu, and M. Meuth. 1986. Structure and sequence of

mutations induced by ionizing radiation at selectable loci in Chinese hamster ovary cells. *J. Mol. Biol.* **192:** 669.
Carver, J.H., G.M. Adair, and D.L. Wandres. 1980. Mutagenicity testing in mammalian cells: Validation of multiple drug resistance markers having practical application for screening potential mutagens. *Mutat. Res.* **72:** 203.
Chasin, L.A. 1974. Mutations affecting adenine phosphoribosyl transferase activity in Chinese hamster cells. *Cell* **2:** 37.
Evans, H.H., J. Mencl, N.-F. Horng, M. Picanti, C. Sanchez, and J. Hozier. 1986. Locus specificity in the mutability of L5178Y mouse lymphoma cells: The role of multilocus deletions. *Proc. Natl. Acad. Sci.* **83:** 4379.
Eves, E.M. and R.A. Farber. 1983. Expression of recessive APRT$^-$ mutations in mouse CAK cells resulting from chromosome loss and duplication. *Somatic Cell Genet.* **9:** 771.
Grosovski, A.J., E.A. Drobetsky, P.J. deJong, and B.W. Glickman. 1986. Southern analysis of genomic alterations in gamma-ray-induced APRT$^-$ hamster cell mutants. *Genetics* **113:** 405.
Hordern, J. and F. Henderson. 1982. Comparison of purine and pyrimidine metabolism in G1 and S phases of HeLa and Chinese hamster ovary cells. *Can. J. Biochem.* **60:** 422.
Jones, G.E. and P.A. Sargent. 1974. Mutants of cultured Chinese hamster cells deficient in adenine phosphoribosyl transferase. *Cell* **2:** 43.
Lowy, I., A. Pellicer, J.F. Jackson, G. Sim, S. Silverstein, and R. Axel. 1980. Isolation of transforming DNA: Cloning of the hamster APRT gene. *Cell* **22:** 817.
Meuth, M. and J.E. Arrand. 1982. Alterations of gene structure in ethyl methanesulfonate-induced mutants of mammalian cells. *Mol. Cell. Biol.* **2:** 1459.
Mitchell, D., R.M. Humphrey, G.M. Adair, L.H. Thompson, and J.M. Clarkson. 1987. The importance of (6-4) photoproducts and cyclobutane dimers for split-dose recovery in UV-irradiated normal and hypersensitive rodent cell lines. *Mutat. Res.* (in press).
Moore, M.M., D. Clive, J.C. Hozier, B.E. Howard, A.G. Batson, N.T. Turner, and J. Sawyer. 1985. Analysis of trifluorothymidine-resistant (TFT^r) mutants of L5178Y/TK +/− mouse lymphoma cells. *Mutat. Res.* **151:** 161.
Nalbantoglu, J. and M. Meuth. 1986. DNA amplification-deletion in a spontaneous mutation of the hamster *aprt* locus: Structure and sequence of the novel joint. *Nucleic Acids Res.* **14:** 8361.
Nalbantoglu, J., O. Goncalves, and M. Meuth. 1983. Structure of mutant alleles at the *aprt* locus of Chinese hamster ovary cells. *J. Mol. Biol.* **187:** 575.
Nalbantoglu, J., G.A. Phear, and M. Meuth. 1986a. Nucleotide sequence of hamster adenine phosphoribosyl transferase (aprt) gene. *Nucleic Acids Res.* **14:** 1914.
Nalbantoglu, J., D. Hartley, G. Phear, G. Tear, and M. Meuth. 1986b. Spontaneous deletion formation at the *aprt* locus of hamster cells: The presence of short sequence homologies and dyad symmetries at deletion termini. *EMBO J.* **5:** 1199.
Simon, A.E. and M.W. Taylor. 1983. High-frequency mutation at the adenine phosphoribosyltransferase locus in Chinese hamster ovary cells due to deletion of the gene. *Proc. Natl. Acad. Sci.* **80:** 810.
Simon, A.E., M.W. Taylor, and W.E.C. Bradley. 1983. Mechanism of mutation at the

aprt locus in Chinese hamster ovary cells: Analysis of heterozygotes and hemizygotes. *Mol. Cell. Biol.* **3:** 1703.

Simon, A.E., M.W. Taylor, W.E.C. Bradley, and L.H. Thompson. 1982. Model involving gene inactivation in the generation of autosomal recessive mutants in mammalian cells in culture. *Mol. Cell. Biol.* **2:** 1126.

Stout, J.T. and C.T. Caskey. 1985. HPRT: Gene structure, expression and mutation. *Annu. Rev. Genet.* **19:** 127.

Taylor, M.W., H.V. Hershey, and A.E. Simon. 1979. An analysis of mutation at the adenine phosphoribosyl transferase locus. *Banbury Rep.* **2:** 211.

Taylor, M.W., A.E. Simon, and R.M. Kothari. 1985. The APRT system. In *Molecular cell genetics* (ed. M. Gottesman), p. 311. Wiley, New York.

Thompson, L.H., S. Fong, and K. Brookman. 1980. Validation of conditions for the efficient detection of HPRT and APRT mutations in suspension-cultured Chinese hamster ovary cells. *Mutat. Res.* **74:** 21.

Yandell, D.W., T.P. Dryja, and J.B. Little. 1986. Somatic mutations at a heterozygous autosomal locus in human cells occur more frequently by allele loss than by intragenic structural alterations. *Somatic Cell Mol. Genet.* **12:** 255.

Mutation at the Human Major Histocompatibility Complex for Genotoxicity Testing

JANICE A. NICKLAS, J. PATRICK O'NEILL, MARK ALLEGRETTA, AND RICHARD J. ALBERTINI
Genetics Laboratory
University of Vermont
Burlington, Vermont 05401

OVERVIEW

The possible uses of the human major histocompatibility complex (HLA) genes in mutagenicity testing are described. The advantages of HLA over other genes include the fact that it is autosomal, that it is polymorphic so that most individuals are heterozygous, that an effective selection system exists for the detection of mutation at just one allele, and that a number of linked genes can be studied simultaneously to detect the extent of deletions. In addition, molecular probes are available for the study of mutants. HLA mutants have already been selected in human cell lines. Techniques are being developed to isolate HLA mutants derived in vivo and to develop an in vitro system to study HLA and hypoxanthine–guanine phosphoribosyl transferase (*hgprt*) mutation simultaneously. The necessity of studying several genes when quantitating mutation and of studying autosomal loci and the ideal nature of HLA as a mutant assay system is discussed.

INTRODUCTION

Mutations at the *hgprt* gene is widely used as a measure of genotoxicity. This includes in vitro and in vivo assays in both human and animal systems (Caskey and Kruh 1979; Albertini et al. 1982, 1985; Morley et al. 1983; Albertini 1985; Jones et al. 1985; Seifert et al. 1987).

The *hgprt* gene is utilized both because of ease of mutant selection (mutants are resistant to purine analogs such as 6-thioguanine [TG]) and because of its X-linkage; only one allele need be mutated in both males (who have one X chromosome) and females (who have only one active X chromosome) to reveal the HGPRT$^-$ phenotype. However, recently, the HGPRT mutant assay systems have come under scrutiny because of this X linkage of *hgprt*. It is thought that *hgprt* gene deletion could cause cell lethality, if the loss of X-chromosomal material is substantial and there is thus a concurrent deletion of linked vital genes. This lethality and resulting loss of deletional HGPRT

Banbury Report 28: Mammalian Cell Mutagenesis

mutants could definitely affect observed mutation frequencies, especially during study of clastogens such as X-irradiation.

The use of autosomal genes for mutation study should solve this problem, as deletion of a region of one chromosome would still leave functioning vital genes on the other chromosome of the pair. Several autosomal mutant assay systems have been developed, e.g., thymidine kinase (TK) (Clive et al. 1972) and adenine phosphoribosyl transferase (APRT) (Chasin 1974; Jones and Sargent 1974). However, these assays require the use of cells that are already functionally hemizygous for the gene (i.e., one allele is already mutant) and thus are not generally applicable for population screening.

The major HLA genes of man are an autosomal system which circumvents the problem of the inability to select a heterozygote mutant. The HLA complex maps to a 3 cM (6000 kbp) region on the short arm of chromosome 6 (Fig. 1) and contains a large number of genes involved as surface molecules in the immune response. The class I genes (*B*, *C*, *A*, and Q_a-Tla-like genes) code for a 44-kD protein which associates with β_2-microglobulin (a 12-kD protein whose gene maps to human chromosome 15). The class II molecules, which consist of an α-chain (34 kD) and a β-chain (28 kD), both coded at HLA, are divided into three groups: *DR*, *DQ* (formerly *MB* or *DC*), and *DP*

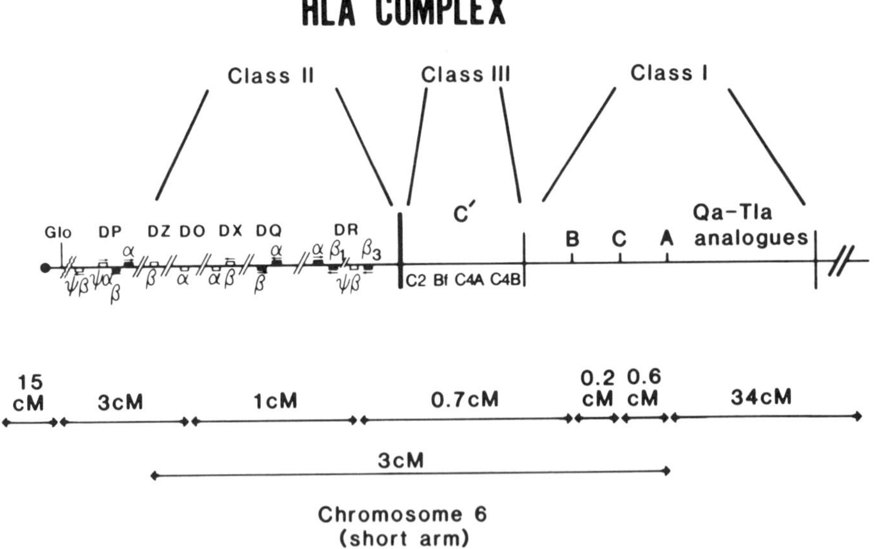

Figure 1
Diagram of the HLA gene complex on the short arm of chromosome 6.

(formerly *SB*). The genes currently known to exist in the HLA complex are one *DR*α, three *DR*β, two *DQ*α, two *DP*β, two *DP*α, one "*DZ*"α, and one "*DO*"β. There are also HLA class III genes that code for various molecules important in the complement pathway.

The genes of the HLA complex have a high level of polymorphism in the population. Twenty-three alleles of *A*, 47 alleles of *B*, 8 alleles of *C*, 14 alleles of *DR*, 3 alleles of *DQ*, and 6 alleles of *DP* are known (Bodmer and Bodmer 1984). Such polymorphism is presumed to exist because of selection for heterozygosity at HLA in the individual. Heterozygosity is hypothesized to increase immune responsiveness and thus fitness.

Sera specific to the different HLA alleles are available and are used to HLA-type individuals. Also, a number of monoclonal antibodies are available to selected specificities (Colombani et al. 1982). These specific antibodies allow independent detection of both alleles of each gene in most individuals. Because most individuals are heterozygous for HLA alleles, mutation to a hemizygous state can easily be detected. In fact, the antibodies can be used to select HLA mutant cells (Pious et al. 1973; Kavathas et al. 1980; Nicklas et al. 1984; Spring et al. 1985). Treatment of T cells with a specific antibody followed by complement will cause lysis of all cells that bind the antibody. Only cells lacking the HLA molecule to which the antibody is directed (i.e., mutants) will escape lysis. These cells can then be grown and studied.

The HLA system should allow detection of any mutants causing loss of a specificity. This includes deletions, frameshifts, point mutations, and chromosomal rearrangements. Having the entire HLA complex to analyze will allow determination of the extent of deletions by detection of loss of neighboring HLA genes concurrently. Another benefit of HLA study is the detailed molecular characterization of the region. Genomic clones and cDNA exist for class I, class II, and complement genes, and thus a detailed analysis can be made of the types of mutation induced by using Southern and/or Northern analysis. In addition, a large number of antibodies to the HLA proteins exist, allowing study of the mutant proteins. It may be possible to study missense mutations in this way. Antibodies to different epitopes on a molecule can be used against mutants to determine if only one part of the molecule has been altered by the mutation. Other types of mutants, such as somatic crossovers causing homozygosity, control mutants affecting expression, and for class I genes, loss of β_2 microglobulin expression could also be detected in this system.

RESULTS

In work done several years ago, one of us (J.A.N.) selected in vitro HLA mutants of lymphoblastoid (B) cell lines (LCLs) (Nicklas et al. 1984, 1985).

Cells of a line called LCL 526 (A2, B44, C5, DR1, DQw1, DPw3/A24, B27, B27, C2, DR4, DQw3, DPw4) from a male donor were mutagenized with 300 rads from a ^{137}Cs source, allowed to recover for 5 days, and then treated with either an anti-A2 (Parham and Brodsky 1981) or an anti-B27 (Grumet et al. 1982) monoclonal antibody (45 min at room temperature) followed by complement (anti-DR, Pel Freez) for 90 minutes at room temperature. The cells were then plated out on soft agar (0.35% low-temperature agarose, FMC Corp.) media over a feeder fibroblast layer. Colonies were picked and retyped for loss of HLA specificities. Treatment with γ-radiation induced mutations in the HLA genes; mutants were selected as they survived treatment with anti-HLA antibodies and complement, whereas all the parental cells were killed. For example, treatment of γ-irradiated cells of LCL 526 with an anti-A2 monoclonal antibody will select A2$^-$ mutants. Some of these, in addition to A2 loss, have C5 loss or A2, C5, B44 loss, and so forth, including loss of the entire haploid genotype (haplotype), presumably due to deletion of part of chromosome 6. This loss can be detected serologically by antibody typing.

Figure 2 is a diagrammatic representation of the mutants generated in these studies. As shown, a variety of mutants were induced, including various null mutants (i.e., B null, A null, DR null) generated by secondary selection on primary mutants. The molecular basis of the induced mutations was studied by Southern analysis using a variety of HLA cDNA probes (*DRβ* [Long et al. 1982], *DRα* [Wake et al. 1982], *DCβ* [Larhammer et al. 1982], *SBα* [Auffray et al. 1984], *SB*B [Roux-Dosseto et al. 1983], B7 [Sood et al. 1981]). Figure 3 shows two representative Southern blots and demonstrates that a number of mutants have lost fragments corresponding to different genes.

The use of specific HLA probes allows definition of gross alterations within the target gene (gene coding for the selected antigen). The additional use of probes for flanking HLA genes demonstrates whether they too are involved in the lesion (e.g., a large deletion). The deletion of all or a portion of a single haplotype can be detected because the polymorphism seen in HLA-serologic types is reflected at the DNA level by restriction-fragment-length polymorphism (RFLP). In Southern blots, the two types of haplotype-loss mutants possible in a heterozygous individual usually show loss of reciprocal fragments, thereby allowing assignment of most fragments to one or the other chromosome. In some instances, fragments cannot be assigned to either chromosome, because they are not lost in any of the mutants. These fragments may lie outside the boundaries of deletions existing in the mutants, may represent identical copies on both chromosomes, or may comigrate with other fragments thereby making interpretation difficult. The new fragments seen in some mutants may represent breakpoint fragments. Some mutants appear to contain shorter deletions than other mutants, retaining fragments

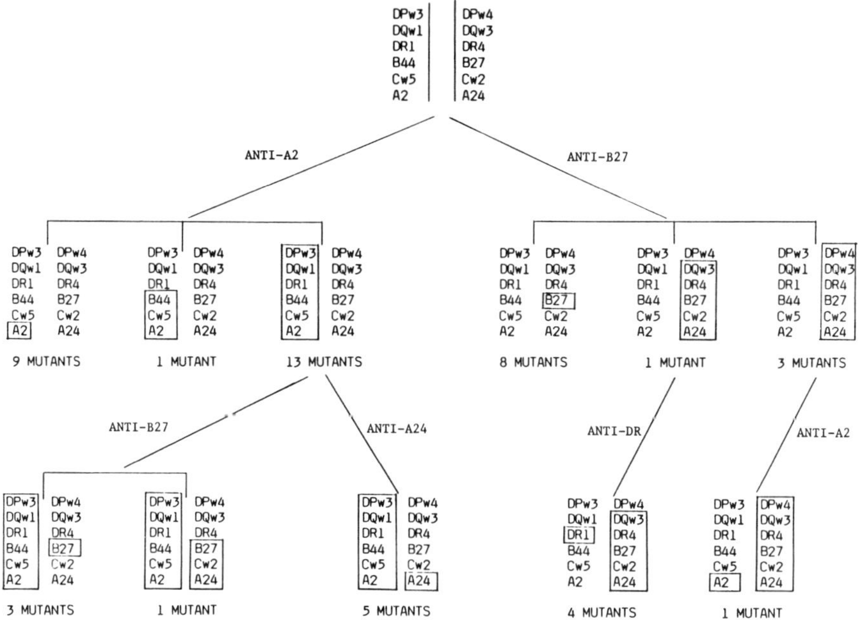

Figure 2

HLA mutants selected in cell line LCL 526 after X-irradiation. LCL 526 is an Epstein-Barr-virus-transformed B-cell line created from cells of a male with postdysenteric Reiter's syndrome (courtesy of J. Taurog and Y. Miyachi, University of Minnesota, Minneapolis). Cells were treated with 300 rads ^{137}Cs irradiation and allowed 5 days expression time; they were then treated with the indicated monoclonal antibody and complement, and plated in soft agar. Putative mutants were picked after 14 days and retested. Primary mutants were again treated with 300 rads and selected with a different monoclonal antibody and complement to obtain secondary "double" mutants. The HLA alleles are listed in two columns to indicate the maternal and paternal chromosome 6, respectively. Boxed alleles indicate serologically detected loss of that HLA specificity in the mutant. Anti-A2 is the monoclonal antibody BB7.2 (Parham and Brodsky 1981); Anti-B27 is the monoclonal B27M2 (Grumet et al. 1982); Anti-A24 is the monoclonal antibody A11.24 (courtesy of F. Carl Grumet); and Anti-DR is the monoclonal L243 (Lampson and Levy 1980).

that other mutants lack. The extent of deletion in the mutants can thus be determined.

Current studies have twofold aims: (1) to develop an HLA assay for in vivo mutation and (2) to develop a joint HLA and HGPRT in vitro system.

Methods to be used for the in vivo system will combine the HGPRT T-cell-cloning assay with HLA-selection methods. Briefly, T lymphocytes are isolated from the blood of an HLA A2 heterozygous individual on Ficoll-hypaque, and phytohemagglutinin (PHA)-stimulated for approximately 40 hours. HLA $A2^-$ cells are selected by treatment with an anti-A2-monoclonal antibody (BB7.2), followed by complement to kill the $A2^+$ cells. The cells are

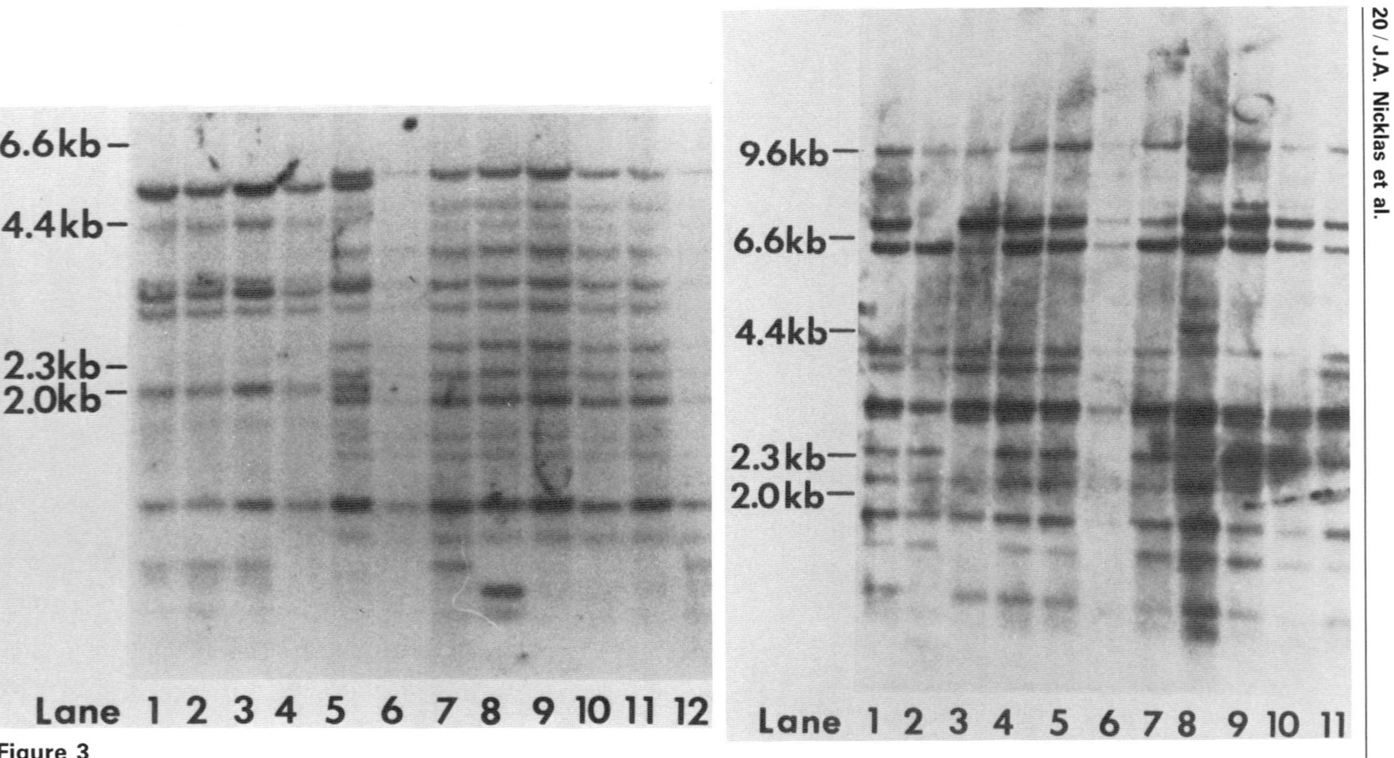

Figure 3

(*Left*) A Southern blot of DNAs cut with *Msp*I and probed with an HLA-DRβ probe. (*1–4*) DNAs from mutants with deletions on the A24, C2, B27, and DR4 haplotypes; (*5*) DNA from the parental 526; (*6–12*) DNAs from mutants with deletions on the A2, C5, B44, and DR1 haplotypes. (*Right*) A Southern blot of DNA cut with *Msp*I and probed with an HLA class I probe: (*1*) 526 parent; (*2*) A2, C5, B44, and DR1 haplotype mutant; (*3*) A24, C2, B27, and DR4 haplotype mutant; (*4*) A2$^-$/A24, C2, B27, and DR4 haplotype mutant; (*5*) A2$^-$ mutant; (*6*) A2$^-$ mutant; (*7*) A2$^-$ mutant; (*8*) A2$^-$ mutant; (*9*) A2$^-$ mutant; (*10*) A2$^-$ mutant; (*11*) A2$^-$ mutant.

then cloned by limiting dilution with feeder cells and medium containing T-cell growth factor. Clones that are recovered following selection are tested for the A2$^-$ phenotype. Few mutants have been isolated in preliminary experiments, as most of the resultant clones did not retest as A2$^-$. Antibody and complement selection is known to be very inefficient. The linkage of the anti-A2 antibody to ricin (Vallera et al. 1983) is being developed for selection to obviate the need for complement and improve the efficiency of selection.

The development of an in vitro HLA and HGPRT system should allow the comparison of mutations in these genes in the same cell line. The cell line 526, already used for HLA mutation studies, is being selected for HGPRT mutants. In a preliminary experiment, several HGPRT$^-$ mutants were obtained. Cells (526) were irradiated with 300 rads (^{137}Cs) and selected at various time-points after irradiation in medium plus 10 μM 6-thioguanine. This demonstrates that HGPRT mutants can be selected in this cell line that has already been used to detect HLA loss mutants. Thus, mutants in both markers have now been induced in a cell line to be used for comparative in vitro studies.

DISCUSSION

We have described our preliminary attempts to use the HLA genes for the detection of in vivo mutants and as part of an HGPRT (X chromosomal)-HLA (autosomal) in vitro mutation assay.

Experiments in a number of laboratories (Pious et al. 1973; Kavathas et al. 1980), including the results shown here, have demonstrated the feasibility of isolating HLA mutants in vitro. The isolation of HGPRT mutants in the same cell line has now been accomplished, as shown here. This should allow the comparison of the frequencies and types of mutations seen at the two loci (X chromosomal vs. autosomal).

The development of the in vivo assay for HLA-loss mutant T cells appears to be more difficult, possibly due to antigen density on, or lytic resistance of, fresh T cells versus the B-cell lines used for in vitro studies. The use of a ricin-conjugated antibody or other variations in technique will help to resolve these difficulties.

The use of HLA as a second system to the HGPRT assays appears to be important for a number of reasons. The first is that study of a single gene (e.g., *hgprt*) does not allow one to make generalizations about mutagenesis in the whole genome. It is not possible to know if the gene under study is representative "average" gene, or if it is "peculiar" or unique in some way. Studies of several genes can allow generalizations to be made, if patterns are seen in types and frequencies of mutation. The second is that the *hgprt* X linkage of the gene may limit the types of mutations seen. Recently, at the

Environmental Mutagenesis Society Annual Meeting in 1986 in Baltimore, concerns were raised about the use of X-linked gene for mutation studies. It is thought that a significant number of deletions and other chromosomal aberrations may not be detected in such screens because of the effect of these lesions on vital flanking genes. If such X-linked flanking genes are deleted or inactivated by translocation, then cell lethality will occur (lowering the apparent mutant frequency) because no functioning gene copy remains. However, for autosomal genes, a deletion or aberration in one chromosome still leaves its homolog with functioning alleles, and there is no resulting cell lethality. Since the large bulk of the human genome is autosomal, it is possible that X-linked-gene-mutation assays, such as HGPRT, will not be representative of genomic mutation due to clastogens.

Because assays involving HGPRT are a major contribution to mutation analysis, this question requires further study. It will be necessary to induce mutations with clastogenic and nonclastogenic mutagens at an autosomal gene and compare the frequencies of mutation seen with those of an X-chromosomal gene. Studies of the spectra of induced mutations are also important. Does an autosomal gene-mutant collection contain more deletions and other aberrations than an X-linked gene-mutant collection? A number of clastogenic agents need to be studied to determine whether any effect is consistent and to define the frequencies and types of mutations that might be missed by reliance on an X-linked-gene-mutation screen. Once such effects are known, they could be perhaps compensated for in future HGPRT assays. The great advantage of HLA is the ability to select against one allele using a specific antibody to the protein produced by that allele. Because of the great polymorphism seen in HLA alleles in the human population, nearly every individual will carry two different alleles of each HLA gene allowing selection of mutants.

The other important advantage is that HLA contains a number of closely linked genes and much is known about their molecular organization. The extent of deletions can be accurately mapped, both by loss of specific HLA proteins as detected by antibodies and by loss of restriction fragments on Southern blots using a variety of available HLA gene probes.

These advantages, in our opinion, make HLA the system of choice for autosomal gene mutagenicity studies. The feasibility of selecting in vitro mutants has already been demonstrated and work to isolate in vivo mutants is in progress.

ACKNOWLEDGMENTS

The early HLA work was supported by National Institutes of Health grants ROI-AI-17687, ROI-AI-19007, and 2-POI-AM-13083 to Dr. F.H. Bach at the University of Minnesota. Current work is supported by an Environmental Protection Agency cooperative agreement (CR813565-01-0) to J.P.O. We thank Inge Gobel for typing this manuscript.

REFERENCES

Albertini, R.J. 1985. Somatic gene mutations in vivo as indicated by the 6-thioguanine-resistant T-lymphocytes in human blood. *Mutat. Res.* **150:** 411.

Albertini, R.J., K.S. Castle, and W.R. Borcherding. 1982. T-cell cloning to detect the mutant 6-thioguanine resistant lymphocytes present in human peripheral blood. *Proc. Natl. Acad. Sci.* **79:** 6617.

Albertini, R.J., J.P. O'Neill, J.A. Nicklas, N.H. Heintz, and P.C. Kelleher. 1985. Alterations of the *hprt* gene in human 6-thioguanine resistant T-lymphocytes arising *in vivo*. *Nature* **316:** 369.

Auffray, C., J.W. Lillie, D. Arnot, D. Grossberger, D. Kappes, and J.L. Strominger. 1984. Isotypic and allotypic variation of human class II histocompatibility antigen α-chain genes. *Nature* **308:** 327.

Bodmer, J. and W. Bodmer. 1984. Histocompatibility 1984. *Immunol. Today* **5:** 251.

Caskey, C.T. and G.D. Kruh. 1979. The HPRT locus. *Cell* **16:** 1.

Chasin, L.A. 1974. Mutations affecting adenine phosphoribosyl transferase activity in Chinese hamster cells. *Cell* **2:** 37.

Clive, D., W.G. Flamm, M.R. Machesko, and N.J. Berheim. 1972. A mutational assay system using the thymidine kinase locus in mouse lymphoma cell. *Mutat. Res.* **16:** 77.

Colombani, J., J. Dausset, V. LePage, L. Degos, J. Kalil, and M. Fellous. 1982. HLA monoclonal antibody registry: A proposal. *Tissue Antigens* **20:** 161.

Grumet, F.C., B.M. Fendly, L. Fish, S. Foung, and E.G. Engleman. 1982. Monoclonal antibody (B27M2) subdividing HLA-B27. *Hum. Immunol.* **5:** 61.

Jones, G.E. and P.A. Sargent. 1974. Mutants of cultured Chinese hamster cells deficient in adenine phosphoribosyl transferase. *Cell* **2:** 43.

Jones, I.M., K. Burkhart-Schultz, and A.V. Carrano. 1985. A method to quantify spontaneous and in vivo induced thioguanine-resistant mouse lymphocytes. *Mutat. Res.* **147:** 97.

Kavathas, P., F.H. Bach, and R. DeMars. 1980. Gamma ray-induced loss of expression of HLA and glyoxalase I alleles in lymphoblastoid cells. *Proc. Natl. Acad. Sci.* **77:** 4251.

Lampson, L.A. and R. Levy. 1980. Two populations of Ia-like molecules on a human B cell line. *J. Immunol.* **125:** 293.

Larhammer, D., L. Schenning, K. Gustafsson, K. Wiman, L. Claesson, L. Rask, and P.A. Peterson. 1982. Complete amino acid sequence of an HLA-DR antigen-like β chain as predicted from the nucleotide sequence: Similarities with immunoglobulins and HLA-A, -B, and -C antigens. *Proc. Natl. Acad. Sci.* **79:** 3687.

Long, E.O., C.T. Wake, M. Strubin, N. Gross, R.S. Accolla, S. Carrel, and B. Mach. 1982. Isolation of distinct cDNA clones encoding HLA-DR_{β} chains by use of an expression assay. *Proc. Natl. Acad. Sci.* **79:** 7465.

Morley, A.A., K.J. Trainor, R. Seshadri, and R.B. Ryall. 1983. Measurement of in vivo mutations in human lymphocytes. *Nature* **302:** 155.

Nicklas, J.A., H.J. Noreen, N. Ohta, F.C. Grumet, and F.H. Bach. 1985. Analysis of new HLA loss in mutants of lymphoblastoid cell lines. *Transplant. Proc.* **17:** 776.

Nicklas, J.A., Y. Miyachi, J.D. Taurog, S. Wee, L. Chen, F.C. Grumet,and F.H. Bach. 1984. HLA-B27 loss variants of a lymphoblastoid cell line: Genetic and cellular characterization. *Hum. Immunol.* **11:** 19.

Parham, P. and F.M. Brodsky. 1981. Partial purification and some properties of BB7.2—A cytotoxic monoclonal antibody with specificity for HLA-A2 and a variant of HLA-A28. *Hum. Immunol.* **3:** 277.

Pious, D., P. Hawley, and G. Forrest. 1973. Isolation and characterization of HL-A variants in cultured human lymphoid cells. *Proc. Natl. Acad. Sci.* **70:** 1397.

Roux-Dosseto, M., C. Auffray, J.W. Lillie, J.M. Boss, D. Cohen, R. DeMars, C. Mawas, J.G. Seidman, and J.L. Strominger. 1983. Genetic mapping of a human class II antigen β-chain cDNA clone to the SB region of the HLA complex. *Proc. Natl. Acad. Sci.* **80:** 6036.

Seifert, A.M., W.E.C. Bradley, and K. Messing. 1987. Exposure of nuclear medicine patients to ionizing radiation is associated with rises in HPRT-mutant frequency in peripheral T-lymphocytes. *Mutat. Res.* (in press).

Sood, A.K., D. Pereira, and S.M. Weissman. 1981. Isolation and partial nucleotide sequence of a cDNA clone for human histocompatibility antigen HLA-B by use of an oligodeoxy nucleotide primer. *Proc. Natl. Acad. Sci.* **78:** 616.

Spring, B., C. Fonatsch, C. Muller, G. Pawelec, J. Kompf, P. Wernet, and A. Ziegler. 1985. Refinement of HLA gene mapping with induced B-cell line mutants. *Immunogenetics* **21:** 277.

Vallera, D.A., R.C. Ash, E.D. Zanjani, J.H. Kersey, T.W. LeBien, P.C.L. Beverley, D.M. Neville, Jr., and R.J. Youle. 1983. Anti-T-cell reagents for human bone marrow transplantation: Ricin linked to three monoclonal antibodies. *Science* **222:** 512.

Wake, C.T., E.O. Long, M. Strubin, N. Gross, R. Accolla, S. Carrel, and B. Mach. 1982. Isolation of cDNA clones encoding HLA-DR α chains. *Proc. Natl. Acad. Sci.* **79:** 6979.

Historical Overview of the Mouse Lymphoma TK$^{+/-}$ Mutagenicity Assay

DONALD CLIVE
Genetic Toxicology Laboratory
Burroughs Wellcome Company
Research Triangle Park, North Carolina 27709

Being asked to do an historical overview of the principal subject of your professional career is a moment of mixed emotions. There is a touch of sadness that your work and, by inference, yourself are of sufficient vintage to be considered proper grist for the historical mill. On the other hand, this is more than compensated for by the sense of satisfaction that your contributions or perspectives warrant such a request and by the awareness that your version of events will be the one on record.

In what follows, I will be concentrating on only a few of the most critical events in the development of the L5178Y/TK$^{+/-}$ → TK$^{-/-}$ mouse lymphoma assay (MLA) as I perceive them. No doubt others would select other highlights for discussion, since there is always a significant subjective component to recent history. In addition, even when supplemented by notebooks and the literature, memory is fallible. Finally, the requirement that it all fit into 15 pages of double-spaced text (including references) subjects the self-indulgence of the author to the tolerance of the audience and editor.

Therefore, having caveatted sufficiently to cover both omissions and commissions, let me proceed.

OVERVIEW

The thymidine kinase heterozygote (TK)$^{+/-}$–3.7.2C cell line currently in use for in vitro mutagenicity studies was not the *tk* heterozygote of choice at the time, was not designed to produce small- and large-colony mutants nor to detect single-gene and viable chromosomal mutations, was a last-ditch compromise to get something functioning at the *tk* locus in mouse lymphoma cells, and illustrates the major role that serendipity can play in the biological sciences. The TK$^{+/-}$ → TK$^{-/-}$ MLA has developed in five stages.

1. Development of methodologies for producing and isolating the three *tk* genotypes ($tk^{+/+}$ and $tk^{-/-}$ homozygotes and the $tk^{+/-}$ heterozygote), and optimizing mutagenicity assay conditions with the first $tk^{+/-}$ heterozygote.

Banbury Report 28: Mammalian Cell Mutagenesis

This appeared to be a noncontroversial equivalent of a single-gene mutation assay such as hypoxanthine-guanine phosphoribosyl transferase (HGPRT) (Clive et al. 1972a,b).

2. Loss of this first $tk^{+/-}$ heterozygote and the recognition that subsequent heterozygotes produced distinctly bimodal distributions of mutant-colony sizes; these were eventually interpreted in terms of single-gene and viable chromosomal mutations.
3. A period of diversification of MLA and controversy over the significance of small-colony mutants (Amacher et al. 1979, 1980; Clive et al. 1979, 1980; Amacher and Paillet 1981).
4. A cytogenetic period, led by Hozier's laboratory, in which the cytogenetic interpretation of small-colony mutants was confirmed (Clive et al. 1980, 1983; Hozier et al. 1981, 1982, 1983, 1985; Sawyer et al. 1985; Appelgate and Hozier, this volume).
5. The current molecular period, again in Hozier's laboratory, in which most mutations in both large- and small-colony mutants are being found to involve large-scale deletions (Evans et al. 1986; Appelgate and Hozier, this volume).

$tk^{+/-}$ – 1.1.1 Heterozygote

Two classes of $tk^{+/-}$ heterozygotes have been recognized in L5178Y mouse lymphoma cells based on spontaneous mutation rates and on the presence or absence of a clearly bimodal distribution of mutant colony sizes. Our first $tk^{+/-}$ heterozygote ($tk^{+/-}$ – 1.1.1, see Table 1) was isolated in the spring of 1971 in Gary Flamm's laboratory at the National Institute of Environmental Health Sciences (NIEHS) by cloning untreated and ultraviolet (UV)-irradiated (100 ergs/mm^2) wild-type (TK$^{+/+}$ – 1) cells in the presence of 5-bromodeoxyuridine (BrdU) (50 μg/ml; 1 × 10^5 cells/ml). Out of a total of 2 × 10^8 UV-irradiated cells cloned in two independent experiments, only one BrdU-resistant colony was found; none was found out of one-fifth as many

Table 1
Derivation of First TK$^{+/+}$, TK$^{+/-}$, and TK$^{-/-}$ Cell Lines

Phenotype change	Genotype change	Mutant frequency	Mutation rate (per generation)
UV-induced			
BrdU^s/THMG^r → BrdU^r/THMG^s	$tk^{+/+} \rightarrow tk^{-/-}$	5×10^{-9}	5×10^{-9}
Spontaneous			
BrdU^s/THMG^r → BrdU^r/THMG^s	$tk^{+/+} \rightarrow tk^{-/-}$	$<2 \times 10^{-8}$	$\ll 2 \times 10^{-8}$
BrdU^r/THMG^s → BrdU^s/THMG^r	$tk^{-/-} \rightarrow tk^{+/-}$	$4–50 \times 10^{-8}$	6×10^{-9}
BrdU^s/THMG^r → BrdU^r/THMG^s	$tk^{+/-} \rightarrow tk^{-/-}$	9×10^{-6}	1×10^{-7}

unirradiated cells. This implied UV-induced and spontaneous mutant frequencies of 5×10^{-9} and less than 2×10^{-8}, respectively.

On the reasonable assumption that such low frequencies reflected diploidy at the *tk* locus, this $TK^{-/-}$ mutant (designated as $TK^{-/-} - 1.1$, following the demonstrated inability of these growing cells to incorporate [^{3}H]thymidine into DNA and, also, the absence of TK activity by direct enzyme assay) was allowed to accumulate spontaneous $TK^{+/-}$ revertants that were then quantitated by cloning in the presence of THMG (3 μg/ml thymidine +5 μg/ml hypoxanthine +0.1 μg/ml methotrexate +7.5 μg/ml glycine). In four experiments spanning 6–48 generations of nonselective growth, 136 presumptive $TK^{+/-}$ revertant colonies were discovered out of 7×10^8 BrdU-resistant cells cloned for spontaneous reverse mutant frequencies, ranging between 4×10^{-8} and 46×10^{-8} mutations/locus and a mean spontaneous mutation rate of 6×10^{-9} mutations/locus/generation.

One of these spontaneously arising-THMG^r-revertant clones (designated as $TK^{+/-}1.1.1$ following the determination of its TK-enzyme activity as close to half that of the wild-type $TK^{+/+}1$ cells) was grown to culture size in the continued presence of THMG. (It should be noted that this heterozygote and its grandparental $TK^{+/+}$ cells were the only TK-competent mouse lymphoma cells that could be continuously maintained in THMG-supplemented medium; all subsequent lines would die out after a few days in such medium. It is not clear if this relates to the differences between this $TK^{+/-}$ cell line and all subsequent ones). These cells were cloned in the presence of BrdU 3 times between 4 and 12 generations after removal of THMG. The spontaneous forward mutant frequency and mutation rate, $TK^{+/-} \rightarrow TK^{-/-}$, so determined were, respectively, 9×10^{-6} mutations/locus and 1×10^{-7} mutations/locus/generation.

Following reconstruction experiments to determine optimal cloning conditions, the ability of the $TK^{+/-}$–1.1.1 cell line to respond to two classic mutagens, X-rays and ethyl methanesulfonate (EMS), and hycanthone

Table 2
Mutagenesis of $TK^{+/-} - 1$ Cell Line

Treatment	Mutant frequency	Comments
None	9×10^{-6}	
X ray (600 rads)	83×10^{-6}	large colonies only
	314×10^{-6}	very small colonies only
Total	397×10^{-6}	all colonies
EMS (400 μg/ml; 10% survival)	1900×10^{-6}	all colonies
Hycanthone (18 μg/ml; 20% survival)	34×10^{-6}	

methanesulfonate (HMS) was determined at this time (Clive et al. 1973); a sampling of these results is shown in Table 2.

Following this promising beginning, these three cell lines were lost to a combination of mycoplasma contamination and liquid nitrogen tank leak.

$tk^{+/-}$–3.7.2 Heterozygote and Small-colony Mutants

A second sample of L5178Y mouse lymphoma cells ($TK^{+/+}$–2) was obtained from J. Spalding (also our source for the $TK^{+/+}-1$ cells) at NIEHS, and the above isolation steps were repeated to isolate a second *tk* heterozygote ($tk^{+/-}$–2.1.1). Unfortunately, this cell line, when cloned in BrdU in a mutagenesis assay, gave a pronounced bimodel distribution of colony sizes. Although small colonies had been seen with the $tk^{+/-}-1.1.1$ heterozygote under these conditions (see Table 2 for X-rays and EMS treatments), those small colonies were quite tiny and fairly easy to regard as an irrelevant artifact of the BrdU selection or cloning conditions. The new $tk^{+/-}$–2.1.1 heterozygote, on the other hand, spawned small colonies of a size that could not be ignored.

Rather than explore these small colonies in any detail, it was decided to try to avoid them by isolating an independent *tk* heterozygote from a third sample of L5178Y cells, $TK^{+/+}$–3. Seven EMS-induced, BrdU-resistant clones were isolated, grown up, confirmed as being TK-deficient, and designated $TK^{-/-}-3.1$ through $TK^{-/-}-3.7$ (Clive and Voytek 1977). THMG-resistant variants were found in six of these seven lines. The largest of these colonies were isolated for each starting cell line, assayed for TK-enzyme activity, and cloned for spontaneous mutability to BrdU-resistance. They were designated as shown in Table 3. All enzyme activities were intermediate

Table 3
Derivation of Third $TK^{+/+}$, $TK^{+/-}$, and $TK^{-/-}$ Cell Lines

Starting cells	EMS-induced $TK^{+/+} \rightarrow TK^{-/-}$ mutant frequency	$TK^{-/-}$ mutant no.	Spontaneous $TK^{-/-} \rightarrow TK^{+/-}$ mutant frequency	$TK^{+/-}$ mutant no.
$TK^{+/+}-3$	6×10^{-8}	$TK^{-/-}-3.1$	3×10^{-8}	$TK^{+/-}-3.1.1$
		$TK^{-/-}-3.2$	4×10^{-8}	$TK^{+/-}-3.2.1$
		$TK^{-/-}-3.3$	0.8×10^{-8}	$TK^{+/-}-3.3.1$
		$TK^{-/-}-3.4$	0.8×10^{-8}	$TK^{+/-}-3.4.1$
		$TK^{-/-}-3.5$	$<0.8 \times 10^{-8}$	
		$TK^{-/-}-3.6$	12×10^{-8}	$TK^{+/-}-3.6.1$
				$TK^{+/-}-3.6.2$
				$TK^{+/-}-3.6.3$
		$TK^{-/-}-3.7$	12×10^{-8}	$TK^{+/-}-3.7.1$
				$TK^{+/-}-3.7.2$

between those of $TK^{+/+}$ and $TK^{-/-}$ (ranging from approximately one third to two thirds the activity of $TK^{+/+}$–3 cells); each presumptive heterozygote spawned a bimodel distribution of BrdU-resistant colonies.

By then it was late 1972, we had no "clean" mutagenesis assay utilizing a heterozygous *tk* locus, success in isolating what we wanted appeared dim, so that pragmatism dictated trying to work with what we had, which was $TK^{+/-}$–3.7.2.

Small-colony ($\sigma TK^{-/-}$) Mutants

The first order of business was determining the nature of the small colonies on the BrdU plates. Including these with the large colony mutants yielded previously unheard of high mutation rates. Were they present in the $TK^{+/-}$ –3.7.2 cells as a subpopulation? Were they real $TK^{-/-}$ mutants or artifacts? If real mutants, were they "leaky" or completely amorphic mutants? If amorphs, was the agar inhibiting their growth? If leaky, was the BrdU inhibitory? Subclones of $TK^{+/-}$–3.7.2 consistently produced the same spectrum of mutant colonies, and THMG cleansing failed to eliminate them; thus, they were arising de novo and spontaneously. Enzyme assays consistently showed zero TK activity in both small- and large-colony variants, thereby ruling out leaky mutants as the explanation. Secondary BrdU effects on the growth of these colonies, as well as agar effects, were not causal when it was found that the cells from small colonies grew slowly for several generations in a nonselective medium in the absence of agar (Clive et al. 1980). This suggested that the small colonies were (1) real mutants that (2) arose spontaneously and (3) possessed a heritable defect in the growth rate of the constituent cells (i.e., a second mutation was present in these small colony-forming cells).

Next, what was the nature of this mutation? By this time, some mutagenicity studies had begun to show specificity in the proportions of small- and large-colony mutants. EMS, for instance, induced up to 80% large-colony mutants, whereas hycanthone and methyl methanesulfonate (MMS) induced up to 80% small-colony mutants. In these studies, small colonies were present in the viable count plates as well as the BrdU plates, but in far lower proportion (e.g., 5–15% on the viable count plates and up to 80% on the mutant plates). If the small-colony mutant phenotype were being induced independently of the *tk* mutation, it should be present at the same frequency in $TK^{-/-}$ and $TK^{+/-}$ populations. However, the small colony phenotype was preferentially associated with mutation at the *tk* locus and therefore was not an independent event. If two mutations in the same cell were not being induced independently of each other, then the simplest genetic mechanism that could explain their association is that the two genes involved were linked and that multilocus mutations (e.g., deletions) which affected the *tk* gene and a gene involved in normal growth had occurred in small-colony mutants. In the mid-to-late 1970s, this translated into chromosomal mutations at this heterozygous *tk* locus but not at the hemizygous *hgprt* locus (Brown and Clive 1977; Clive et al. 1979; Clive and Moore-Brown 1979; Moore-Brown and Clive 1979).

By 1975, I had left NIEHS and was working at Burroughs Wellcome Company. Three things happened over a short period of time that considerably improved both the MLA and our understanding of these small-colony mutants. First, precise, sensitive, and reproducible colony sizing on an automatic colony counter (supplemented with colony-size discrimination capability) became possible, permitting better quantitation of small-colony ($\sigma TK^{-/-}$) and large-colony ($\lambda TK^{-/-}$) mutants. Second, trifluorothymidine (TFT), the first of a series of Burroughs Wellcome antiviral drugs, while being tested for mammalian cell mutagenesis in MLA, was serendipitously found to arrest TK-competent cells within a single cell division (unlike BrdU, which required three or more cell divisions, thereby producing a hazy "lawn" throughout the plates and compromising mutant counts, mutant isolation, and the confidence with which one could regard the small-colony mutants as "real." This arresting property singled out TFT as a far superior selective agent for $TK^{-/-}$ mutants than BrdU (Clive et al. 1979; Moore-Brown et al. 1981). Third, John Hozier was "discovered" by Martha Moore and JoEllen Lewtas. This last discovery led to the modern version of the MLA as a combination single-gene, multigene, and viable chromosomal mutation assay (Clive et al. 1980, 1983; Hozier et al. 1981, 1982, 1983; Clive 1983; Moore et al. 1985a,b). These aspects of MLA will be dealt with in detail by others (Evans et al.; Moore et al.; Applegate and Hozier; all this volume) and need not be discussed further in this overview.

Diversification and Controversy

Numerous investigators around the world have encountered difficulties with the original MLA. Two major variations based on cell-growth difficulties have been established, and a third variation based on scoring only large-colony mutants has been used briefly in a few laboratories.

Starting in the late 1970s, D.E. Amacher announced (1) that $TK^{+/-}$ -3.7.2 cells grew poorly in Fischer's medium and that RPMI 1640 was a superior culture medium for these cells (Amacher et al. 1978), and (2) that $\sigma TK^{-/-}$ mutants were not true, stable mutants at all but were system artifacts related to delayed cytotoxicity, a position expressed as recently as 1984 (Amacher and Paillet 1981; Amacher 1984). The first of these claims has been seen in our laboratory only after prepared Fischer's medium was deliberately stored in a lighted refrigerator for several hours. RPMI medium appears to be more robust under such storage conditions (either by the distributor or in the laboratory); it is the better medium for this reason and because it also permits faster growth of $TK^{+/-}$–3.7.2 cells both in suspension culture and under cloning conditions. Today, a number of laboratories routinely use RPMI 1640 medium with the MLA.

However, a great drawback to this medium relates to the other claim by Amacher and Paillet (1981) of the nonmutant stature of the small colonies growing in the presence of TFT. This claim was based in part on analyzing a bimodal distribution as if it were unimodal. As a result, our previously reported loss of small-colony mutants with posttreatment time was misinterpreted as a posttreatment time-dependent increase in colony size, and hence as a recovery from delayed cytotoxicity, rather than as a dilution effect resulting from the slower suspension growth of these mutant cells during extended-expression times. Amacher's interpretation was supported by his observation that a high proportion of the small colonies arising in TFT-supplemented plates from untreated cultures was sensitive to TFT following isolation, growth, and rechallenge with TFT. This observation has been confirmed elsewhere for RPMI medium but was found not to be a problem with Fischer's medium (Moore and Howard 1982). Most spontaneous, large TFT-resistant colonies in RPMI medium, and all such from Fischer's medium, are true mutants, and all mutagen-*induced* small or large TFT-resistant colonies appear to be true mutants regardless of culture medium.

As a result, RPMI 1640 medium quickly became accepted as an alternative, perhaps superior medium, and, in a few laboratories, small-colony mutants began to be ignored. This last trend quickly dissipated when the previously mentioned cytogenetic results began to appear from John Hozier's laboratory.

Recent attempts to explore the nature of these RPMI-specific, nonmutant, small TFT-resistant colonies in our own laboratory have been disappointing due to an inability to generate them at the high frequencies reported by other investigators. At present, it is not known why this should be, but, if it holds up, we would have no remaining concerns over switching to this richer medium.

A second variation in the MLA was developed in Jane Cole's laboratory in the United Kingdom. Here, it was noted that a continuous range of mutant colony sizes was showing up in TFT-supplemented plates rather than our reported bimodal distribution. This problem proved intractable and led to the development of the so-called "fluctuation" version. In this, small numbers of cells were cloned in each of many microwells containing TFT-supplemented liquid medium, rather than in soft-agar-cloning medium. I am not sufficiently aware of details to comment further on this version nor on the likely problems leading up to it, except that the work was done by highly competent individuals who did not easily give up on trying to get the original system to work.

The two soft agar versions of the MLA were validated around the turn of the decade, one from the point of view of large- and small-colony mutants (Clive et al. 1979) and the other restricted to large-colony mutants only (Amacher et al. 1979, 1980). In retrospect, both of these efforts suffered from the same major problems that have plagued the use of genotoxicity assays for predicting carcinogenicity from the mid-1970s, namely, the near absence of convincing noncarcinogens. For this reason, both of these validation efforts failed to show up the low specificity of this assay that was reported nearly half a decade later by Shelby and Stasiewicz (1984).

The major controversy, however, was not in how to run the MLA but in the interpretation of what we were calling small-colony or chromosomal mutations. As mentioned earlier, Amacher led the school that believed that these were not true mutants but represented the effects of delayed cytotoxicity. This viewpoint should not be judged harshly on the basis of what we have learned in the ensuing half decade. When we first encountered these small colonies in the early 1970s, our initial reaction was the same as Amacher's in 1981 (Amacher and Paillet 1981). However, since Fischer's medium allowed for extremely clean selection, we were fortunate in not having to account for the nonmutants that grew in RPMI selection medium.

Lessons in Mammalian Cell Mutagenesis

A number of general principles of mammalian cell mutagenesis have emerged from the decade and a half of working with this system. First and foremost, it has been observed in a number of laboratories that the majority of compounds that are mutagenic in this system induce primarily small-colony mutants, whereas very few compounds (e.g., several ethylating agents, ICR 170, and some methotrexate) induce predominantly large-colony mutants. This translates into the following three principles: (1) Viable specific-locus mutations can occur by chromosomal mechanisms in addition to the traditional intragenic events. (2) The majority of specific-locus mutations in mammalian cells may be of chromosomal or at least multigenic dimensions. (3) Clastogenicity is probably the most general genetic event resulting from treatment of mammalian cells with genotoxic agents.

It eventually became apparent that the MLA was capable of identifying too many compounds as genotoxins. It was *desirable* that certain Ames-negative carcinogens (e.g., procarbazine) were potent mutagens in the MLA. It was *acceptable* that several "classic" noncarcinogenic clastogens (e.g., caffeine and methotrexate), which were negative in other pure gene mutational assays, were positive in the MLA, since these were already known to be clastogenic. However, it was *unsuitable* that the MLA identified a large proportion of novel compounds as being genotoxic, whereas the Ames assay was usually negative. When this pattern became apparent, the MLA soon acquired the reputation of being too sensitive to predict genotoxic/carcinogenic potential reliably. Within our own testing experience, it was apparent that MLA positives usually correlated with in vitro clastogenicity; other investigators noted similar strong correlations with sister chromatid exchanges, and few, if any, unique MLA-positive results existed once adequate testing was performed. In addition, I am not aware of any compound, aside from certain pyrimidine analogs that differentially select for preexisting $TK^{-/-}$ cells, for which the observed MLA-positive result did not arise from a de novo induction of stable $TK^{-/-}$ mutants.

Out of these observations grew the area of work in which I am most involved at the present (Clive 1987a,b), and the final principle of mammalian cell mutagenesis—a high proportion of compounds are positive in one or more of the most widely used in vitro genotoxicity assays; in some cases, this genotoxicity relates to in vitro testing conditions that are irrelevant to the whole animal or to reasonable conditions of exposure. Why should this be? We should not forget that the genetic machinery in higher eukaryotes performs highly complex operations on about 1m of DNA contained within each approximately 20-μm diameter cell nucleus. Its fidelity depends on the integrity of several mutually interacting homeostatic mechanisms. Some of these mechanisms are sensitive to significant departures from physiological conditions (e.g., transient hypoxia, low-pH, and high-osmolarity, resulting in genotoxicity) (Cifone 1985; Gaulden 1986; Rice et al. 1986; Galloway et al. 1987).

All of these nonphysiological conditions can occur at some (usually high) concentrations of certain test compounds; such high-dose-related genetic damage is not likely to be relevant to genetic risk in the whole animal. These and other as yet unidentified in vitro artifacts may be responsible for some of the embarassing mutagenic activities detected among noncarcinogens (Shelby and Stasiewicz 1984). It is well known that most chemicals are cytotoxic at some concentration; indeed, most in vitro short term tests specify testing at up to cytotoxic concentrations. Since at least some cytotoxic mechanisms have genotoxic consequences, this indicates that in vitro genotoxicity is a natural and expected property of high concentrations of many chemicals. In those instances, in vitro genotoxicity may have no more relevance to human or whole-animal risk than does in vitro cytotoxicity. The obvious solution might appear to be to reduce the severity of in vitro treatment in order to stay below the level of significant cytotoxicity. However, in vitro treatment conditions cannot be greatly decreased in severity without compromising the detection of certain rodent carcinogens and increasing the incidence of false negatives to an unacceptable level. It is critical, therefore, that each compound tested be evaluated for test conditions that may lead to false positive results, particularly when positive responses are observed only at high concentration and/or high cytotoxicity.

In summary, although in vitro genotoxicity assays have proven disappointing in rapidly and accurately detecting potential rodent carcinogens (Tennant et al. 1987), the MLA appears to have emerged as a versatile and informative system for understanding the full complexity of mammalian cell mutagenesis. The reasons for this will become clear in subsequent presentations (Moore et al.; Appelgate and Hozier; both this volume).

REFERENCES

Amacher, D.E. 1984. The L5178Y/TK gene mutation assay system. *Chem. Mutagens* **9:** 183.

Amacher, D.E. and S.C. Paillet. 1981. Trifluorothymidine-resistance and colony size in L5178Y/TK$^{+/-}$ cells treated with methyl methanesulfonate. *J. Cell. Physiol.* **106:** 349.

Amacher, D.E., S.C. Paillet, and V.A. Ray. 1978. 2-Acetylaminofluorene mutagenicity at the TK locus in L5178Y cells pretreated with sodium phenobarbital or Aroclor 1254. In *Abstract from the 9th Annual Environmental Mutagen Society Meeting*, p. 55.

———. 1979. Point mutations at the thymidine kinase locus in L5178Y mouse lymphoma cells. II. Application to genetic toxicological testing. *Mutat. Res.* **64:** 391.

Amacher, D.E., S.C. Paillet, G.N. Turner, V.A. Ray, and D.S. Salsburg. 1980. Point mutations at the thymidine kinase locus in L5178Y mouse lymphoma cells. II. Test validation and interpretation. *Mutat. Res.* **72:** 447.

Brown, M.M.M. and D. Clive. 1977. The utilization of trifluorothymidine as a selective agent for TK$^{-/-}$ mutants in L5178Y mouse lymphoma cells. In *Abstract from the 8th Annual Environmental Mutagen Society Meeting,* p. 69.

Cifone, M.A. 1985. Relationship between increases in the mutant frequency in L5178Y TK$^{+/-}$ mouse lymphoma cells at low pH and metabolic activation. *Environ. Mutagen.* (suppl. 3) **7:** 27.

Clive, D. 1983. Viable chromosomal mutations affecting the TK locus in L5178Y/TK$^{+/-}$ mouse lymphoma cells: The other half of the assay. *Ann. N.Y. Acad. Sci.* **407:** 253.

———. 1987a. Genetic toxicology: From theory to practice. *J. Drug. Dev. Clin. Res.* **1:** 11.

———. 1987b. Genetic toxicology: Can we design predictive in vivo assays? *Mutat. Res.* (in press).

Clive, D. and M.M. Moore-Brown. 1979. The L5178Y/TK$^{+/-}$ mutagen assay system: Mutant analysis. *Banbury Rep.* **2:** 421.

Clive, D. and P. Voytek. 1977. Evidence for chemically-induced structural gene mutations at the thymidine kinase locus in cultured L5178Y mouse lymphoma cells. *Mutat. Res.* **44:** 269.

Clive, D., A.G. Batson, and N.T. Turner. 1980. The ability of L5178Y/TK$^{+/-}$ mouse lymphoma cells to detect single gene and viable chromosome mutations: Evaluation and relevance to mutagen and carcinogen screening. In *The predictive value of short-term screening tests in carcinogenicity evaluation* (ed. G.M. Williams et al.), p. 103. Elsevier, New York.

Clive, D., W.G. Flamm, and M.R. Machesko. 1972a. Mutagenicity of hycanthone in mammalian cells. *Mutat. Res.* **14:** 262.

Clive, D., W.G. Flamm, and J.B. Patterson. 1972b. A mutational assay system using the thymidine kinase locus in mouse lymphoma cells. *Mutat. Res.* **16:** 77.

Clive, D., J. Hozier, and M.M. Moore. 1983. "Single-gene" and viable chromosome mutations affecting the TK locus in L5178Y mouse lymphoma cells. *Ann. N.Y. Acad. Sci.* **407:** 420.

Clive, D., W.G. Flamm, M.R. Machesko, and N.J. Bernheim. 1973. Specific-locus assay systems for mouse lymphoma cells. *Chem. Mutagens* **3:** 79.

Clive, D., K.O. Johnson, J.F.S. Spector, A.G. Batson, and M.M.M. Brown. 1979. Validation and characterization of the L5178Y/TK$^{+/-}$ mouse lymphoma mutagen assay system. *Mutat. Res.* **59:** 61.

Evans, H.H., J. Mencl, M.-F. Horng, M. Ricanati, C. Sanchez, and J. Hozier. 1986. Locus specificity in the mutability of L5178Y mouse lymphoma cells: The role of multilocus lesions. *Proc. Natl. Acad. Sci.* **83:** 4379.

Galloway, S.M., D.A. Deasy, C.L. Bean, M.J. Armstrong, A.R. Kraynak, and M.O. Bradley. 1987. Effects of high osmotic strength on chromosome aberrations, sister chromatid exchanges and DNA strand breaks, and the relation to toxicity. *Mutat. Res.* (in press).

Gaulden, M.E. 1986. Sodium chloride-induced chromosome stickiness: An examination in living cells of its relation to chromosome breakage. *Environ. Mutagen.* (suppl. 6) **8:** 30 (Abstr.).

Hozier, J., J. Sawyer, D. Clive, and M. Moore. 1982. Cytogenetic distinction between the TK^{+} and TK^{-} chromosomes in the L5178Y/TK$^{+/-}$–3.7.2C mouse-lymphoma cell line. *Mutat. Res.* **105:** 451.

———. 1983. Cytogenetic analysis of small-colony L5178Y TK$^{-/-}$ mutants early in their clonal history. *Ann. N.Y. Acad. Sci.* **407:** 423.

———. 1985. Chromosome 11 aberrations in small colony L5178Y TK$^{-/-}$ mutants early in their clonal history. *Mutat. Res.* **147:** 237.

Hozier, J., J. Sawyer, M. Moore, B. Howard, and D. Clive. 1981. Cytogenetic analysis of the L5178Y/TK$^{+/-}$ → TK$^{-/-}$ mouse lymphoma mutagenesis assay system. *Mutat. Res.* **84:** 169.

Moore, M.M. and B.E. Howard. 1982. Quantitation of small colony trifluorothymidine-resistant mutants of L5178Y/TK$^{+/-}$ mouse lymphoma cells in RPMI-1640 medium. *Mutat. Res.* **104:** 287.

Moore, M.M., D. Clive, B.E. Howard, A.G. Batson, and N.T. Turner. 1985a. In situ analysis of trifluorothymidine-resistant (TFT^{r}) mutants of L5178Y/TK$^{+/-}$ mouse lymphoma cells. *Mutat. Res.* **151:** 147.

Moore, M.M., D. Clive, J.C. Hozier, B.E. Howard, A.G. Batson, N.T. Turner, and J. Sawyer. 1985b. Analysis of trifluorothymidine-resistant (TFT^{r}) mutants of L5178Y/TK$^{+/-}$ mouse lymphoma cells. *Mutat. Res.* **151:** 161.

Moore-Brown, M.M. and D. Clive. 1979. The L5178Y/TK$^{+/-}$ mutagen assay system: In situ results. *Banbury Rep.* **2:** 71.

Moore-Brown, M.M., D. Clive, B.E. Howard, A.G. Batson, and K.O. Johnson. 1981. The utilization of trifluorothymidine (TFT) to select for thymidine kinase-deficient (TK$^{-/-}$) mutants from L5178Y/TK$^{+/-}$ mouse lymphoma cells. *Mutat. Res.* **85:** 363.

Rice, G.C., C. Hoy, and R.T Schimke. 1986. Transient hyposa enhances the frequency of dihydrofolate reductase gene amplification in Chinese hamster ovary cells. *Proc. Natl. Acad. Sci.* **83:** 5978.

Sawyer, J., M.M. Moore, D. Clive, and J. Hozier. 1985. Cytogenetic characterization of the L5178Y/TK$^{+/-}$–3.7.2C mouse lymphoma cell line. *Mutat. Res.* **147:** 243.

Shelby, M.D. and S. Stasiewicz. 1984. Chemicals showing no evidence of carcinogenicity in long-term, two-species rodent studies: The need for short-term test data. *Environ. Mutagen.* **6:** 871.

Tennant, R.W., B.H. Margolin, M.D. Shelby, E. Zeiger, J.K. Haseman, J. Spalding, W. Caspary, M. Resnick, S. Stasiewicz, B. Anderson, and R. Minor. 1987. Prediction of chemical carcinogenicity in rodent from in vitro genetic toxicity assays. *Science* **236:** 933.

The Use of the *hgprt* versus *gpt* Locus for Quantitative Mammalian Cell Mutagenesis

ABRAHAM W. HSIE
Division of Environmental Toxicology
Department of Preventive Medicine and
Community Health
The University of Texas Medical Branch
Galveston, Texas 77550

OVERVIEW

The Chinese hamster ovary (CHO) cell clone K_1-BH_4 has been used for about 10 years to quantify environmental-agent-induced toxicity and mutation at the hypoxanthine-guanine phosphoribosyl transferase (*hpgrt*) locus on the X chromosome (the CHO/HGPRT assay). The CHO/HGPRT assay fulfills the cellular, genetic, and biochemical criteria for a specific-locus mutational assay. The quantitative nature of this assay has been utilized to study the structure-activity (mutagenicity and cytotoxicity) relationships of various classes of chemicals. In our studies with physical agents, we found that this assay can quantify the mutagenic effects of UV light but not X-irradiation. Recently, we transformed an HGPRT-deficient CHO cell (X3/5) with the plasmid vector pSV2*gpt* and isolated a transformant, AS52. AS52 cells carry a single copy of the xanthine-guanine phosphoribosyl transferase (*gpt*) gene (the bacterial equivalent of the mammalian gene *hgprt*) stably integrated into, presumably, one of the host autosomes. Our studies demonstrated that AS52 cells are hypersensitive to physical and chemical agents that exert their biological effects via reactive oxygen species that ultimately lead to oxidative DNA damage. The apparent hypermutability of AS52 cells probably results from a higher recovery of multilocus deletion mutants in AS52 cells than in K_1-BH_4 cells, rather than a higher yield of induced mutants.

INTRODUCTION

It has been 8 years since we wrote a volume to assess the progress of mammalian cell mutagenesis (Hsie et al. 1979a). That volume was titled as *Mammalian Cell Mutagenesis: The Maturation of the Test Systems*, since mammalian cell mutational assays had developed (matured) to the point where they were ready to be used for studies of mechanisms of gene mutation and screening for mutagenic activity of environmental chemicals. During the past 8 years, notable progress has been made in this field.

This paper reports on aspects of our studies using CHO cells in mutagenesis in my research and on mutagenicity testing. I discuss the development of the mutational assay at the *hgprt* locus in CHO cells (CHO/HGPRT assay) (Hsie et al. 1975a, 1978, 1981), the utility of the CHO/HGPRT assay including its strength and weakness, and the development of the CHO-AS52/GPT system (Tindall et al. 1984, 1986) and its use in studying oxidative DNA damage and, possibly, the recovery of mutants resulting from multilocus deletion (Hsie et al. 1986; Stankowski and Hsie 1986; Stankowski et al. 1986).

RESULTS

The CHO/HGPRT Assay: The Strength

After studying the role of cyclic nucleotides in growth differentiation and malignancy, "reverse transformation," in mammalian cells (Hsie and Puck 1971; Hsie et al. 1971; Puck et al. 1972), I realized that the auxotrophic system of Chinese hamster cells offered a unique means to induce some nutrient-requiring mutants with definable biochemical alterations (Puck and Kao 1967; Chu and Malling 1968); however, quantification of auxotrophic mutation was hampered by the findings, that only a small fraction of the recovered colonies were bona fide mutants. The HGPRT system (Chu and Malling 1968) appeared to be highly promising if 6-thioguanine (TG), instead of 8-azaguanine, was used as a selective agent (Chasin 1973).

Our first experiment demonstrated that mutation to TG resistance (TG^r) as induced by ethyl methanesulfonate (EMS) is quantifiable. Treatment of CHO cell clone K_1-BH_4 (referred to as CHO-K_1-BH_4 cells or K_1-BH_4 cells), with EMS (25–800 μg/ml) for 16 hours caused a concentration-dependent exponential lethality after a distinct shoulder region, where there was no appreciable loss of cell survival. Mutation induction occurred over the entire concentration range (Hsie et al. 1975a). This experiment was conducted using a protocol based on the best available information and technology in 1974. During the next 5 years, various experiments were performed to document the choice of the standard experimental conditions necessary for quantifying mutation at the *hgprt* locus was correct (O'Neill et al. 1977a,b; Hsie et al. 1978, 1979b, 1981; O'Neill and Hsie 1979a,b).

It is noteworthy that we demonstrated a linear dose-response for mutation is induced by EMS in the near-diploid subclones of K_1-BH_4 and 51-11 but not in the near-tetraploid 5T-111 subclone (Chasin 1973; Hsie et al. 1977). A single-hit event is necessary to induce a specific-locus mutation in the diploid cells, since these cells contain a single functional *hgprt* gene in an active X chromosome. In the tetraploid cells with two functional *hgprt* genes localized on two active X chromosomes, a two-hit event, which is expected to occur at a frequency near background level even at high mutagen concentrations, is required to induce a mutation.

Utility of the CHO/HGPRT Assay: A Weakness

The quantitative nature of the CHO/HGPRT assay has been used for studies of structure-activity (mutagenicity and cytotoxicity) relationship of various classes of chemicals as tabulated in the Gene-Tox work group reports of the U.S. Environmental Protection Agency (EPA). So far, the mutagenic activity of over 100 chemicals has been reported and evaluated (Hsie 1981; Hsie et al. 1981; Li et al. 1987).

In our first experiment with physical agents, we found that irradiation with UV light caused a dose-dependent increase of cytotoxicity and mutagenicity (Hsie et al. 1975b). We found that X-irradiation is weakly mutagenic, causing a dose-dependent increase of mutation to TG^r in the range of 200–800 rads (delivered at 100 rads/minute). The shape of the dose-response curve could not be adequately defined from ten repeated experiments due to the weak mutagenicity of this agent in the CHO/HGPRT assay (O'Neill et al. 1977a).

Development of the CHO-AS52/GPT System: A Complementation

Demonstration that a plasmid vector such as pSV2*gpt* could be introduced into an appropriate mammalian cell and expressed in the new host offered the possibility of studying mutagenesis at the molecular level (Mulligan and Berg 1981). We transformed an X-ray-induced, HGPRT-deficient X3/5 subclone of K_1-BH_4 cells with pSV2*gpt* and produced transformants containing various copies of the *gpt* gene transformant (referred to as AS52) that carries a single copy of the *gpt* gene stably integrated into the high-molecular-weight DNA of the host (Tindall et al. 1984, 1986).

When irradiated with X-rays, we found that although X-irradiation is equitoxic to AS52 cells and to the parental CHO-K_1-BH_4 cells, X-rays are approximately ten times more mutagenic to AS52 cells than to K_1-BH_4 cells. Thus, AS52 cells are hypersensitive to X-ray-induced mutations. Using Southern blot analyses, we found that one X-irradiation-induced and one spontaneously arising TG^r clone exhibited a deletion in the structural gene of *gpt* (Tindall et al. 1984).

Later, we expanded the X-ray experiments and extended the study to include EMS, ICR-191, and UV. We found that EMS, ICR-191, and UV produced no detectable alteration of the *gpt* and *hgprt* gene in AS52 cells and K_1-BH_4 cells, respectively, as determined by Southern blot analyses. These agents produce dose-response curves for cytotoxicity and mutagenicity in AS52 cells similar to those of the parental K_1-BH_4 cells (Stankowski and Hsie 1986; Stankowski et al. 1986). However, X-rays produced mostly deletions of the *gpt* and *hgprt* gene. Again, we found that AS52 cells were much more sensitive than K_1-BH_4 cells to the mutagenic effects of X-irradiation.

Evidence for Reactive Oxygen Species Inducing Mutations in Mammalian Cells

Since AS52 cells can sensitively quantify the mutagenic effects of X-rays, AS52 cells were used as a biological dosimeter to study the role of reactive oxygen species in mutation induction.

To ensure that the hypermutability of AS52 cells is not unique to X-rays, we proceeded to study the effects of neutrons (at a dose-rate of 2.33–11.7 rads/minute) and found that the pattern of X-ray-induced differential mutagenic response between K_1-BH_4 cells and AS52 cells was produced also by neutrons. Like X-rays, neutrons resulted in nearly identical lethality to both cell types; however, neutrons are ten times more effective in inducing HGPRT mutation in AS52 cells than in K_1-BH_4 cells (Fig. 1) (Hsie et al. 1986).

Reactive oxygen species are being implicated in the toxic action of numerous chemicals (Dormandy 1983). If reactive oxygen species were to mediate the mutagenic effects of radiation such as X-rays and neutrons, then radiomimetic chemicals such as streptonigrin and bleomycin, which are known to produce superoxide and hydroxy radicals, would be expected to be equitoxic to both cell types and more mutagenic to AS52 cells than to K_1-BH_4 cells. Likewise, oxidizing compounds such as potassium superoxide and hydrogen peroxide, which are reactive oxygen species, should be more mutagenic to AS52 cells and equitoxic to both cell types. Our experiments fulfilled such expectations (Fig. 1). Early experiments showed that agents such as EMS, ICR-191, and UV light, which do not produce reactive oxygen species, do not elicit differential mutagenic response in both cell types (Stankowski and Hsie 1986; Stankowski et al. 1986) (Table 1). Taken together, these experiments support the view that reactive oxygen species induce gene mutations in mammalian cells.

DISCUSSION

Demonstration that reactive oxygen species induce differential mutagenicity in K_1-BH_4 cells and AS52 cells has several implications.

First, the differential mutagenic effect is specific to agents that produce reactive oxygen species.

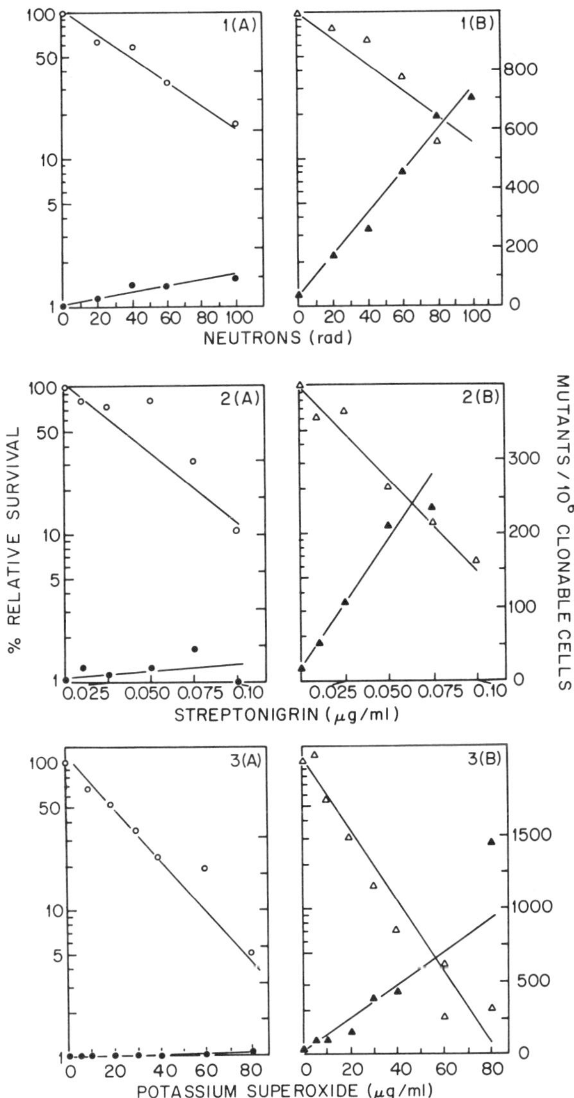

Figure 1

The cytotoxic (○, △) and mutagenic (●, ▲) effects of neutron-irradiations (*1*), of streptonigrin (*2*) and of potassium superoxide (*3*) to CHO-K_1-BH_4 cells (*A*) and to AS52 cells (*B*). (Adapted from Hsie et al. 1986.)

Second, the high mutagenic response of AS52 cells to radiation and radiomimetic chemicals suggests that these cells are sensitive enough to be used to define the dose-response relationship from experiments using a larger

Table 1
Interrelationships of Reactive Oxygen Species, Hypermutability, and Oxidative DNA Damage/Deletion Mutation

Agents	Major mode of actions: production of reactive oxygen species	hypermutability to AS52 cells	induction of oxidative DNA damage and/or deletion mutation
Physical			
X-rays	+	+	+
neutrons	+	+	+
UV light	−	−	−
Chemical			
radiomimetic			
streptonigrin	+	+	+
bleomycin	+	+	+
oxidizing			
potassium superoxide	+	+	+
hydrogen peroxide	+	+	+
missense mutagenic			
EMS	−	−	−
ICR-191	−	−	−

Data from Hsie et al. (1986).

number of radiation doses and different dose rates. This is an important concern from the viewpoint of radiation protection because the existing mammalian cell mutational assays ". . . have neither the resolution nor the precision to make clear whether, in some cases, a zero slope is obtained as opposed to a positive initial slope . . ." (NCRP 1980).

Third, since X-ray induces predominantly deletion mutations (Stankowski and Hsie 1986), the low mutagenic activity of X-irradiation in K_1-BH_4 cells, where the *hgprt* gene exists in a hemizygous state, is likely due to lethal events associated with multilocus deletions that reduce the number of recoverable TG^r mutants. In AS52 cells, the *gpt* gene may be stably integrated into an autosome. The genes localized in that autosome probably exist in the diploid state. Even a multilocus deletion induced by X-rays could produce a hemizygous state in the essential genes that may flank the *gpt* gene (in contrast to a loss of these essential genes with a deletion larger than the *hgprt* gene in the K_1-BH_4 cells), resulting in a higher probability of recovering induced TG^r mutants. By inference from experiments with X-rays, the low mutagenic response in K_1-BH_4 cells to other agents that produce reactive oxygen species may also be due to low mutant recovery, rather than to reduced mutation induction. This leads to the prediction that agents that are known to produce reactive oxygen species are expected to produce predominantly deletion mutation and/or chromosomal aberration.

Fourth, demonstration that mutants induced by radiation and reactive oxygen producers may not survive to be scored as mutants in the K_1-BH_4 assays implies that assessment of genetic risk for these types of agents using hemizygous genetic markers is likely to be underestimated. Thus, there may be a need to reevaluate assessment of genetic risk relative to ionizing radiation.

Fifth, it is known that many naturally occurring chemicals can be reduced by intracellular reductases and, in turn, reduce oxygen to reactive oxygen species via a redox cycling mechanism (Kappus and Sies 1981). These natural compounds together with background radiations and man-made reactive oxygen producers may ultimately cause spontaneous (deletion) mutations and possibly also cancer and aging.

Our discussion of the relationship between mutation induction and mutant recovery relied heavily on the data using Southern blot analysis that has a sensitivity of detecting a restriction fragment of 100–200 bp (Tindall et al. 1984; Stankowski and Hsie 1986; Stankowski et al. 1986). It would be more convincing if a large number of TG^r mutants induced by various mutagenic agents were analyzed at the level of nucleotide sequence, using recently developed techniques (Myers et al. 1985; R.A. Gibbs and C.T. Caskey, in prep.). We plan to perform such analyses in collaboration with R. Gibbs and C.T. Caskey of Baylor College of Medicine.

SUMMARY

Five years ago, we observed the differential mutagenic effect of X-irradiation on K_1-BH_4 and AS52 cells. The concept that mutant recovery is the basis for the differential mutagenic effect strengthened gradually during the intervening years. Since the differential mutagenicity was demonstrated using neutrons, streptonigrin, bleomycin, potassium superoxide, and hydrogen peroxide in 1985, I believe that it is timely to address the role of reactive oxygen species in inducing gene mutation and to emphasize the relationship of mutation induction and mutant recovery.

I predicted that those agents that elicit differential mutagenicity in the K_1-BH_4 and AS52 assays would produce the so-called chromosomal type mutation, resulting in small thymidine kinase-deficient mutant colonies in the mouse lymphoma L5178 cell mutational assay (Clive et al. 1983). This prediction has been fulfilled in L5178 cells (Clive et al. 1983; Evans et al. 1986; Moore et al. 1986) and in human lymphoblast cells (Yandell et al. 1986).

This concept on the relationship of mutation induction and mutant recovery in mammalian cells is reminiscent of the observations made about 20 years ago in mice (Russell 1971; Rinchik et al. 1986) and in *Neurospora* (Webber and de Serres 1965; Brockman et al. 1984). It is gratifying to see that what is true for a simple CHO cell is also true for a sophisticated mouse.

ACKNOWLEDGMENTS

I thank R.L. Schenley and L.C. Waters of Oak Ridge National Laboratory (ORNL) for reviewing the draft, J.F. Young (ORNL) for preparing the manuscript as well as F.T. Kenney (ORNL) and M.A. Waters (EPA) for hosting me at ORNL and the EPA, respectively. Support as a Distinguished Visiting Scientist of the EPA was under interagency agreement DOE-40-1601-85 and with the Department of Energy under contract (DE-AC05-840R2140) with Martin Marietta Energy Systems, Inc. Although the research described in this paper has been funded by the EPA, it has not been subjected to EPA review and therefore does not necessarily reflect the views of the EPA, and no official endorsement should be inferred.

REFERENCES

Brockman, H.E., F.J. de Serres, T.M. Ong, D.M. DeMarini, A.J. Katz, A.J.F. Griffith, and R.S. Stafford. 1984. Mutation tests in *Neurospora crassa*. *Mutat. Res.* **133:** 87.

Chasin, L.A. 1973. The effect of ploidy on chemical mutagenesis in cultured Chinese hamster cells. *J. Cell. Physiol.* **82:** 299.

Chu, E.H.Y. and H.V. Malling. 1968. Chemical induction of specific locus mutations in Chinese hamster cells *in vitro*. *Proc. Natl. Acad. Sci.* **61:** 1306.

Clive, D., R. McCuen, J.F.S. Spector, C. Piper, and K.H. Mavournin. 1983. Specific gene mutations in L5178Y cells in culture. *Mutat. Res.* **115:** 225.

Dormandy, T.L. 1983. An approach to free radicals. *Lancet* **II** (8357)**:** 1010.

Evans, H.H., J. Mencl, M.F. Horng, M. Ricanati, C. Sanchez, and J. Hozier. 1986. Locus specificity in the mutability of L5178Y mouse lymphoma cells: The role of multilocus lesions. *Proc. Natl. Acad. Sci.* **83:** 4379.

Hsie, A.W. 1981. Structure-mutagenicity analysis with the CHO/HGPRT system. *Food Cosmet. Toxicol.* **19:** 617.

Hsie, A.W. and T. Puck. 1971. Morphological transformation of Chinese hamster cells by dibutyryl adenosine cyclic 3′,5′-monophosphate and testosterone. *Proc. Natl. Acad. Sci.* **68:** 358.

Hsie, A.W., C. Jones, and T.T. Puck. 1971. Further changes in differentiation state accompanying the conversion of Chinese hamster cells to fibroblastic form by dibutyryl adenosine cyclic 3′,5′-monophosphate and hormones. *Proc. Natl. Acad. Sci.* **68:** 1648.

Hsie, A.W., J.P. O'Neill, and V.K. McElheny, eds. 1979a. The maturation of test systems. *Banbury Rep.* **2**.

Hsie, A.W., P.A. Brimer, R. Machanoff, and M.H. Hsie. 1977. Further evidence for the genetic origin of mutations in mammalian somatic cells: The effects of ploidy level and selection stringency on dose-dependent chemical mutagenesis to purine analogue resistance in Chinese hamster ovary cells. *Mutat. Res.* **45:** 271.

Hsie, A.W., P.A. Brimer, T.J. Mitchell, and D.G. Gosslee. 1975a. The dose-response relationship for ethyl methanesulfonate-induced mutations at the hypoxanthine-guanine phosphoribosyltransferase locus in Chinese hamster ovary cells. *Somatic Cell Genet.* **1:** 247.

———. 1975b. The dose-response relationship for ultraviolet light-induced mutations at the hypoxanthine-guanine phosphoribosyltransferase locus in Chinese hamster ovary cells. *Somatic Cell Genet.* **1:** 383.

Hsie, A.W., J.P. O'Neill, J.R. San Sebastian, and P.A. Brimer. 1979b. The CHO/HGPRT mutation assay: Progress with quantitative mutagenesis and mutagen screening. *Banbury Rep.* **2:** 407.

Hsie, A.W., D.A. Casciano, D.B. Couch, D.R. Krahn, J.P. O'Neill, and B.L. Whitfield. 1981. The use of Chinese hamster ovary cells to quantify specific locus mutation and to determine mutagenicity of chemicals. *Mutat. Res.* **86:** 193.

Hsie, A.W., L. Recio, D.S. Katz, C.Q. Lee, M. Wagner, and R.L. Schenley. 1986. Evidence for reactive oxygen species inducing mutations in mammalian cells. *Proc. Natl. Acad. Sci.* **83:** 9616.

Hsie, A.W., J.P. O'Neill, D.B. Couch, J.R. San Sebastian, P.A. Brimer, R. Machanoff, J.C. Riddle, A.P. Li, J.C. Fuscoe, M.L. Forbes, and M.H. Hsie. 1978. Quantitative analysis of radiation- and chemical-induced cellular lethality and mutagenesis in Chinese hamster ovary cells. *Radiat. Res.* **76:** 471.

Kappus, H. and H. Sies. 1981. Toxic drug effects associated with oxygen metabolism: Redox cycling and lipid peroxidation. *Experientia* **37:** 1233.

Li, A.P., R.S. Gupta, R.H. Heflich, and J.S. Wassom. 1987. A review and analysis of the Chinese hamster ovary cells/hypoxanthine-guanine phosphoribosyltransferase assay to determine the mutagenicity of chemical agents. *Mutat. Res.* (in press).

Moore, M.M., A. Amtower, G.H.S. Strauss, and C. Doerr. 1986. Genotoxicity of gamma-irradiation in L5178 mouse lymphoma cells. *Mutat. Res.* **174:** 149.

Mulligan, R.C. and P. Berg. 1981. Selection for animal cells that express the *Escherichia coli* gene coding for xanthine-guanine phosphoribosyltransferase. *Proc. Natl. Acad. Sci.* **78:** 2072.

Myers, R.M., Z. Larin, and T. Maniatis. 1985. Detection of single base substitutions by ribonuclease cleavage at mismatches in RNA:DNA duplex. *Science* **230:** 1242.

National Council on Radiation Protection and Measurements (NCRP). 1980. Influence of dose and its distribution in time and dose-response relationship for low-LET radiations. *NCRP Rep.* **64:** 80.

O'Neill, J.P. and A.W. Hsie. 1979a. The CHO/HGPRT mutagenicity assay: Experimental procedure. *Banbury Rep.* **2:** 55.

———. 1979b. The CHO/HGPRT mutagenicity assay: Adaptation for mutagen screening. *Banbury Rep.* **2:** 311.

O'Neill, J.P., P.A. Brimer, R. Machanoff, G.P. Hirsch, and A.W. Hsie. 1977a. A quantitative assay of mutation induction at the hypoxanthine-guanine phosphoribosyltransferase locus in Chinese hamster ovary cells (CHO/HGPRT system): Development and definition of the system. *Mutat. Res.* **45:** 91.

O'Neill, J.P., D.B. Couch, R. Machanoff, J.R. San Sebastian, P.A. Brimer, and A.W. Hsie. 1977b. A quantitative assay of mutation induction at the hypoxanthine-guanine phosphoribosyltransferase locus in Chinese hamster ovary cells (CHO/HGPRT system): Utilization with a variety of mutagenic agents. *Mutat. Res.* **45:** 103.

Puck, T.T. and F.T. Kao. 1967. Treatment of 5-bromodeoxyuridine and visible light for isolation of nutritionally deficient mutants. *Proc. Natl. Acad. Sci.* **58:** 1227.

Puck, T.T., C.A. Waldren, and A.W. Hsie. 1972. Membrane dynamics in the action of dibutyryl adenosine 3′,5′-cyclic monophosphate and testosterone on mammalian cells. *Proc. Natl. Acad. Sci.* **69:** 1943.

Rinchik, E.M., L.B. Russell, N.G. Copeland, and N.A. Jenkins. 1986. Molecular genetic analysis of the dilute-short-ear (d-se) region of the mouse. *Genetics* **112:** 321.

Russell, L.B. 1971. Definition of functional unit in a small chromosomal segment of the mouse and its use in interpreting the nature of radiation-induced mutations. *Mutat. Res.* **11:** 107.

Stankowski, L.F., Jr. and A.W. Hsie. 1986. Quantitative and molecular analyses of radiation-induced mutation in AS52 cells. *Radiat. Res.* **105:** 37.

Stankowski, L.F., Jr., K.R. Tindall, and A.W. Hsie. 1986. Quantitative and molecular analyses of ethyl methanesulfonate- and ICR-induced mutation in pSV*gpt*-transformed CHO cells. *Mutat. Res.* **160:** 133.

Tindall, K.R., L.F. Stankowski, Jr., R. Machanoff, and A.W. Hsie. 1984. Detection of deletion mutations in pSV*gpt* transformed cells. *Mol. Cell. Biol.* **4:** 1411.

———. 1986. Analyses of mutation in pSV*gpt*-transformed CHO cells. *Mutat. Res.* **160:** 121.

Webber, B.B. and F.J. de Serres. 1965. Induction kinetics and genetic analysis of x-ray-induced mutations in the AD-3 region of *Neurospora crassa*. *Proc. Natl. Acad. Sci.* **53:** 430.

Yandell, D.W., T.P. Pryja, and J.B. Little. 1986. Somatic mutations at a heterozygous autosomal locus in human cells occur more frequently by allele loss than by intragenic structural alterations. *Somatic Cell Mol. Genet.* **12:** 255.

Comments

Glickman: Have you used pseudogene probes like DQ to look at your blots? Have you looked for conversion in your system?

Nicklas: No. They are generally over maybe five amino acids, and I don't think that's something you can pick up.

Glickman: So they are called conversion events, but they are actually a very small sequence of the regions?

Nicklas: (Nods affirmatively)

Tindall: Gerry, are you cloning deletion endpoints or the region in the Z4 chromosome that yields the high-frequency deletions.

Adair: Obviously, that is an important thing to do. Another thing we are very interested in trying, is to do in situ hybridizations like John Hozier. I think that will be important in order to confirm what we see in these gene-mapping studies. All of our evidence so far has been based on analyses of somatic cell hybrids constructed using heterozygous or hemizygous cell lines.

Essentially, what we have done has been to utilize two *aprt*-linked electrophoretic markers, *mdh*2 and *ldhA*. In CHO cells, *mdh*2 is hemizygous, with a single allele on chromosome Z7, whereas alleles for *ldhA* are expressed on both the Z3 and the Z4 chromosomes. We have isolated a cell line that is doubly marked; it has an altered electrophoretic mobility mutation for the Z7-*mdh*2 locus, as well as a heterozygous electrophoretic mobility shift mutation for the Z4-*ldhA* locus. This doubly marked line was the parental line that we used to isolate our series of 16 *aprt* functional heterozygotes. So, if you hybridize each of those lines with an $aprt^{-/-}$ cell line and then look for segregation of the chromosome carrying the functional *aprt* gene, you can ask, "Does the *aprt* gene cosegregate with the MDH2 marker or the LDHA marker?" That was our basic approach.

Glickman: My question is for Dr. Hsie. You described oxygen effects that are clearly very important, but you also have indicated, or made the suggestion, that perhaps they would make good dosimetry methods for a variety of reasons. Have you begun to think about the molecular side of the dosimetry?

Hsie: Oxygen chemistry has become so sophisticated that the chemists will never want to pin down anything for you. You are lucky to show that this particular oxygen species responded positively. In a way, I think that will be the goal eventually—to correlate quantitatively and also qualitatively, if possible, which oxygen radicals or hyperoxide radicals cause which type of damage. There is a growing evidence for 8-hydroxy-deoxy guanine as a modified base resulting from the reaction between reactive oxygen species producer(s) and cellular DNA.

Hutchinson: The interesting feature is that Abe (Hsie) shows that the superoxide radical, for example, is much more mutagenic in AS52 cells than in CHO-K1 cells. That is unexpected because superoxide would be expected to form only simple products in DNA that would give point mutations, as do agents such as UV and EMS. These latter agents have similar mutagenic effects on AS52 and on CHO-K1 cells. Abe is just showing us how little we know at the molecular level. This will have to be followed up.

Hsie: In response to the question raised by Barry Glickman and Frank Hutchinson, I don't have a quantitative answer, but the concerns are there. I think AS52 cells appear to offer a potential for a very useful biological dosimeter, at least at this stage of the game, to qualitatively detect these reactive oxygen species that produce reactions leading to the observed genetic damage.

Hutchinson: There is at least some evidence that the sizes of the deletions occurring spontaneously are quite different from those induced by X-rays. If that is true, then that would tend to suggest a different mechanism.

Hsie: Yes, you are absolutely right. In fact, the data Tindall analyzed in our place were only of a couple of mutants. Stankowski extended his data to cover about 200. Still, I must emphasize first that they were using a Southern blot; a lot of small deletions will never be seen. In our future collaboration with Richard Gibbs and Tom Caskey using their rapid DNA sequence technique, Frank's (Hutchinson) questions will hopefully be answered.

de Serres: The whole problem in working with spontaneous mutations is to try to decide whether the mutants that you're collecting are a consequence of normal DNA replication and repair or whether they are due to artifacts in the medium. Very early, we were able to show in *Neurospora*, for example, that if you autoclave the medium too long and then use it right away, your spontaneous background *ad-3* mutant frequency increases. Now, obviously, those are not spontaneous mutations, but they are induced by hydrolysis products in the medium.

Also, when you go from prokaryotes to eukaryotes, because of the obvious differences in chromosome structure, again it's apples and oranges. The ground rules for prokaryotes are probably completely different from eukaryotes. In this particular forum we are going to try to resolve differences between spontaneous and induced specific-locus mutations in eukaryotes. I think we have a big enough problem just going from lower eukaryotes such as with *Neurospora* to where some of you are, especially Lee (Russell), with regard to reconciling what we think we know of spontaneous and induced specific-locus mutations in higher eukaryotic organisms. I think there are fundamental differences with regard to mutagenesis at the molecular level in prokaryotes versus eukaryotes. I think that it is very difficult to develop ground rules that will apply to all model systems.

Hsie: One of the things which I talked with Herman Brockman is perhaps whether a particular locus we are looking at is really hemizygous for the particular chromosome. If you only delete that particular gene, actually that chromosome is still heterozygous with a hemizygous gene. Let us talk about minus and zero. When the genes do not exist to begin with, you call that 0. When you have it and then lose it, you call that −.

The second thing is when you deal with the V79 cell, it is derived from the male Chinese hamster lung tissue. Functionally and structurally, there is only one *hgprt* gene in V79 cells. However, consider the female origin of the CHO cell. It should have two chromosomes, hence, two *hgprt* genes, but only one *hgprt* gene is active. We should consider to use +/0 to describe the situation of *hgprt* gene in V79 cells and +/−, indicating inactive, to describe the situation in CHO cells.

Chu: You have compared the X-ray response in the induction of gene mutation and cell killing in two situations. One is the cell with the resident *hprt* gene and the other is the cell into which the *gpt* sequence was introduced and incorporated. Survival of the two types of cells should be the same because they are of the same host cell. Yet, you showed that for mutation induction there are two curves. One is a quadratic response for *gpt*, and the other is a more linear response for the resident gene. I wonder what is the reason for the two curves?

Hsie: When Stankowski analyzed the molecular basis of the mutants induced by X-rays, he picked the mutant clones randomly. One might expect that at a low dose most mutants are of base damage type; at a high dose, it may be predominantly deletion mutations. That might be expected because, purely from a dose/response term, you can see how much smoother it is at the beginning, but you can get all kinds of scattering at high doses. Southern blot is enough to answer such questions. As I said earlier in answering Frank Hutchinson's comments, I feel that data really need to be analyzed using a good DNA-sequencing technique of the particular gene. Probably, Jack Little would go even further and analyze the flanking DNA sequences of the particular locus that is being selected.

de Serres: One of the problems we had organizing this particular conference was one of terminology. Everybody has been using the same terms to refer to different events. One of the ground rules that I think we need to establish very early on here is what we are talking about when we use particular terminology.

Glickman: What we have tried to do, because we have had to face this problem, is to make a definition that I would like to add here, because it's very complementary. We decided to say that if you can't see it on a Southern blot, it's a point mutant. Then you sequence it, and you discover it's a 17-base-pair deletion; it's still a point mutant. We have defined that in some papers with a footnote saying, "Today our definition of point mutants includes things that are too small to be differentiated on a Southern blot."

Hutchinson: How would you classify a base-substitution mutation in an enhancer?

de Serres: I would call it a point mutation.

Glickman: Then you have to say gene, regulatory regions, and so forth, and so you are already outside the gene as defined. That was my point. You have to be broader than intragenic, because we now know that a gene is defined as something that produces a protein. That is an insufficient definition of the genetic target with which we are trying to deal.

Clive: I would like to stop at point one. The reason I used the term "gene mutations" is that John Hozier's lab appears to be showing that most of our large-colony mutants appear to be complete deletions of the gene, so they are not intragenic nor are they point mutations.

Hsie: That's multilocus.

Hozier: It's possibly multilocus mutations. We don't know the extent of the genetic damage in most of these mutants yet.

Chasin: We know they're large deletions. Why not just leave the term at that?

de Serres: Then they're multilocus deletions. The term multilocus deletions should be reserved for this kind of situation, where you have simultaneous deletion of the gene of interest, as well as other genes in the immediately adjacent area.

Glickman: There's a problem here. Suppose you have a 500-base-pair deletion, and it happens to be all within *aprt*? That's fine. I'm calling those deletions.

de Serres: They're intralocal deletions.

Glickman: Just straight deletions. Suppose I have a 50,000-base-pair deletion in *aprt*; it's now a multilocus deletion. If it happened to be in *hprt*, where you've got 44,000 base pairs and all that intragenic garbage, it's now an intragenic deletion by your definition, but in my case, it's all at the locus.

de Serres: But what you're saying is that unless you have the molecular data you can't give anything a name. I am trying to establish ground rules for referring to specific-locus mutations in order that we can communicate effectively. Then, when we have the molecular data, we will know to what extent the initial classification works. If all our $tk^{-/-}$ colonies turn out to be point mutations, okay! Then we're led to one conclusion, but if, indeed, some of the large colonies can result from multilocus deletion, then obviously we've got some problems. We don't know mechanistically in most loci how to relate these terms to what's going on at the molecular level. Until we can do that, we have to have some kind of mutually agreed upon terminology so that we can communicate more effectively.

Hutchinson: I would like to strongly support Barry's point of using operational definitions. You can think in terms of three categories. You could talk about point mutations, which don't change the Southern blot. As a second category, there are deletions in which at least part of the gene for which you are probing has been left out of the genome. The third term is gross rearrangements, in which the connectivity of the genome has been changed.

Auletta: But you do need to distinguish between intragenic deletions and deletions that affect more than one gene.

Hutchinson: At the moment, it's not clear that the technology is up to doing this.

Glickman: But the phenotypes can do it.

Hutchinson: Sure. They can, but they might not.

Little: I was going to say that, when you talk about multilocus deletions, you may well mean loss of genetic material at multiple loci. However, that may not be purely deletional; that might be a recombinational event. So, in a sense, deletion is an improper term.

Hsie: Also, if you delete a nonessential flanking gene of your genetic locus of interest, you may not be sure that you have lost that nonessential gene.

Differential Recovery of Mutants at Different Loci

Specific-locus Studies with Two-component Heterokaryons of *Neurospora crassa* Predict Differential Recovery of Mutants in Mammalian Cells

FREDERICK J. DE SERRES
Center for Life Sciences and Toxicology
Research Triangle Institute
Research Triangle Park, North Carolina 27709

OVERVIEW

Studies on specific-locus mutations in various eukaryotic organisms have shown that they result from both gene mutations and multilocus deletions. By using two-component heterokaryons of *Neurospora crassa*, it is possible to recover both classes of specific locus mutations. In such heterokaryons, heterozygous in the adenine-3 (*ad-3*)-region for the two closely linked loci *ad-3A* and *ad-3B*, it is possible to recover gene mutations at each locus as well as multilocus deletions covering each locus or both loci simultaneously. X-irradiation produces both classes of specific locus mutations; the relative frequency of each class is dependent on the X-ray dose. Gene mutations increase linearly, and multilocus deletions increase as the square of the X-ray dose. Homology tests with genetic markers with supplementable or nonsupplementable loss of function in the genetic regions immediately adjacent to the *ad-3* region have shown, for example, that multilocus deletions in the *ad-3* region may also cover the histidine-3 (*his-3*) and lysine-4 (*lys-4*) loci located to the left, other markers in the "X" region between the *ad-3A* and *ad-3B* loci, and the nicotinic acid (*nic-2*) locus located to the right. The spectrum of specific locus mutations in the *ad-3* region obtained after X-ray treatment of two-component heterokaryons of *Neurospora* is identical with that obtained in the *d* (dilute) *se* (short-ear) region of the mouse. Growth-rate studies on X-ray-induced specific locus mutations at the *ad-3A* or *ad-3B* loci have shown that gene mutations at each locus are completely recessive, and that heterokaryons between them grow at wild-type rate. However, heterokaryons between multilocus deletions at either the *ad-3A* locus or the *ad-3B* locus, alone or in combination, show heterozygous effects resulting in a marked reduction in growth rate. In addition, experiments with various environmental chemicals to induce specific locus mutations in the *ad-3* region

Banbury Report 28: Mammalian Cell Mutagenesis

of *Neurospora* have shown that in some cases, only gene mutations are induced; in all other cases, both classes of specific locus mutations are recovered. The relative frequencies of *ad-3* mutants in each class are mutagen-dependent.

INTRODUCTION

As a result of the pioneering work of H.J. Muller (1927), there has been concern about the genetic effects of ionizing radiation on man. This early work showed that X-rays could produce genetic damage in the germ cells of *Drosophila* that was transmitted to the progeny of the irradiated parents. In the late 1960s and early 1970s this concern was extended to environmental chemicals as a result of exploratory work, primarily in *Salmonella* and *Escherichia coli*, which showed that there were mutagens in just about every class of chemical agents including: drugs, cosmetics, foods, beverages, food additives, pesticides, and industrial and household chemicals (see de Serres 1976). This concern over the genetic effects of chemicals gave rise to the new field of genetic toxicology that focuses on the rapid identification of chemicals with mutagenic activity, as well as the assessment of their potential for causing adverse human health effects in terms of cancer, birth defects, and genetic disease.

With regard to adverse human health effects, there are three main types of genetic damage of concern: (1) the production of specific locus mutations, (2) chromosome aberrations, and (3) aneuploidy. In eukaryotic organisms, specific locus mutations result from both gene mutations and chromosomal events affecting multiple loci (multilocus deletions). Single gene mutations result from subtle changes within the gene ranging from single-base-pair substitutions and frameshift mutations (point mutations) to intralocal deletions covering many base pairs. In these cases the gene is altered but intact. In the case of multilocus deletions, the gene in question is functionally inactivated (usually by physical removal from the chromosome by chromosome breakage events), usually along with other genes in the immediately adjacent regions. X-irradiation produces both classes of specific locus mutations; the relative frequency of each class depends on the X-ray dose (Webber and de Serres 1965).

Various committees, such as the Biological Effects of Ionizing Radiation (BEIR) Committee (BEIR III 1980) and the Committee on Chemical Environmental Mutagens (CCEM) Report (1983), have focused on the risk of the production of gene mutations. In these reports and others that have been concerned with genetic risk assessment, whereas attention has been focused on the production of gene mutations, the genetic risk of successful transmission of multilocus deletions has been essentially ignored. There is compelling evidence for heterozygous effects of X-ray-induced specific locus mutations, resulting from multilocus deletion in *Neurospora* (de Serres 1965), in *Drosophila* (Temin 1978; Shukla and Auerbach 1981), and in the mouse (Russell et al. 1966; Russell 1971).

As a result of screening tests on environmental chemicals, we now know that many of these chemicals produce high frequencies of mutations affecting multiple loci in *Neurospora* at the *ad-3A* and *ad-3B* loci (de Serres, this paper; F.J. de Serres and J.E. Brockman, unpubl.) and in small colonies (presumptive multilocus deletions) in strain L5178Y of mouse lymphoma cells in culture at the $tk^{+/-}$ locus (Moore et al. 1985a,b and this volume).

The specific locus test in the mouse (Russell 1954) is an excellent example of a model system that was developed and used for many years to detect, recover, and analyze the genetic damage produced by ionizing radiation and chemical mutagens. The seven loci at which mutations can be detected with this system are those that affect morphological characters in the mouse. Particularly useful loci are dilute (*d*) and short ear (*se*), which are very closely linked. As a result of this close linkage, *d*, *se*, and *d se* mutants are recovered. Gene mutations can be expected to give rise to mutants, which are phenotypically *d* or *se*, and multilocus deletions to mutants that are phenotypically *d*, *se*, and *d se*. It is this latter class that is due exclusively to multilocus deletion (Russell 1971). One of the problems with this whole-animal system is the resources required to perform extensive experiments. Although a highly relevant assay with regard to genetic risk assessment for the human population, these resource requirements impose serious limitations on the types and number of experiments that can be carried out. As a result, other investigators have tried to develop model systems with eukaryotic organisms that could be used for more extensive analysis of the parameters associated with radiation and chemically induced specific locus mutations as well as to collect large samples of specific locus mutants for genetic and biochemical analysis.

One of the main objectives of this volume is to evaluate the general utility of various specific locus assays in eukaryotic organisms with regard to their ability to provide data that can be used in genetic risk assessment for the human population. To do this assessment successfully, data on the full range of genetic damage that will produce specific locus mutations are required. There is increasing evidence that the spectrum of genetic damage detected at different genetic loci in mammalian cells in culture is markedly dependent on the recovery procedures and on whether the locus is heterozygous or hemizygous (Evans et al. 1986 and this volume; Moore et al., this volume).

Recovery of Specific Locus Mutations in the *ad-3* Region of Two-component Heterokaryons of *N. crassa* Resulting from Gene Mutations and Multilocus Deletion

It has been useful to contrast the various specific locus assays developed with mammalian cells in culture (Evans et al. 1986) with the specific locus system developed with a two-component heterokaryon of *Neurospora* that is heterozygous in the *ad-3* region (de Serres and Osterbind 1962; de Serres and Malling 1971; de Serres 1981).

Early studies (de Serres 1956) showed that the *ad-3A* and *ad-3B* loci (which control sequential steps in the purine biosynthetic pathway and can be recovered on the basis of the accumulation of a reddish-purple pigment in the vacuoles of the mycelium) are closely linked on linkage group I. Because of this close linkage, it seemed possible to develop an assay in *Neurospora* that would be able to recover the same array of specific locus mutations as can be recovered in the mouse with the closely linked *d* and *se* loci. Since *Neurospora* is a haploid organism, mutations in the *ad-3* region resulting from gene mutation should grow on adenine-supplemented minimal medium (reparable *ad-3* mutations), whereas those mutations resulting from multilocus deletion should be nonsupplemental on any medium (irreparable *ad-3* mutations) and haplolethal (see Atwood 1955). Thus, by using a two-component heterokaryon, heterozygous in the *ad-3* region (Table 1), it was possible to recover reparable mutations resulting from gene mutation at the *ad-3A* or *ad-3B* locus and irreparable mutations resulting from multilocus deletion of *ad-3A* alone, *ad-3B* alone, or both loci, *ad-3A ad-3B* (de Serres and Osterbind 1962; de Serres 1964). These data showed that it was possible to mimic a higher eukaryotic, diploid organism in *Neurospora* by using a two-component heterokaryon, where two genetically different haploid nuclei are together in the same cytoplasm.

Subsequent studies with X-rays (Webber and de Serres 1965) were performed to study the induction kinetics of these two different classes of specific locus mutations in the *ad-3* region. This series of experiments showed that gene mutations increase linearly with X-ray dose and that multilocus deletions increase as the square of X-ray dose. Even more important was the conclusion from these experiments that the spectrum of X-ray induced specific locus mutations is dose-dependent. Prior to these studies, there were striking differences in the literature with regard to the nature of X-ray-induced specific locus mutations in different organisms (Giles et al. 1955; de Serres 1958). Since most specific locus mutations in other organisms were recovered from experiments where high X-ray exposures were used to enhance mutant recovery, the *Neurospora* data provided an explanation for why the majority of specific locus mutations recovered with other organisms were found to be due to multilocus deletion.

Table 1
Genetic Composition of Each Component of a Two-Component Heterokaryon Used to Recover Specific Locus Mutants in the *ad-3* Region (Heterokaryon 12)

Component number	Strain number	Linkage group									
				I				III	IV	V	VI
I	74-OR60-29A	A	*his-2*	*ad-3A*	*ad-3B*	*nic-2*	+	*ad-2*	+	*inos*	+
II	74-OR31-16A	A	+	+	+	+	*al-2*	+	*cot-1*	+	*pan-2*

Homology Tests on Multilocus Deletions with Genetic Markers in the Immediately Adjacent Genetic Regions

Genetic analysis of the extent and type of functional inactivation in irreparable *ad-3* mutants was performed to determine whether the postulated multilocus deletions would cover other loci in the immediately adjacent genetic regions (de Serres 1968, 1969; de Serres and Brockman 1968). Some mutants originally classified as irreparable *ad-3* mutants turned out to be, on more detailed genetic analysis, gene mutations at the *ad-3A* or *ad-3B* locus with a separate site of closely linked recessive lethal damage in the immediately adjacent genetic regions ($ad\text{-}3^R + RL^{CL}$). These mutants proved invaluable in demonstrating that the functional inactivation in other mutants that were

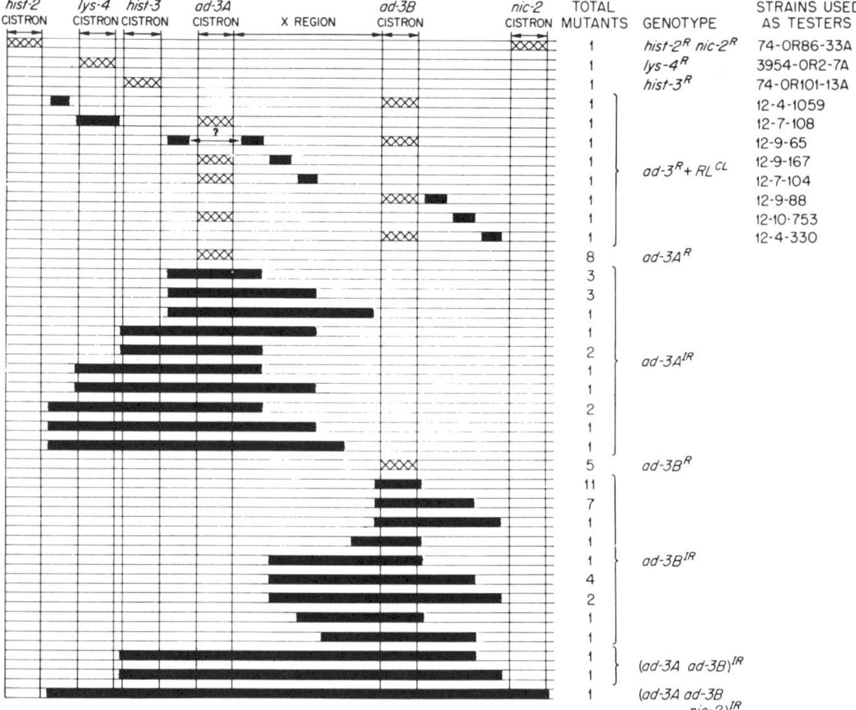

Figure 1

Complementation map of the *ad-3* and immediately adjacent regions showing the extent and type of functional inactivation in various types of *ad-3* mutations induced in a dikaryon in relation to markers at other loci in the immediately adjacent regions. (XXXX) Recessive lethal damage supplementable on supplemented basal medium. (■) Recessive lethal damage nonsupplementable on supplemented basal medium. Irregular ends indicate unknown limits. (Reprinted, with permission, from de Serres 1969).

also scored as multilocus deletions in the *ad-3* region actually did cover other genetic loci located to the left of *ad-3A*, between *ad-3A* and *ad-3B* in the X region, and to the right of *ad-3B* (de Serres 1969). As shown in Figure 1, *ad-3* mutants in this latter category map as a series of overlapping chromosome deletions as has been shown in studies of X-ray-induced specific locus mutations by other investigators in *Drosophila* (Temin 1978; Shukla and Auerbach 1981) and in the mouse (Russell et al. 1966; Russell 1971). In the latter two instances, new molecular studies using techniques for DNA cloning and sequencing have actually demonstrated the loss of an extensive number of base pairs in the DNA in specific locus mutations resulting from multilocus deletion in both organisms (Rinchik et al. 1986; Reardon et al. 1987).

Heterozygous Effects of Multilocus Deletions in *Neurospora*

Studies by L.B. Russell on *d* and *se* mutants resulting from multilocus deletion (deficiencies) in the mouse (Russell et al. 1966; Russell 1971) have shown striking heterozygous effects in terms of reduced transmission through the germ line, early death, sterility, and reduced size (runting).

In *Neurospora*, a study of the heterozygous effects of multilocus deletions in the *ad-3* region was made by studying their effects on the growth rate on forced two-component heterokaryons (dikaryons). These studies (de Serres 1965; de Serres and Miller 1987) showed heterozygous effects of multilocus deletions alone and even more striking effects in combination (Fig. 2 and 3). These data show that this class of gene mutation in the *ad-3* region is not completely recessive, an observation that when coupled with the data on

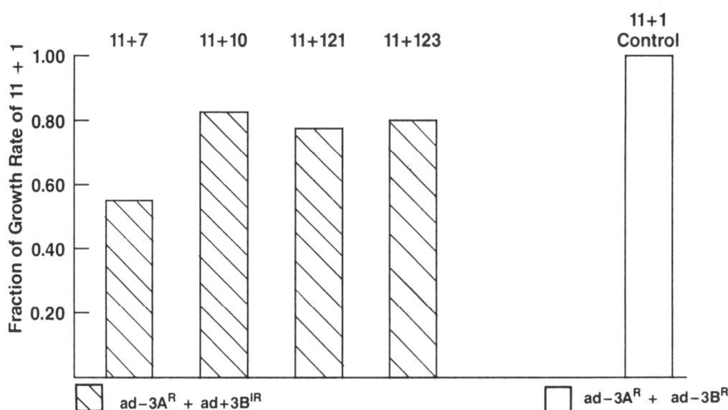

Figure 2

Comparison of the heterozygous effects of single multilocus deletions in forced dikaryons of *N. crassa*.

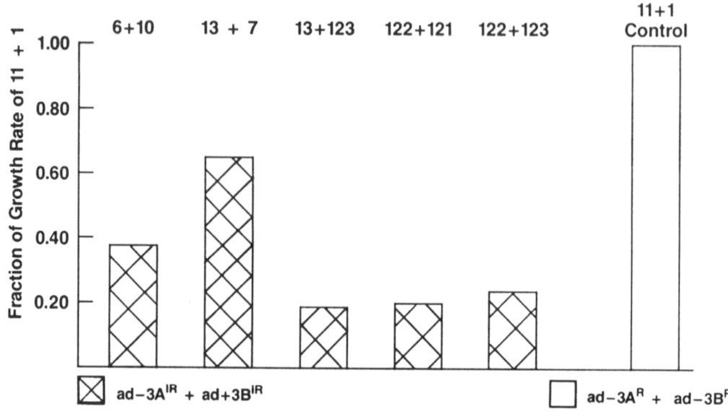

Figure 3

Comparison of heterozygous effects of double multilocus deletions in forced dikaryons of *N. crassa*.

multilocus deletions in the *d se* region in the mouse has important implications for genetic risk assessment.

Comparison of the Relative Frequencies of Gene Mutations and Multilocus Deletions among *ad-3* Mutations Induced by Different Chemical Mutagens

The genetic characterization of *ad-3* mutations recovered from forward-mutation experiments on two-component heterokaryons of *Neurospora* (de Serres and Malling 1971; de Serres 1981) makes it possible to determine the relative frequencies of gene mutations at each locus (*ad-3*), as well as the relative frequencies of multilocus deletions covering each or both loci (*ad-*3^{IR}) (Fig. 4). The genotypes of gene mutations are: *ad-3*A^R and *ad-3*B^R. The genotypes of multilocus deletions are: *ad-3*A^{IR}, *ad-3*B^{IR}, (*ad-3B nic-2*)^{IR}, (*ad-3A ad-3B*)^{IR}, (*ad-3A ad-3B nic-2*)^{IR} (see Figure 5). More complex genotypes have also been found: *ad-3*A^R *ad-3*B^{IR} and *ad-3*A^{IR} *ad-3*B^R; but like the *ad-3*A^R *ad-3*B^R double mutant they usually occur at very low frequencies (F.J. de Serres, unpubl.).

The initial genetic analysis of chemically induced *ad-3* mutations was complicated by inadequate testers for the trikaryon tests used to characterize genetically the newly induced mutants (Brockman and Goven 1965). Subsequent studies (de Serres and Brockman 1968; Brockman et al. 1969; Malling and de Serres 1970) showed that *ad-3* mutations originally classified in these experiments, where 2-methoxy-6-chloro-9-(3-[ethyl-2-chloroethyl]aminopropylamino)acridine-dihyrochloride (ICR-170), nitrous acid (NA), or *N*-methyl-*N′*-nitro-*N*-nitroso-guanidine (MNNG) were used as mutagens, were actually gene mutations with a closely linked site of recessive

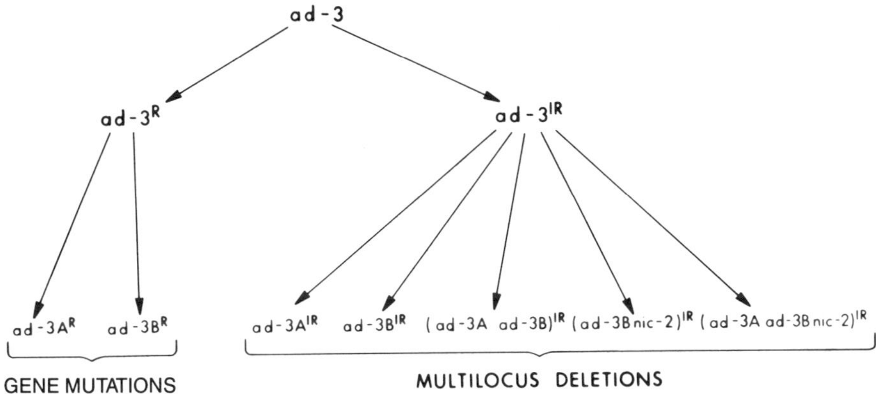

Figure 4

Subclasses of *ad-3* mutants revealed by genetic analysis (modified from de Serres 1983).

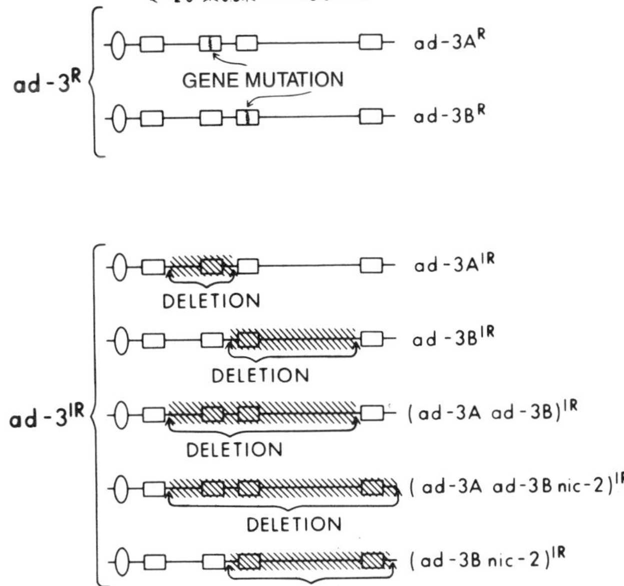

Figure 5

Extent of genetic damage in gene mutations and multilocus deletions in the *ad-3* region of *N. crassa* (modified from de Serres 1983).

lethal damage (*ad-3*^{R} + RL^{CL}) rather than *ad-3*^{IR}. With the use of these *ad-3* mutants with closely-linked sites of recessive lethal mutation as testers in the genetic characterization of *ad-3* mutations induced in two-component heterokaryons, more reliable classifications and more detailed genetic mapping were possible.

The differences in the relative frequencies of gene mutations (*ad-3*^{R}) and multilocus deletions (*ad-3*^{IR}) found after treatment of a two-component heterokaryon (heterokaryon 12, H-12) are shown in Tables 2 and 3. These studies has shown that for many chemicals (Table 2) the *ad-3* mutants recovered are predominantly or exclusively gene mutations. The small percentages of *ad-3* mutants classified as *ad-3*^{IR} are either *ad-3* mutants of spontaneous origin or gene mutations with a closely linked site of recessive lethal damage *ad-3*^{R} + RL^{CL}. Homology tests on all such mutants listed as *ad-3*^{IR} in Table 2 should help us distinguish between these two alternatives. In Table 3 is a listing of those chemicals that induce *ad-3* mutants by both gene mutation and multilocus deletion. The relative frequencies of *ad-3* mutants in each class is markedly mutagen-dependent.

DISCUSSION

The study of specific locus mutations in the *ad-3* region of a two-component heterokaryon of *N. crassa* has shown that both gene mutations and multilocus deletions can be recovered as predicted by specific locus studies by other investigators in *Drosophila* (Temin 1978; Shukla and Auerbach 1981) and in the mouse in the *d se* region (Russell et al. 1966; Russell 1971). The *Neurospora* studies have shown that after X-irradiation, these two different classes of mutations show different induction kinetics, with gene mutations increasing linearly with dose and multilocus deletions increasing as the square of the dose (Webber and de Serres 1965). Genetic analysis of the multilocus deletions by a series of homology tests with genetic markers in the immediately adjacent genetic regions has shown that in the main part they map as a series of overlapping chromosomal or interstitial deletions as predicted from studies on other organisms. The fact that many environmental chemicals produce specific locus mutations by both gene mutation and multilocus deletion predicts that a substantial fraction of specific locus mutations at particular loci will not be recovered in mammalian cells in culture at those loci that are hemizygous, since they are haplolethal. Full recovery can only be expected at those loci, such as the *tk* locus in mouse lymphoma cells or human cells in culture, that are heterozygous. Thus, with such chemical mutagens as hycanthone methanesulfonate, the specific locus mutations that can be recovered at hemizygous loci are only a small fraction of those actually induced.

Table 2
Chemicals that Induce *ad-3* Mutations Predominantly or Exclusively by Point Mutations

Experiment number	Mutagen	Relative frequency[a]		Reference
		ad-3^{R}	*ad-3*^{IR}	
12-4	NA	417 (99.8)	1 (0.2)	Brockman et al. (1969)
12-9	ICR-170	187 (100)	0 (0)	de Serres and Brockman (1968)
12-27	MNNG	953 (99.4)	6 (0.6)	Malling and de Serres (1970)
12-21	HA	202 (99.0)	2 (1.0)	Malling and de Serres (1971)
12-28	DEN	92 (98.9)	1 (1.1)	Malling and de Serres (1972)
12-196	4NQO	184 (100)	0 (0)	Ong et al. (1975)
12-197	4HAQO	210 (100)	0 (0)	Ong et al. (1975)
12-314	NDZ	199 (100)	0 (0)	Ong et al. (1979)
12-315	MTZ	139 (98.1)	3 (1.9)	Ong et al. (1979)
12-267	AF-2	262 (99.2)	2 (0.8)	Ong and de Serres (1981)
12-264	SQ18506	212 (98.6)	3 (1.4)	Ong and de Serres (1981)
12-265	FANFT	213 (98.6)	3 (1.4)	Ong and de Serres (1981)
12-163	ENU	218 (99.1)	2 (0.9)	de Serres (1983)

[a]Number of each genotype and percentage (in parentheses).

Abbreviations: NA, nitrous acid; ICR-170, 2-methyl-6-chloro-9-[3-(ethyl-2-chloroethyl) aminopropylamino] acridine dihydrochloride; MNNG, *N*-methyl-*N'*-nitro-*N*-nitrosoguanidine; HA, hydroxylamine; DEN, diethylnitrosamine; 4NQO, 4-nitroquinoline 1-oxide; 4HAQO, 4-hydroxyaminoquinoline 1-oxide; NDZ, niridazole; MTZ, metronidazole; AF-2, 2(2-furyl)-3-(5-nitro-2-furyl) acrylamide; SQ18506, trans-5-amino-3-[2-(5-nitro-2-furyl)-vinyl]-1,2,4-oxadiazole; FANFT, 2-formylamino-4-(5-nitro-2-furyl) thiazole; ENU, ethylnitrosourea.

Table 3
Chemicals that Induce *ad-3* Mutations by Both Point Mutation and Multilocus Deletion

Experiment number	Mutagen	Relative frequency[a]		Reference
		ad-3^{R}	*ad-3*^{IR}	
12-64	MMS	501 (94.5)	29 (5.5)	Malling and de Serres (1973)
12-180	EI	180 (97.3)	5 (2.7)	Ong and de Serres (1973)
12-189	PMMT	170 (94.4)	10 (5.6)	Ong and de Serres (1973)
12-194	PDMT	45 (80.3)	11 (19.6)	Ong and de Serres (1973)
12-199	DEP	201 (93.1)	15 (6.9)	Ong and de Serres (1975)
12-217	DEO	162 (70.4)	68 (29.6)	Ong and de Serres (1975)
12-243	HYC	19 (17.8)	88 (82.2)	Ong and de Serres (1980)
12-250	LUC	11 (16.9)	54 (83.1)	Ong and de Serres (1980)
12-248	IA-3	15 (18.5)	66 (81.5)	Ong and de Serres (1980)
12-254	IA-4	16 (12.9)	108 (87.1)	Ong and de Serres (1980)
12-244	IA-5	11 (12.0)	81 (88.0)	Ong and de Serres (1980)
12-225	EDB	327 (84.3)	61 (15.7)	de Serres and Malling (1983)

[a]Number of each genotype and percentage (in parenthesis).

Abbreviations: MMS, methylmethanesulfonate; EI, ethylenimine; PMMT, 1-phenyl-3-monomethyltriazene; PDMT, 1-phenyl-3-dimethyltriazene; DEP, 1,2,4,5,-diepoxypentane; DEO, 1,2,7,8,diepoxyoctane; HYC, hycanthone methanesulfonate; LUC, leucanthone hydrochloride; IA-3, hycanthone methanesulfonate indazole analog; IA-4, leucanthone methanesulfonate indazole analog; IA-5, hycanthone methanesulfonate indazole analog; EDB, ethylene dibromide.

REFERENCES

Atwood, K.C. 1955. Indispensable gene functions in *Neurospora*. *Proc. Natl. Acad. Sci.* **39:** 1027.

BEIR III. 1980. National Academy of Sciences/National Research Council, Advisory Committee on the Biological Effects of Ionizing Radiations (BEIR). *The effects on populations of exposure to low levels of ionizing radiations*. National Academy of Sciences Press, Washington, D.C.

Brockman, H.E. and W. Goben. 1965. Mutagenicity of a monofunctional alkylating agent derivative of acridine in *Neurospora*. *Science* **147:** 750.

Brockman, H.E., F.J. de Serres, and W.E. Barnett. 1969. Analysis of *ad-3* mutants induced by nitrous acid in a heterokaryon of *Neurospora crassa*. *Mutat. Res.* **7:** 307.

CCEM Report. 1983. National Academy of Sciences/National Research Council, Committee on Chemical Environmental Mutagens (CCEM). *Identifying and estimating genetic impact of chemical mutagens*. National Academy of Sciences Press, Washington, D.C.

de Serres, F.J. 1956. Studies with purple adenine mutants in *Neurospora crassa*. I. Structural and functional complexity in the *ad-3* region. *Genetics* **41:** 668.

———. 1958. Studies with purple adenine mutants in *Neurospora crassa*. III. Reversion of X-ray-induced mutants. *Genetics* **43:** 187.

———. 1964. A genetic analysis of the structure of the *ad-3* region of *Neurospora crassa* by means of irreparable recessive lethal mutations. *Genetics* **50:** 21.

———. 1965. Impaired complementation between nonallelic mutations in *Neurospora*. *Natl. Cancer Inst. Monogr.* **18:** 33.

———. 1968. Genetic analysis of the extent and type of functional inactivation in irreparable recessive lethal mutations in the *ad-3* region of *Neurospora crassa*. *Genetics* **58:** 69.

———. 1969. Comparison of the complementation and genetic maps of closely linked nonallelic markers on linkage group I on *Neurospora crassa*. *Mutat. Res.* **8:** 43.

———. 1976. Prospects for a revolution in the methods of toxicological evaluation. *Mutat. Res.* **38:** 165.

———. 1981. Induction and genetic characterization of specific locus mutations in the *ad-3* region in two-component heterokaryons of *Neurospora crassa*. In *Short-term tests for chemical carcinogens* (ed. H.F. Stich and R.H.C. San), p. 175. Springer-Verlag, New York.

———. 1983. The use of *Neurospora* in the evaluation of the mutagenic activity of environmental chemicals. *Environ. Mutagen.* **5:** 341.

de Serres, F.J. and H.E. Brockman. 1968. Homology tests on presumed multilocus deletions in the *ad-3* region of *Neurospora crassa* induced by the acridine mustard ICR-170. *Genetics* **58:** 79.

de Serres, F.J. and H.V. Malling. 1971. Measurement of recessive lethal damage over the entire genome and at two specific loci in the *ad-3* region of a two-component heterokaryon of *Neurospora crassa*. In *Chemical mutagens: Principles and methods for their detection* (ed. A. Hollaender), vol. 2, p. 311. Plenum Press, New York.

———. 1983. The role of *Neurospora* in evaluating chemicals for mutagenic activity. *Ann. N.Y. Acad. Sci.* **407:** 177.

de Serres, F.J. and I.R. Miller. 1987. Genetic modification of the heterozygous effects of multilocus deletions in heterokaryons of *Neurospora crassa*. *Mutat. Res.* (in press).

de Serres, F.J. and R.S. Osterbind. 1962. Estimation of the relative frequencies of X-ray-induced viable and recessive lethal mutations in the *ad-3* region of *Neurospora crassa*. *Genetics* **47:** 793.

Evans, H.H., J. Mencl, N.-F. Horng, M. Picanti, C. Sanchez, and J. Hozier. 1986. Locus specificity in the mutability of L5178Y mouse lymphoma cells: The role of multilocus deletions. *Proc. Natl. Acad. Sci.* **83:** 4379.

Giles, N.H., F.J. de Serres, and C.W.H. Partridge. 1955. Comparative studies of X-ray-induced forward and reverse mutation. *Ann. N.Y. Acad. Sci.* **59:** 536.

Malling, H.V. and F.J. de Serres. 1979. Genetic effect of N-methyl-N′-nitro-N-nitrosoquanidine in *Neurospora crassa*. *Mol. Gen. Genet.* **106:** 195.

———. 1971. Hydroxylamine-induced purple adenine (*ad-3*) mutants in *Neurospora crassa*. I. Characterization of mutants by genetic tests. *Mutat. Res.* **12:** 35.

———. 1972. Genetic characterization of diethylnitrosamine-induced purple adenine (*ad-3*) mutants in *Neurospora crassa*. *Cancer Res.* **32:** 1273.

———. 1973. Genetic analysis of purple adenine (*ad-3*) mutants induced by methyl methanesulfonate in *Neurospora crassa*. *Mutat. Res.* **18:** 1.

Moore, M.M., D. Clive, B.E. Howard, A.G. Bateson, and N.T. Turner. 1985a. In situ analysis of trifluorothymidine (TFT^r) mutants of L5178Y/$TK^{+/+}$ mouse lymphoma cells. *Mutat. Res.* **151:** 147.

Moore, M.M., D. Clive, C. Hozier, B.E. Howard, A.G. Batson, N.T. Turner, and J. Sawyer. 1985b. Analysis of trifluorothymidine-resistant (TFT^r) mutants of L5178/ $TK^{+/+}$ mouse lymphoma cells. *Mutat. Res.* **151:** 161.

Muller, H.J. 1927. Artificial transmutation of the gene. *Science* **66:** 84.

Ong, T.-M. and F.J. de Serres. 1973. Genetic characterization of *ad-3* mutants induced by chemical carcinogens, 1-phenyl-3-monomethyltriazene and 1-phenyl-3,3-dimethyltriazene, in *Neurospora crassa*. *Mutat. Res.* **20:** 17.

———. 1975. Mutagenic evaluation of antischistosomal drugs and their derivatives in *Neurospora crassa*. *J. Toxicol. Environ. Health* **1:** 271.

———. 1980. Genetic analysis of *ad-3* mutants induced by hycanthone, lucanthone, and their indazole analogs in *Neurospora crassa*. *J. Environ. Pathol. Toxicol.* **4:** 1.

———. 1981. Genetic analysis of *ad-3* mutants induced by AF-2 and two other nitrofurans in *Neurospora crassa*. *Environ. Mutagen.* **3:** 151.

Ong, T.-M., B.E. Matter, and F.J. de Serres. 1975. Genetic characterization of *ad-3* mutants induced by 4-nitroquinoline a-oxide and 4-hydroxyaminoquinoline 1-oxide in *Neurospora crassa*. *Cancer Res.* **35:** 291.

Ong, T.-M., B. Slade, and F.J. de Serres. 1979. Mutagenicity and mutagenetic specificity of metronidazole and niridazole in *Neurospora crassa*. *J. Environ. Pathol. Toxicol.* **2:** 1109.

Reardon, J.T., C.A. Liljestrand-Golden, R.L. Dusenbery, and P.D. Smith. 1987. Molecular analysis of diepoxybutane-induced mutations at the *Rosy* locus of *Drosophila melanogaster*. *Genetics* **115:** 323.

Rinchik, E.M., L. Russell, N.G. Copeland, and N.A. Jenkins. 1986. Molecular genetic analysis of the dilute short-ear region (*d se*) of the mouse. *Genetics* **112:** 321.

Russell, L.B. 1971. Definition of functional units in a small chromosomal segment of the mouse and its use in interpreting the nature of radiation-induced mutations. *Mutat. Res.* **11:** 107.

Russell, L.B., M.S. Steele, and H.M. Thompson, Jr. 1966. Deficiencies in the mouse. In Biology Division Semiannual Progress Report (ORNL-3999), p. 90. Oak Ridge National Laboratory, Oak Ridge, Tennessee.

Russell, W.L. 1954. Genetic effects of radiation in mammals. In *Radiation biology* (ed. A. Hollaender), vol. 1, p. 825. McGraw-Hill, New York.

Shukla, P.T. and C. Auerbach. 1981. Genetical tests for the frequency of small deletions among EMS-induced point mutations in *Drosophila*. *Mutat. Res.* **83:** 81.

Temin, R. 1978. Partial dominance of EMS-induced mutations affecting viability in *Drosophila melanogaster*. *Genetics* **89:** 315.

Webber, B.B. and F.J. de Serres. 1965. Induction kinetics and genetic analysis of X-ray-induced mutations in the *ad-3* region of *Neurospora crassa*. *Proc. Natl. Acad. Sci.* **53:** 430.

Characterization of the AS52 Cell Line for Use in Mammalian Cell Mutagenesis Studies

LEON F. STANKOWSKI, JR.* AND KENNETH R. TINDALL†
*Pharmakon Research International, Inc.
Waverly, Pennsylvania 18471
†Laboratory of Genetics
National Institute of Environmental Health Sciences
Research Triangle Park, North Carolina 27709

OVERVIEW

A genetically engineered cell line has been evaluated and compared to the Chinese hamster ovary/hypoxanthine-guanine phosphoribosyl transferase (CHO/HPRT) assay for its ability to detect mutations induced by a variety of chemical and physical agents. AS52 cells contain a single, functional, stably integrated copy of the *Escherichia coli* xanthine-guanine phosphoribosyl transferase (XPRT) gene (*gpt*). Mutations at the *gpt* locus can be detected as 6-thioguanine resistant (6-TG^r) colonies under conditions identical to those for detecting mutations at the *hprt* locus in CHO-K1-BH4 cells. Similar cytotoxic responses were observed in the two cell lines for each agent examined. The AS52 and CHO-K1-BH4 cell lines respond with approximately the same efficiency to mutation induction by classical point mutagens. In contrast, mutant induction by X-rays and a number of other clastogens or agents thought to cause deletions was much greater at the *gpt* locus (ratios of XPRT:HPRT mutants ranged from 2.8 to 89). Ethanol, formaldehyde, and methanol reproducibly induced mutations at the *gpt* locus in AS52 cells but yielded equivocal results or were nonmutagenic at the *hprt* locus. Cytotoxic treatments with acetone, DMSO, HCl, and sucrose, however, were nonmutagenic at both loci. These data support the use of the AS52 cell line as a more sensitive indicator of deletion mutations in routine mutagenesis assays.

INTRODUCTION

The *hprt* locus has been used extensively in the study of mammalian cell mutagenesis (Caskey and Kruh 1979). A variety of mammalian cell systems utilize this locus for the detection and quantification of mutations, as well as molecular analyses of mutant loci. The *hprt* locus is present in mammalian cells on the X chromosome in a functionally hemizygous state; it codes for the purine salvage pathway enzyme, HPRT. Forward mutations, *hprt*$^+$ to *hprt*$^-$, can be selected by virtue of resistance to TG^r whereas reverse mutations, *hprt*$^-$ to *hprt*$^+$, can be selected by growth in the presence of aminopterin or azaserine.

The protein coding sequences of *hprt* are comprised of 654 bp of DNA (Konecki et al. 1982); however, the presence of eight intervening sequences enlarges the total size of the gene to approximately 40 kb pairs (Melton et al. 1984). Although gross rearrangements of *hprt* sequences have been observed for a number of spontaneous and induced TG^r mutants (Fuscoe et al. 1983; Albertini et al. 1985; Turner et al. 1985; Stankowski and Hsie 1986; Stankowski et al. 1986), molecular cloning and direct DNA sequence analysis of mutant genes has been extremely difficult. To overcome this limitation, an *hprt*$^-$ Chinese hamster ovary (CHO) cell line (derived from the CHO-K1-BH4 cell line) has been genetically engineered by transfecting with the *E. coli* *gpt* gene and isolating a cell line that contains a single copy genomic integration (Tindall et al. 1984). The *gpt* gene is the functional equivalent of *hprt*, and *gpt*$^-$ TG^r mutants can be readily isolated. This cell line has been designated AS52, and appears to be quite useful for mammalian cell mutagenesis studies. Results for the evaluation of twenty agents (Table 1) for mutagenicity in the AS52 and CHO-K1-BH4 cell lines are presented.

Table 1
List of Agents Evaluated

Agent	Abbreviation
Acetone	ACE
Actinomycin D	ACD
Adriamycin	ADR
Bleomycin	BLM
Cis-platinum	*cis*P
Cyclophosphamide	CP
Dimethylnitrosamine	DMN
Dimethyl sulfoxide	DMSO
Daunomycin	DAU
Ethylmethane sulfonate	EMS
Ethanol	EtOH
Formaldehyde	FORM
Hydrochloric acid	HCl
ICR-170	—
ICR-191	—
Methanol	MeOH
Mitomycin C	MMC
Sucrose	SUCR
UV-light (254 nm)	UV
X-rays	—

RESULTS

The AS52 cell line was constructed from a CHO cell line containing a deletion mutation that affects most of the *hprt* locus. This cell line carries a single functional copy of the *E. coli* xanthine-guanine phosphoribosyl transferase (XPRT) gene (*gpt*) stably integrated into its genome (Tindall and Hsie 1983; Tindall et al. 1984, 1986; Stankowski and Hsie 1986; Stankowski et al. 1986). Because of the functional similarity of these two purine salvage enzymes, conditions for the selection of forward and reverse mutations of *hprt* in CHO-K1-BH4 cells are readily applicable to the selection of mutations at the *gpt* locus in AS52 cells.

The *gpt* gene within the AS52 cell line is more stable than other pSV2*gpt*-transformed cell lines previously described (Stankowski et al. 1986; Tindall et al. 1986). The spontaneous mutation rate for the *gpt* locus in AS52 cells has been estimated to be 2×10^{-6} mutants/cell/generation. This compares favorably with the spontaneous rate of 9×10^{-7} mutants/cell/generation established for the *hprt* locus in CHO-K1-BH4 cells (O'Neill et al. 1983). The stability of the *gpt* gene in AS52 cells is further demonstrated by the pooled results observed for negative control cultures over a one-year period. In 85 independent cultures from 21 independent experiments, the average spontaneous mutant frequency was 13.24 ± 9.08 (1 SD) $\times 10^{-6}$ following 7–8 days of subculture for phenotypic expression in nonselective medium.

The phenotypic expression time of EMS-, UV-, or X-ray-induced XPRT mutants was determined to be 7 days (Stankowski and Hsie 1986; Stankowski et al. 1986). Recovery of mutants induced by all three agents reached a maximum following 7 days of subculture in nonselective medium and remained constant for up to 17 days examined. As previously reported for expression of HPRT mutants in CHO-K1-BH4 cells (O'Neill et al. 1977; O'Neill and Hsie 1979), the phenotypic expression time for XPRT mutants was independent of the type or dose of mutagen used. Coincidentally, the phenotypic expression times for XPRT mutants in AS52 cells and HPRT mutants in CHO-K1-BH4 cells are approximately equal.

Because of the slightly higher spontaneous mutation rate, AS52 cells are routinely subcultured in the mycophenolic acid (MPA) selective medium described previously (Tindall et al. 1984). AS52 cells were initially grown in this selective medium until just prior to treatment. Recently, we have begun to remove the MPA selective medium and switch to nonselective medium approximately 24 hours prior to treatment to guard against possible MPA effects on nucleotide pools. However, comparison of both conditions suggest that MPA selection immediately prior to treatment has no effect on induced mutant frequency (L.F. Stankowski, unpubl.).

Cytotoxicity and mutation induction by 20 different agents have been compared in the two cell lines. Cytotoxic responses of the two cell lines following treatment with all 20 agents exhibited relatively minor differences (data not shown). For almost all agents evaluated (Table 1), the slopes of the cytotoxicity curves were slightly different between the two cell lines ($p < 0.05$). However, D_{37} values for 19 of these agents vary at most by a factor of two between cell lines. Usually, the differences are approximately 10% or less; although, the difference is approximately threefold for DMN + S9. An important aspect to these analyses is that neither cell line is consistently more nor less sensitive to the cytotoxic effects of all of the agents examined. These results indicate that the differences observed are agent specific (not cell line specific) most likely reflecting clonal variations between the two cell lines other than drug resistance mutations or differences in DNA repair.

In contrast to the cytotoxicity results, much larger differences in mutation to TG^r were observed between the *gpt* and *hprt* loci for some of the agents evaluated (Table 2). The results of the mutagenesis studies appear to fall into three agent-specific groups (A, B, and C, see Table 2). Group A includes agents that are not mutagenic at either locus (ACE, DMSO, HCl, and SUCR). Group B includes agents that are approximately equally mutagenic at either locus (DMN, EMS, ICR-170, ICR-191, and UV). For group B, the ratios of XPRT:HPRT induced mutants per unit dose ranged from 0.55 to 1.6. Group B agents are classical point mutagens thought to induce missense or nonsense mutations (Ward 1975; Cerutti 1976; Coulondre and Miller 1977; Miller and Schmeissner 1979; Fuscoe et al. 1982; Hutchinson et al. 1983; Wallace 1983). Group C includes agents that are much more mutagenic at the *gpt* locus than at *hprt* (ACD, ADR, BLM, cis-P, CP, DAU, EtOH, FORM, MeOH, MMC, and X-rays). EtOH, FORM, and MeOH belong to this latter group since they are mutagenic only at the *gpt* locus. For agents in group C, the XPRT:HPRT mutant ratios ranged from 2.8 to 89. All agents in group C have been demonstrated to be clastogenic in vivo and/or in vitro (Schinzel and Schmid 1976; Obe and Ristow 1979; Preston et al. 1981; Heddle et al. 1983; Sorsa et al. 1985).

Interestingly, EtOH(+S9), FORM(±S9), and MeOH(+S9) were mutagenic at the *gpt* locus, but not at the *hprt* locus. FORM is a potent clastogen in both cell lines (J.R. San Sebastian and L.F. Stankowski, unpubl.) but was strongly mutagenic only in AS52 cells, inducing primarily deletions (Tindall and Stankowski, this volume). EtOH and MeOH were also mutagenic only at the *gpt* locus and only in the presence of S9. EtOH and MeOH can be metabolized by S9 to acetaldehyde and FORM, respectively (Obe and Ristow 1979). These metabolites are most likely the ultimate mutagens detected in AS52 cells. This hypothesis is supported by the observation that formaldehyde dehydrogenase eliminates MeOH-induced mutagenesis in the AS52 cell line (C.S. Aaron and L.F. Stankowski, unpubl.).

Table 2
Comparison of Mutation Induction Responses at the *gpt* and *hprt* Loci in AS52 and CHO-K1-BH4 Cells

	TG^r Mutants/10^6 Clonable Cells[a]					
	AS52 (XPRT)		CHO-K1-BH4 (HPRT)		Ratio (XPRT : HPRT)	
Agent (unit dose)	−S9	+S9	−S9	+S9	−S9	+S9
(a) *Group A*						
ACE (μg/ml)	Neg[b]	Neg	Neg	Neg	N.A.[c]	N.A.
DMSO (mg/ml)	Neg	Neg	Neg	Neg	N.A.	N.A.
HCl (N)	Neg	Neg	Neg	Neg	N.A.	N.A.
SUCR (mg/ml)	Neg	Neg	Neg	Neg	N.A.	N.A.
(b) *Group B*						
DMN (μg/ml)	—	1.8	—	3.3	—	0.55
EMS (μg/ml)	0.82	—	1.2	—	0.7	—
ICR-170 (ng/ml)	1.3	—	2.1	—	0.62	—
ICR-191 (μg/ml)	0.54	—	0.34	—	1.6	—
UV (J/m^2)	74	—	54	—	1.4	—
(c) *Group C*						
ACD (ng/ml)	0.85	0.79	0.092	0.12	9.2	6.6
ADR (ng/ml)	0.52–18.0[d]	0.15–4.6	1.1	0.59	0.47–16.0	0.25–7.8
BLM (μg/ml)	3.0	0.88	0.34	0.32	8.8	2.8
*cis*P (μg/ml)	140	330	15	23	9.3	14
CP (μg/ml)	—	8.2–170	—	1.9	—	4.3–89
DAU (ng/ml)	3.8	2.5	0.38	0.79	10	3.2
EtOH (mg/ml)	Neg	0.25	Neg	Neg	N.A.	N.A.
FORM (μg/ml)	1.5	2.4	Neg	Neg	N.A.	N.A.
MeOH (mg/ml)	Neg	4.8	Neg	Neg	N.A.	N.A.
MMC (μg/ml)	640	75	130	18	4.9	4.2
X-rays (rad)	0.44–1.4	—	0.072	—	6.1–19	—

Selected data from Stankowski and Hsie (1986), Stankowski et al. (1986, 1988).

[a]Induced mutant frequencies are derived from the slopes of the lines fitted to the data by the least-squares method of linear regression analysis. Pooled results from 2–4 independent experiments for each agent are presented.

[b]Negative response: no dose dependent nor statistically significant induced mutational response.

[c]Not applicable.

[d]ADR, CP, and X-rays produce an apparent dose square response at the *gpt* locus; induced mutant frequencies are therefore estimated from the slopes between adjacent data points and a range is given.

DISCUSSION

Data presented here indicate that the AS52 cell line is comparable to the more commonly used CHO-K1-BH4 cell line (CHO/HPRT assay) with regard to (1) the low spontaneous mutation frequency, (2) the lack of mutagenic response following treatment with cytotoxic nonmutagens, and (3) the mutant induction response following treatment with classical point mutagens. However, the AS52 cell line appears to detect some mutagens that are either nonresponsive or poorly mutagenic in the CHO/HPRT assay. We suggest this response is due to the unique site of integration of the *gpt* gene (most likely autosomal) that allows the recovery of deletions that include adjacent sequences essential for cell viability (multilocus deletions) due to the presence of such sequences on the homologous chromosome. The hemizygous nature of the *hprt* gene precludes the isolation of deletions that include both the *hprt* gene and any adjacent essential sequences. Thus, the induced mutant frequency following treatment with agents that induce deletions will be lower in CHO-K1-BH4 cells (i.e., at the *hprt* locus) because only a subset of the deletions induced will be recovered.

Molecular analyses of isolated mutants indicate that the differential mutant frequency response for *gpt* and *hprt* is associated with deletions (Tindall and Stankowski, this volume). These data are consistant with the recovery of multilocus deletions at *gpt* that are lethal at *hprt*. Similar mutant induction responses for *gpt* and *hprt* appear to correlate with the induction of point mutations or small deletions (Stankowski et al. 1986; Tindall and Stankowski, this volume). Thus, comparative mutagenesis studies using AS52 or CHO-K1-BH4 cells can provide insight regarding mutational mechanisms. For example, the addition of S9 substantially reduces the XPRT:HPRT mutant ratio for BLM (Table 2). However, this decrease is due solely to a lower yield of XPRT mutants. The addition of S9 has no affect upon the induction of HPRT mutants by BLM. These data suggest at least two pathways for mutation induction by BLM. The substantial reduction of the induced mutant frequency at XPRT by the addition of S9 indicates that S9 eliminates or largely reduces the mutational pathway leading to the generation of multilocus deletions. The recovery of HPRT mutants suggests a second mutational pathway for BLM. Assuming multilocus deletions are not recovered by *hprt*, BLM appears to be inducing point mutations or small deletions at *hprt* or, in the presence of S9, at *gpt*. S9 may metabolize BLM to alternate metabolites incapable of inducing large deletions or may simply act as a scavenger of the oxygen radicals known to be produced by BLM (Sausville et al. 1976; Lown and Sim 1977). In contrast, MMC-induced mutations are reduced equally by S9 in the two cell lines. The MMC $\pm$ S9 induced XPRT:HPRT ratios (Table 2) are similar, suggesting a common pathway for

MMC mutagenesis at both loci. Thus, the AS52 cell line may be useful in detecting mutations induced by clastogens as well as providing mechanistic insight regarding the mutation induction response of the AS52 or the CHO-K1-BH4 cell lines.

These data suggest that the *gpt* locus is capable of detecting specific mutational events not detected at the *hprt* locus. Reference to the AS52 cell line as being "hypersensitive" to oxygen radicals (Hsie et al. 1986) is unfortunately misleading. Although the AS52 cell line can detect mutations induced by oxygen radicals, its differential response relative to the *hprt* locus is likely due to the ability to recover large deletions. We suggest that any DNA damage leading to deletions of the target locus and adjacent essential sequences will yield the same results. More accurately, the *hprt* locus in CHO-K1-BH4 cells may be hyposensitive to radiomimetic agents or clastogens due to the inability to recover some deletions. Thus, the induced mutant frequency in CHO-K1-BH4 cells will under represent the number of deletions that have occurred at *hprt*. We hypothesize that the more responsive AS52 cell line allows the recovery of more of the induced deletion mutations at the *gpt* locus, resulting in a higher mutant frequency response.

A lack of mutagenicity for several innocuous or toxic but nonmutagenic chemicals indicates that mutation of the *gpt* locus is both a sensitive and a specific indicator for mutagens. The ease with which the *gpt* gene can be manipulated as well as the ability to recover mutations that are apparently lethal to CHO-K1-BH4 cells supports the use of the AS52 cell line in the routine detection of mutagens and in molecular studies of the mechanistic pathways by which these agents exert their effects.

ACKNOWLEDGMENTS

The expert technical assistance of William G. Tuman, Edmund G. Godek, Gerald J. Kasper, and Thomas F. Golden is greatly appreciated, as is the assistance of Connie Stankowski in the preparation of this manuscript. This research was sponsored by Pharmakon Research International, Inc. and the National Institute of Environmental Health Sciences.

REFERENCES

Albertini, R.J., J.P. O'Neill, J.A. Nicklas, N.H. Heintz, and P.C. Kelleher. 1985. Alteration of the *hprt* gene in human *in vivo*-derived 6-thioguanine-resistant T lymphocytes. *Nature* **316:** 369.

Caskey, C.T. and G.D. Kruh. 1979. The HPRT locus. *Cell* **16:** 1.

Cerutti, P.A. 1976. Base damage induced by ionizing radiation. In *Photochemistry and photobiology of nucleic acids* (ed. S.Y. Wang), vol. 2, p. 375. Academic Press, New York.

Coulondre, C. and J.H. Miller. 1977. Genetic studies of the *lac* repressor. III. Additional correlation of mutational sites with specific amino acid residues. *J. Mol. Biol.* **117:** 525.

Fuscoe, J.C., R.G. Fenwick, Jr., D.H. Ledbetter, and C.T. Caskey. 1983. Deletion and amplification of the HPRT locus in Chinese hamster cells. *Mol. Cell. Biol.* **3:** 1086.

Fuscoe, J.C., J.P. O'Neill, R. Machanoff, and A.W. Hsie. 1982. Quantification and analysis of reverse mutations at the *hgprt* locus in Chinese hamster ovary cells. *Mutat. Res.* **96:** 15.

Hsie, A.W., L. Recio, D.S. Katz, C.Q. Lee, M. Wagner, and R.L. Schenley. 1986. Evidence for reactive oxygen species inducing mutations in mammalian cells. *Proc. Natl. Acad. Sci.* **83:** 9616.

Heddle, J.A., M. Hite, B. Kirkhart, K. Mavourin, J.T. MacGregor, G.W. Newell, and M.F. Salamone. 1983. The induction of micronuclei as a measure of genotoxicity: A report of the U.S. Environmental Protection Agency Gene-Tox Program. *Mutat. Res.* **123:** 61.

Hutchinson, F., T.R. Skopek, and R.D. Wood. 1983. Mechanisms of mutagenesis for lambda phage. *UCLA Symp. Mol. Cell. Biol. New Ser.* **11:** 501.

Konecki, D.S., J. Brennand, J.C. Fuscoe, C.T. Caskey, and A.C. Chinault. 1982. Hypoxanthine-guanine phosphoribosyl transferase genes of mouse and Chinese hamster: Construction and sequence analysis of cDNA recombinants. *Nucleic Acids Res.* **10:** 6763.

Lown, J.W. and S.K. Sim. 1977. The mechanism of the bleomycin-induced cleavage of DNA. *Biochem. Biophys. Res. Commun.* **77:** 1150.

Melton, D.W., D.S. Konecki, J. Brennand, and C.T. Caskey. 1984. Structure, expression, and mutation of the hypoxanthine phosphoribosyl transferase gene. *Proc. Natl. Acad. Sci.* **81:** 2147.

Miller, J.H. and U. Schmeissner. 1979. Genetic studies of the *lac* repressor. X. Analysis of missense mutations in the *lacI* gene. *J. Mol. Biol.* **131:** 223.

Obe, G. and H. Ristow. 1979. Mutagenic, cancerogenic and teratogenic effects of alcohol. *Mutat. Res.* **65:** 229.

O'Neill, J.P. and A.W. Hsie. 1979. Phenotypic expression time of mutagen induced 6-thioguanine resistance in Chinese hamster ovary cells (CHO/HGPRT system). *Mutat. Res.* **59:** 109.

O'Neill, J.P., M.W. Heartlein, and R.J. Preston. 1983. Sister-chromatid exchanges and gene mutations are induced by the replication of 5-bromo- and 5-chlorodeoxyuridine substituted DNA. *Mutat. Res.* **109:** 259.

O'Neill, J.P., P.A. Brimer, R. Machanoff, G.P. Hirch, and A.W. Hsie. 1977. A quantitative assay of mutation induction at the hypoxanthine-guanine phosphoribosyl transferase locus in Chinese hamster ovary cells. (CHO/HGPRT system). Development and definition of the system. *Mutat. Res.* **45:** 91.

Preston, R.J., W. Au, M.A. Bender, J.G. Brewen, A.V. Carrano, J.A. Heddle, A.F. McFee, S. Wolff, and J.S. Wassom. 1981. Mammalian *in vivo* and *in vitro* cytogenetic assays: A report of the U.S. EPA's Gene-Tox Program. *Mutat. Res.* **87:** 143.

Sausville, E.A., J. Peisach, and S.B. Horwitz. 1976. A role for ferrous ion and oxygen in the degradation of DNA by bleomycin. *Biochem. Biophys. Res. Commun.* **73:** 814.

Schinzel, A. and W. Schmid. 1976. Lymphocyte chromosome studies in humans exposed to chemical mutagens. The validity of the method in 67 patients undergoing cytostatic therapy. *Mutat. Res.* **40:** 139.

Sorsa, M., K. Hemminki, and H. Vaino. 1985. Occupational exposure to anticancer drugs—Potential and real hazards. *Mutat. Res.* **154:** 135.

Stankowski, L.F., Jr. and A.W. Hsie. 1986. Quantitative and molecular analyses of radiation-induced mutation in AS52 cells. *Radiat. Res.* **105:** 37.

Stankowski, L.F., Jr., K.R. Tindall, and A.W. Hsie. 1986. Quantitative and molecular analyses of ethyl methanesulfonate- and ICR 191-induced mutation in AS52 cells. *Mutat. Res.* **160:** 133.

Stankowski, L.F., Jr., K.R. Tindall, E.G. Godek, R.J. Matthews, and R.W. Naismith. 1988. Comparison of the cytotoxicity and mutagenicity of seven anticancer drugs in AS52 and CHO-K1-BH4 cells. *Mutat. Res.* (in press).

Tindall, K.R. and A.W. Hsie. 1983. Detection of deletion mutations at the *gpt* locus in pSV2*gpt* transformed CHO cells. *UCLA Symp. Mol. Cell. Biol. New Ser.* **11:** 625.

Tindall, K.R., L.F Stankowski, Jr., R. Machanoff, and A.W. Hsie. 1984. Detection of deletion mutations in pSV2*gpt* transformed cells. *Mol. Cell. Biol.* **4:** 1411.

———. 1986. Analysis of mutation in pSV2*gpt* transformed Chinese hamster ovary cells. *Mutat. Res.* **160:** 121.

Turner, D.R., A.A. Morley, M. Haliandros, R. Kutlaca, and B.J. Sanderson. 1985. *In vivo* somatic mutations in human lymphocytes frequently result from major gene deletions. *Nature* **315:** 343.

Wallace, S.S. 1983. Detection and repair of DNA base damages produced by ionizing radiation. *Environ. Mutagen.* **5:** 769.

Ward, J.F. 1975. Molecular mechanism of radiation-induced damage to nucleic acids. *Adv. Radiat. Biol.* **5:** 181.

Poor Recovery of Mutants Harboring Multilocus Lesions in Hemizygous Chromosomal Regions

HELEN H. EVANS, JAROSLAV MENCL, AND MIN-FEN HORNG
Department of Radiology
Case Western Reserve University
Cleveland, Ohio 44106

OVERVIEW

Previous studies have shown that the mutagenicity of clastogens is much lower when measured by the inactivation of the hypoxanthine-guanine phosphoribosyl transferase (*hprt*) gene on the X chromosome than when measured by the inactivation of the thymidine kinase (*tk*) gene on an autosome in $tk^{+/-}$ heterozygotes. We have measured mutant frequencies induced by clastogens and base-change mutagens in two strains of L5178Y mouse lymphoma cells. Strain LY-R16 is a $tk^{+/-}$ heterozygote, whereas strain LY-R83 possesses only one chromosome 11—the site of the *tk* gene in the mouse. The mutagenicity of clastogens at the *tk* locus was markedly higher in strain LY-R16 than in the monosomic strain LY-R83. These differences in clastogen-induced mutant frequencies between the two strains were much greater than in the case of ethyl methanesulfonate (EMS)- and ultraviolet (UV)-induced mutant frequencies. The very low clastogen-induced mutability of strain LY-R83 at the *tk* locus was similar to the very low clastogen-induced mutability of both strains at the *hprt* and Na^+/K^+ ATPase loci. These results demonstrate that multilocus lesions in target genes, located in hemizygous chromosomal regions, lead to poor recovery of the mutant cells.

INTRODUCTION

The measurement of the mutagenicity of various chemical and physical agents in cultured mammalian cells has generally relied on the determination of the frequency of forward mutations that inactivate genes that code for nonessential products. Target genes of choice for the measurement of forward mutations are those in which activation of one allele results in a measurable phenotypic change. These mutagenicity assays have therefore usually employed cells harboring heterozygous target genes (Clive et al. 1979; Carver et al. 1980), target genes in hemizygous regions (Adair et al. 1983; Nalbantoglu et al. 1983), or codominant target genes (Baker et al. 1974). One of the most commonly used target genes in these determinations has been the *hprt* gene, located on the X chromosome. This chromosome is monosomic in male cells and is almost entirely functionally hemizygous in female cells (Lyon 1961; Mohandas et al. 1979; Liskay and Evans 1980). The measurement of the inactivation of genes that code for nonessential products has been thought to cover all types of mutations; i.e., base changes, frame-shifts, deletions, and rearrangements. However, the mutagenicity of individual agents has been found to vary markedly depending on the location of the target gene, particularly in the case of agents that yield a high proportion of gross chromosomal changes. These clastogenic agents have generally shown a much higher mutagenicity at the heterozygous loci than at genes located in hemizygous regions (Clive et al. 1979). The experiments of Webber and de Serres (1965) indicated that multilocus lesions were lethal in haploid strains of *Neurospora crassa*, since only mutants with intragenic alterations were recovered in these strains. Chu (1971) found that X-radiation produces mutations at the *hprt* locus and suggested that radiation-induced multilocus lesions in surviving $HPRT^-$ mutants could not extend to essential genes on either side of the target *hprt* locus in this hemizygous region. Thus, determination of the mutagenicity of clastogens in which the target gene is the *hprt* locus, or any other gene located in a hemizygous region, would not be expected to include mutants harboring multilocus lesions and therefore would be deceptively low.

In the present work, we have compared the mutagenicity of clastogens and base-change mutagens in two strains of mouse L5178Y-R cells. Strain LY-R16 was found to be heterozygous at the *tk* locus on chromosome 11, and strain LY-R83 was found to possess only one chromosome 11. The clastogen-induced mutation frequencies at the *tk* locus in the monosomic strain were markedly lower than in the heterozygous strain. The low mutation frequencies at the *tk* locus in the monosomic strain were similar to the low clastogen-induced frequencies at the *hprt* and Na^+/K^+ ATPase loci in both strains. These results support the suggestion that the measurement of mutation frequencies at target genes located in hemizygous regions does not include mutants harboring multilocus lesions, since these cells are not likely to survive.

RESULTS

Presumptive $TK^{+/-}$ heterozygous strains of LY5178Y cells were isolated from strain L5178Y-TKR-5 according to the method of Clive and Voytek (1977). L5178Y-TKR-5 is resistant to bromodeoxyuridine (BrdU), lacks TK activity, is a presumptive homozygous $TK^{-/-}$ strain (obtained from J.Z. Beer, Food and Drug Administration, Rockville, Maryland). The cells were grown for 17 days in culture and were plated and recloned in Fischer's medium supplemented with THMG (*t*hymidine [0.12 mM], *h*ypoxanthine [0.04 mM], *m*ethotrexate [0.15 μg/ml], and glycine [0.1 mM]). Two strains (LY-R16 and LY-R83) were isolated that possessed approximately 50% of the TK activity of the $TK^{+/+}$ strain, L5178Y-R. These apparent heterozygotes were similar to the homozygous LY-R $TK^{+/+}$ strain in their response to the cytotoxic effects of various agents and in their mutability at the *hprt* and Na^+/K^+ ATPase loci (Evans et al. 1986).

Cells were treated with radiation or chemical agents in a dose range usually producing 10–80% survival. Following treatment, aliquots were removed for the measurement of survival in terms of colony-forming ability in soft agar medium, and the remainder of the treated (and control) cells were incubated at 37°C to allow expression of mutation. Expression times were 7, 3, and 2 days for measurement of mutations at the *hprt*, Na^+/K^+ APTase, and *tk* loci, respectively. Following expression, cells were plated in soft agar medium in the presence and absence of a selective agent for the measurement of mutant frequency. Selective agents were thioguanine (5 μg/ml), ouabain (Oua) (0.5 mM), and trifluorothymidine (TFT) (4 μg/ml) for measurement of mutations at the *hprt*, Na^+/K^+ ATPase, and *tk* loci, respectively.

Mutant frequencies at the *hprt* locus following treatment with X-radiation, UV radiation, and EMS are shown in Figure 1. The two strains were similar to each other in their mutability by these three agents at the *hprt* locus. However, the mutagenicity of X-radiation at the *hprt* locus was at least 10 times lower than the mutagenicity of EMS or UV radiation (Fig. 1). The two strains were also similar to each other in UV radiation- and EMS-induced mutant frequency at the Na^+/K^+ ATPase locus (Fig. 2). (No ouabain-resistant [Oua^r] mutants have been found to be produced by X-radiation [Baker et al. 1974; Arlett et al. 1975]). The two strains were similar to each other in their mutability at the *tk* locus, following treatment with UV radia-

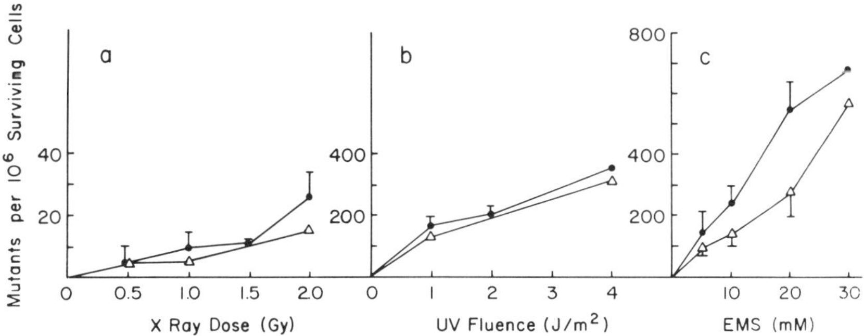

Figure 1

Mutant frequencies induced at the *hprt* locus by X radiation (*a*), UV radiation (*b*), and EMS (*c*). (●) Strain-LY-R16; (△) LY-R83. The graphs were redrawn from data reported by Evans et al. (1986). Vertical lines show the standard error of means derived from replicate experiments.

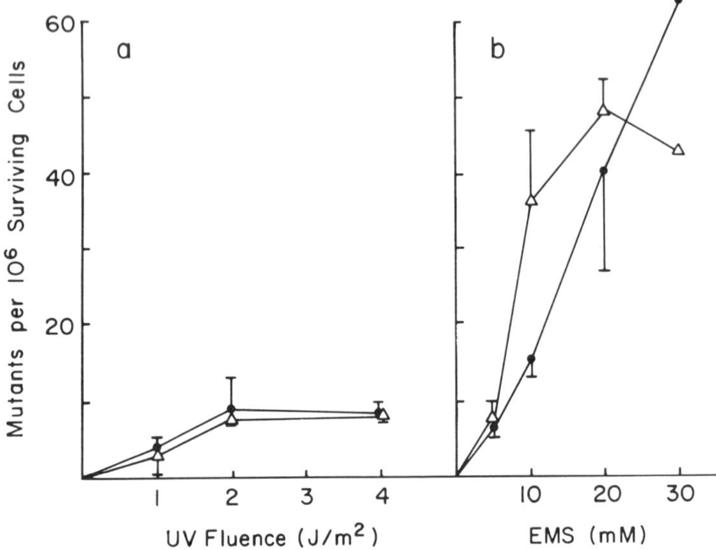

Figure 2

Mutant frequencies induced at the Na^+/K^+ ATPase locus by UV radiation (*a*) and EMS (*b*). (●) Strain LY-R16; (△) LY-R83. The graphs were redrawn from data reported by Evans et al. (1986). Vertical lines show the standard error of means derived from replicate experiments.

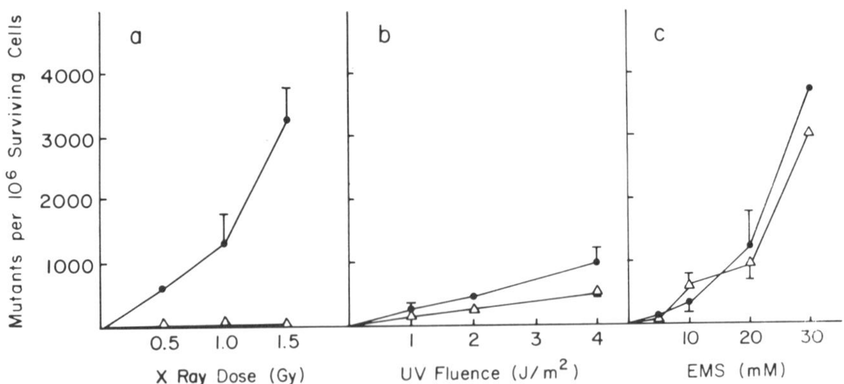

Figure 3

Mutant frequencies induced at the *tk* locus by X radiation (*a*), UV radiation (*b*), and EMS (*c*). (●) Strain LY-R16; (△) LY-R83. (Graphs redrawn from data reported by Evans et al. 1986.) Vertical lines show the standard error of means derived from replicate experiments.

tion and EMS (Fig. 3). However, following treatment with ionizing radiation, the mutability of strain LY-R83 was markedly lower than that of LY-R16. The X-radiation-induced mutability of strain LY-R83 at the *tk* locus was comparable to that observed at the *hprt* locus (Fig. 1). Since UV radiation and EMS are known to produce a predominance of point mutations, and X-radiation results in a large proportion of multilocus lesions and chromosome rearrangements (e.g., see Abrahamson and Wolff 1976; Thacker et al. 1978), the mutability of the two strains was compared following treatment with other clastogenic compounds. The mutability of strain LY-R83 was very low in comparison to strain LY-R16 following treatment with bleomycin (Bm), hydrogen peroxide (HP), and tertiary butyl hydroperoxide (BHP) (Fig. 4), as well as following treatment with the clastogenic, intercolating topoisomerase II inhibitors, 4′-(9-acridinylamino)methanesulfon-*m*-anisidide (*m*-AMSA) and ellipticine (Fig. 5). The sensitivity of the two strains to the cytotoxic effects of these agents is shown in Table 1 in terms of the dose of each agent required to reduce survival to 37%. The two strains were generally similar to each other in their sensitivity to the cytotoxic effects of these agents, except that strain LY-R16 was more sensitive than strain LY-R83 to HP and ellipiticine and more resistant than strain LY-R83 to Bm. The differences in the sensitivity of the two strains to the cytotoxic effects of these compounds were much less than the differences in their mutability by these agents. The mutant frequencies at the D37 dose of the various agents for the two strains of the three loci are also shown in Table 1. It may be noted that the mutant frequency of strain LY-R83 at the *tk* locus was similar to the mutant frequency

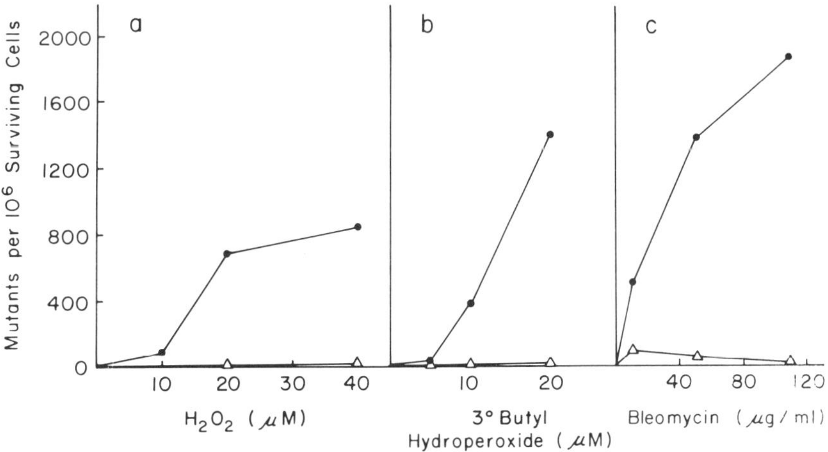

Figure 4

Mutant frequencies induced at the *tk* locus by H_2O_2 (*a*), tertiary butyl hydroperoxide (*b*), and bleomycin (*c*). (●) Strain LY-R16; (△) LY-R83.

Table 1
Sensitivity of Strains LY-R83 and LY-R16 to the Cytotoxic and Mutagenic Effects of Clastogens and Base-change Mutagens

Agent	Dose yielding 37% survival (D37)		Mutant frequency[a] at the D37 dose: *tk* locus		ATPase locus		*hprt* locus	
	R16	R83	R16	R83	R16	R83	R16	R83
EMS	28 mM	22 mM	3100	1250	57	47	660	340
UV-radiation	2 J/m^2	2 J/m^2	450	250	9	8	200	190
X-radiation	1.22 Gy	1.17 Gy	2250	50			10	8
H_2O_2	18 μM	44 μM	500	16				
3°-BHP	27 μM	28 μM	2050	18				
Bleomycin	19 μg/ml	7 μg/ml	730	75				
Ellipticine	240 ng/ml	730 ng/ml	420	18				
m-AMSA	15 ng/ml	19 ng/ml	3150	45				

[a]Mutant frequencies are expressed as mutants/10^6 surviving cells.

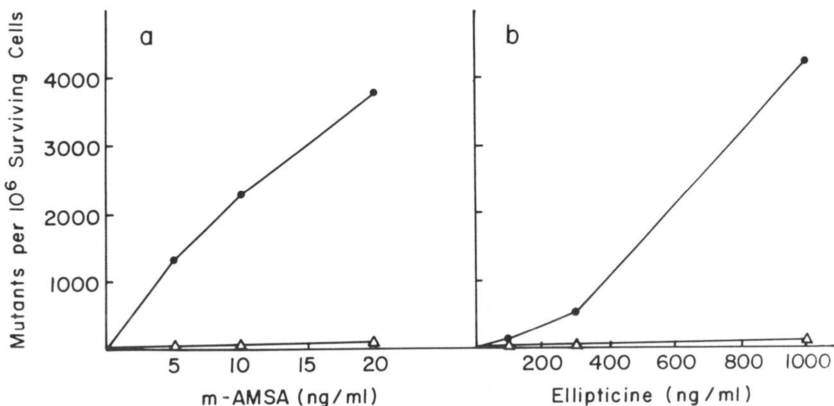

Figure 5

Mutant frequencies induced at the *tk* locus by *m*-AMSA (*a*) and ellipticine (*b*). (●) Strain LY-R16; (△) LY-R83.

at the Na^+/K^+ ATPase and *hprt* loci following treatment with all types of clastogens. In contrast, the clastogen-induced mutability of strain LY-R16 at the *tk* locus was more than 100 times higher than at the *hprt* locus following treatment with X-radiation. Although EMS- and UV-induced mutant frequencies were lower in strain LY-R83 than in strain LY-R16 at all three loci, the differences were much less marked than in the case of clastogen-induced frequencies. EMS- and UV-induced frequencies were higher at the *tk* locus than at the *hprt* locus, which in turn were higher than at the Na^+/K^+ ATPase locus.

Banded karyotype analysis, carried out by C. Sanchez and J. Hozier at the Florida Institute of Technology, according to methods described by Hozier et al. (1981), revealed that each strain possessed only one intact X chromosome (the site of the *hprt* gene). The analysis also showed that whereas strain LY-R16 possessed two copies of chromosome 11 (the site of the *tk* gene), strain LY-R83 possessed only one intact chromosome 11 (Fig. 6).

DISCUSSION

Our results demonstrate that the poor mutagenicity of clastogens at the *hprt* locus parallels their poor mutagenicity at the *tk* locus in an LY-R strain monosomic for chromosome 11. In contrast, clastogens were highly mutagenic at the *tk* locus in the $TK^{+/-}$ heterozygous LY-R strain possessing two copies of chromosome 11. These results agree with the suggestion that clastogens produce multilocus lesions that inactivate not only the target gene,

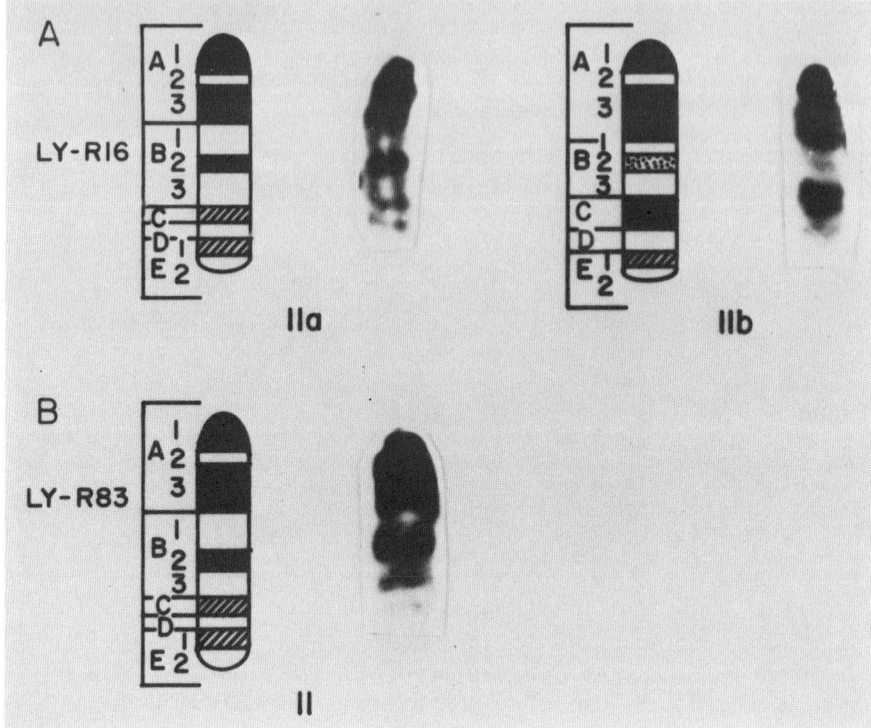

Figure 6
Banded karotype analysis of chromosome(s) 11 in strains LY-R16 (*a*) and LY-R83 (*b*). (Redrawn from Evans et al. 1986.)

but also neighboring genes that are necessary for growth and cell division. When such multilocus lesions occur in hemizygous chromosomal regions, no homologous chromosome is present to supply an active copy of the essential gene, and the mutant cell does not survive (Fig. 7). When multilocus lesions occur on autosomes, a homologous chromosome is present to supply an active copy of the essential gene so that the mutant survives, in some cases as a slow-growing colony (Clive et al. 1979; Moore et al. 1985) (Fig. 7b).

Mutations in the gene for Na^+/K^+ ATPase causing inactivation of the ouabain-binding site of the enzyme, can produce Oua^r ATPase molecules that in turn result in Oua^r cells. However, if the mutation inactivates the enzyme as well as the oubain-binding site, no active Oua^r enzyme is present, and the cell fails to exhibit ouabain resistance. Since the active site of the enzyme and the ouabain-binding site are within the same gene, the induction of Oua^r

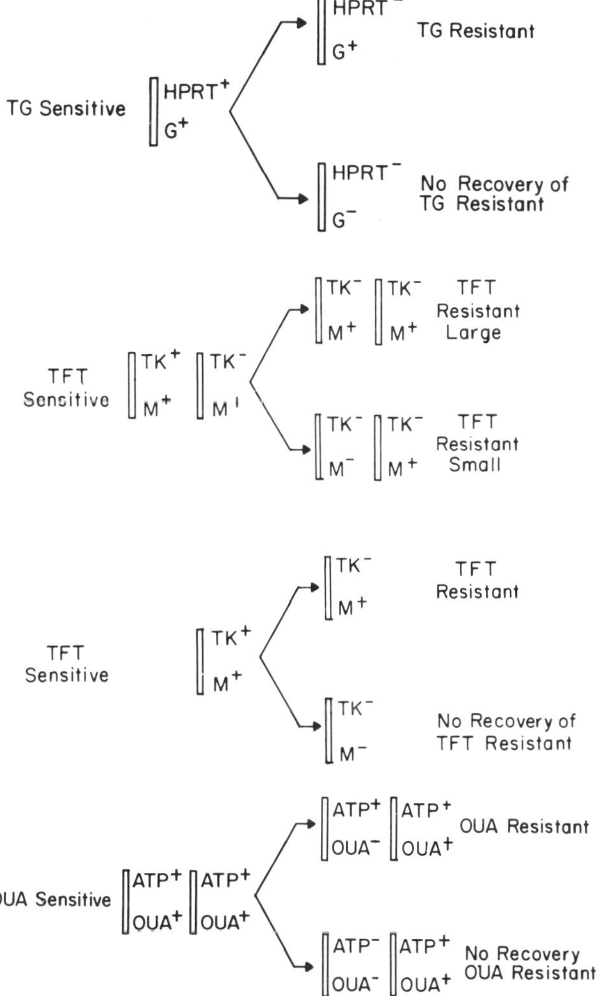

Figure 7

Possible mutational events occurring at the *hprt* locus (TG sensitive), and *tk* locus (TFT sensitive); and the Na^+/K^+ ATPase locus (OUA sensitive). *G* and *M* are hypothetical essential genes neighboring the *hprt* and *tk* genes, respectively.

occurs only after treatment of cells with agents that produce point mutations (Fig. 7c) (Arlett et al. 1975; Thacker et al. 1978).

Results of other investigators have also indicated that high mutant frequencies can be induced by clastogens, when the target gene is located on an autosome. Clive et al. (1979) reported the high mutability of clastogens at the *tk* locus versus the *hprt* locus in mouse L5178Y strain $TK^{+/-}$-3.7.2C, and D.W. Yandell et al. (in prep.) have reported similar results for the human lymphoblastoid line TK6. W.E.C. Bradley et al. (in prep.) and G.M. Adair et al. (in prep.) have found that the mutagenicity of clastogens at the adenine phosphoribosyl transferase (*aprt*) locus is higher when this gene is present in a heterozygous condition rather than in a hemizgyous condition.

Clastogen-induced mutant frequencies of transfected genes have also been shown to be markedly higher than in the case of host genes located in hemizygous chromosomal regions. Thus, Tindall et al. (1984), Stankowski and Hsie (1986), and Hsie et al. (1986) have reported that clastogen-induced mutant frequencies at the bacteria guanine phosphoribosyl transferase (*gpt*) gene, inserted into a CHO chromosome, are markedly higher than at the resident *hprt* gene on the X chromosome. Waldren et al. (1979) have reported high clastogen-induced mutant frequencies at genes located on human chromosome 11 in transfected Chinese hamster cells. Although the hybrid cells harbor only one copy of the human chromosome, none of the human genes are essential for the growth and multiplication of the hamster cells. Hybrid cell lines harboring foreign target genes that are not essential for the growth of the host cell and/or are not located in host hemizygous regions are thus advantageous in the determination of mutant frequencies. These assays eliminate the problem of deceptively low clastogen-induced mutant frequencies obtained at hemizgyous loci, such as the *hprt* locus on the X chromosome, as well as the necessity of isolation heterozygous strains for the measurement of mutation at autosomal sites.

Finally, it may be noted that the disparity between clastogen-induced frequencies at the *hprt* and *tk* loci would be expected to vary from strain to strain depending on the arrangement of the genes on the X chromosome, as well as upon the ability of the various strains to repair DNA lesions (de Serres et al. 1983; Evans et al. 1986).

ACKNOWLEDGMENTS

This research was supported by U.S. Public Health Service grants CA-15901 and CA-23427 awarded by the National Cancer Institute. We thank Dr. T.E. Evans, Case Western Reserve University, for his help in reviewing the manuscript.

REFERENCES

Abrahamson, S. and S. Wolff. 1976. Re-analysis of radiation-induced specific locus mutations in the mouse. *Nature* **264:** 715.

Adair, G.M., R.L. Stallings, R.S. Nairn, and M.J. Siciliano. 1983. High frequency structural gene deletion as the basis for functional hemizygosity of the adenine phosphoribosyl-transferase locus in Chinese hamster ovary cells. *Proc. Natl. Acad. Sci.* **80:** 5961.

Arlett, C.F., D. Turnbull, S.A. Harcourt, A.R. Lehmann, and C.M. Collella. 1975. A comparison of the 8-azaguanine and ouabain-resistance systems for the selection of induced mutant Chinese hamster cells. *Mutat. Res.* **33:** 261.

Baker, R.M., D.M. Brunette, R. Mankovitz, L.H. Thompson, G.F. Whitmore, L. Siminovitch, and J.E. Till. 1974. Ouabain-resistant mutants of mouse and hamster cells in culture. *Cell* **1:** 9.

Carver, J.R., G.M. Adair, and D.L. Wondres. 1980. Mutagenicity testing in mammalian cells. II. Validation of multiple drug-resistance markers having practical application for screening potential mutagens. *Mutat. Res.* **72:** 207.

Clive, D. and P. Voytek. 1977. Evidence for chemically-induced structural gene mutations at the thymidine kinase locus in cultured L5178Y mouse lymphoma cells. *Mutat. Res.* **44:** 269.

Clive, D., K.O. Johnson, J.F.S. Spector, A.G. Batson, and M.M.M. Brown. 1979. Validation and characterization of the L5178Y/TK$^{+/-}$ mouse lymphoma mutagen assay system. *Mutat. Res.* **59:**161.

Chu, E.H.Y. 1971. Mammalian cell genetics. III. Characterization of X ray induced mutations in Chinese hamster cell cultures. *Mutat. Res.* **11:** 23.

de Serres, F.J., H. Inoue, and M.E. Schupbach. 1983. Mutagenesis at the *ad-3A* and *ad-3B* loci in haploid UV sensitive strains of *Neurospora crassa.* VI. Genetic characterization of *ad-3* mutants provides evidence for qualitative differences in the spectrum of genetic alterations between wild-type and nucleotide excision-repair deficient strains. *Mutat. Res.* **108:** 93.

Evans, H.H., J. Mencl, M.F. Horng, M. Ricanati, C. Sanchez, and J. Hozier. 1986. Locus specificity in the mutability of L5178Y mouse lymphoma cells: The role of multilocus lesions. *Proc. Natl. Acad. Sci.* **83:** 4379.

Hozier, J., J. Sawyer, M. Moore, B. Howard, and D. Clive. 1981. Cytogenetic analysis of the L5178Y/TK$^{+/-}$TK$^{-/-}$mouse-lymphoma mutagenesis assay system. *Mutat. Res.* **84:** 169.

Hsie, A.W., L. Recio, D.S. Katz, C.Q. Lee, M. Wagner, and R.L. Schenley. 1986. Evidence for reactive oxygen species inducing mutations in mammalian cells. *Proc. Natl. Acad. Sci.* **83:** 9616.

Liskay, R.M. and R.J. Evans. 1980. Inactive X chromosome DNA does not function in DNA mediated cell transformation for the phosphoribosyltransferase gene. *Proc. Natl. Acad. Sci.* **77:** 4895.

Lyon, M.J. 1961. Gene action in the X chromosome of the mouse. *Nature* **190:** 372.

Mohandas, T., L.J. Shapiro, R.S. Sparkes, and M.C. Sparkes. 1979. Regional assignment of the steroid-sulfatase-X-linked ichthyosis locus: Implications for a noninactivated region on the short arm of the human X chromosome. *Proc. Natl. Acad. Sci.* **76:** 5779.

Moore, M.M., D. Clive, J.C. Hozier, B.E. Howard, A.G. Batson, N.T. Turner, and J. Sawyer. 1985. Analysis of trifluorothymidine-resistant mutants of L5178Y/TK$^{+/-}$ mouse lymphoma cells. *Mutat. Res.* **151:** 161.

Nalbantoglu, J., O. Goncalves, and M. Meuth. 1983. Structure of the mutant alleles at the *aprt* locus of Chinese hamster ovary cells. *J. Mol. Biol.* **167:** 575.

Stankowski, L.F. and A.W. Hsie. 1986. Quantitative and molecular analysis of radiation-induced mutation in A552 cells. *Radiat. Res.* **105:** 37.

Thacker, J., M.A. Stephens, and A. Stretch. 1978. Mutation to oubain-resistance in Chinese hamster cells. Induction by ethylmethanesulfphonate and lack of induction by ionising radiation. *Mutat. Res.* **51:** 255.

Tindall, K.R., L.F. Stankowski, R. Machanoff, and A.W. Hsie. 1984. Detection of deletion mutations in pSV2*gpt*-transformed cells. *Mol. Cell. Biol.* **4:** 1411.

Waldren, C., C. Jones, and T.T. Puck. 1979. Measurement of mutagenesis in mammalian cells. *Proc. Natl. Acad. Sci.* **76:** 1358.

Webber, B.B. and F.J. de Serres. 1965. Induction kinetics and genetic analysis of X-ray-induced mutations in the *ad-3* region of *Neurospora crassa. Proc. Natl. Acad. Sci.* **53:** 430.

Differential Recovery of Induced Mutants at the *tk* and *hgprt* Loci in Mammalian Cells

MARTHA M. MOORE,* KAREN H. BROCK,† DAVID M. DEMARINI,* AND CAROLYN L. DOERR†

*Genetic Toxicology Division
Health Effects Research Laboratory
U.S. Environmental Protection Agency
Research Triangle Park
North Carolina 27711

†Environmental Health Research and Testing, Inc.
Research Triangle Park
North Carolina 27709

OVERVIEW

Human genetic disease is known to result from both point mutations and chromosomal aberrations. Therefore, it is critical that short-term in vitro mammalian tests be evaluated as to their capabilities for detecting both types of lesions. Research to date indicates that L5178Y/TK$^{+/-}$−3.7.2C mouse lymphoma cells permit the measurement of forward mutation at the heterozygous thymidine kinase (*tk*) locus. Extensive studies indicate that these TK-deficient mutants result from both single-gene mutations (large-colony mutants) and mutations involving the expression of multiple loci (small-colony mutants). Comparative mutagenicity studies reveal a quantitative difference in the ability of the *tk* and hypoxanthine-guanine phosphoribosyl transferase (*hgprt*) loci to permit recovery of mutations affecting the expression of multiple loci. Thus, the observed HGPRT mutant frequency will underestimate the genotoxicity of compounds acting as clastogens.

INTRODUCTION

The discovery by Muller that X-rays were mutagenic in *Drosophila melanogaster* ushered in the field of mutagenesis (Muller 1927). In this first

This manuscript has been reviewed by the Health Effects Research Laboratory, U.S. Environmental Protection Agency, and approved for publication. Approval does not signify that the contents necessarily reflect the views and policies of the Agency, nor does mention of trade names or commercial products constitute endorsement or recommendation for use.

report, Muller observed that: "In addition to gene mutations, it was found that there is also caused by X-ray treatment a high proportion of rearrangements of the linear order of the genes." Thus, Muller showed that X-rays appeared to induce two classes of mutation: gene mutations and chromosomal rearrangements.

In 1951, W.L. Russell reported the development of a mouse-specific-locus mutagenesis assay that permitted detection of germ-cell mutations at seven loci (Russell 1952). Russell observed that X-rays induced both single-locus mutations and mutations affecting two closely linked loci. He surmised that some of these double mutants were probably not caused by two independent mutations but, rather, arose from chromosomal rearrangements involving the two loci.

Using the mouse model, de Serres incorporated two closely linked genes adenine-3A and adenine-3B (*ad-3A* and *ad-3B*) into a *Neurospora crassa* heterokaryon (de Serres and Osterbind 1962; de Serres and Malling 1971; de Serres, this volume). The *ad-3* forward-mutation assay in two-component heterokaryons of *Neurospora* permitted the recovery of both single-gene mutations and functional multilocus deletions. Using this system, Webber and de Serres (1965) showed that most of the specific locus mutations in the *ad-3* region induced by X-rays at low doses were point mutations, whereas the majority of the mutants at high doses were multilocus deletions. Using reversion techniques, Malling and de Serres (1973) showed that X-ray-induced point mutations resulted from single-base pair changes (95%) or small intralocal deletions (5%).

Chu and Malling (1968) reported the first quantitative specific-locus mutagenesis in cultured mammalian cells. Subsequently, Chu (1971) reported on the reversion analysis of X-ray-induced HGPRT mutants in V79 cells. Chu concluded that both point mutations and nonrevertible chromosomal deletions were being detected. A year later, Clive et al. (1972) reported the development of a mutagenesis assay using the *tk* locus in L5178Y mouse lymphoma cells (Clive, this volume). This locus was different from the *hgprt* locus because it was located on an autosome and, to make mutagenesis experiments practical, required the development of a cell line heterozygous for the *tk* locus. An interesting feature of this assay was the induction of small- and large-colony TK mutants, the quantitation of which was made possible by the development of trifluorothymidine (TFT) as a selective agent (Brown and Clive 1978; Moore-Brown et al. 1981) and the use of an automatic colony counter (Clive et al. 1979; Moore-Brown and Clive 1979). Small colonies were found to be associated in these mouse lymphoma cells with TK mutants but not with HGPRT mutants; in addition, most compounds produced higher TK mutant frequencies than HGPRT mutant frequencies (Brown and Clive 1978; Clive et al. 1979). These observations led to the proposal, first presented at the 8th Annual Environmental Mutagen Society meeting in 1977, "that mutations at the heterozygous *tk* locus represent both chromosomal and genic events, whereas *hgprt* mutations occur primarily at the gene level" (Brown and Clive 1978).

Because human genetic disease is known to result from both point mutations and chromosomal aberrations, it is critical that short-term tests be understood as to their capabilities of detecting both types of lesions. We present evidence that the *tk* locus of L5178Y/TK$^{+/-}$ −3.7.2C cells can be used to evaluate both single-gene mutation and mutation involving the expression of multiple loci (defined here as chromosomal mutation). We also compare mutant recovery at the *tk* and *hgprt* loci.

RESULTS

The analysis of the ability of the *tk* locus to permit the recovery of both single-gene mutation and chromosomal mutation revolves around an analysis of the small- and large-colony TFT-resistant mutants. Several approaches have been used: (1) phenotypic analysis of mutants, (2) banded karyotype analysis of mutants, (3) relative capabilities of particular agents to induce small and large colonies, (4) comparison of mutagenesis in a TK heterozygous versus a TK hemizygous cell line, (5) comparison of mutant frequencies at the *tk* and *hgprt* loci, and (6) evaluation of expression times of both loci.

Phenotypic Analysis of Mutants

A large number of both small- and large-colony mutants has been isolated, clonally expanded in nonselective (no TFT) medium, and tested for their phenotypic properties (Moore et al. 1985b). Briefly, 150 of 152 mutants from mutagen-treated cultures and 163 of 168 spontaneous mutants were TFT resistant when rechallenged approximately 1 week after isolation (3 weeks after induction). All of the 41 mutants assayed for enzyme activity were TK deficient. The small-colony phenotype was found to correlate with slow cellular growth rates (doubling times $>$ 12 hr), rather than from effects of the TFT selection or mutagen toxicity. More recently, Meyer et al. (1986) reported that 97 mutants isolated from treated cultures and 69 mutants from untreated cultures were all (except for one large-colony mutant from an untreated culture) TFT resistant upon rechallenge with TFT.

Banded Karyotype Analysis

The cytogenetic characterization of L5178Y/TK$^{+/-}$–3.7.2C mouse lymphoma cell line has been published by Sawyer et al. (1985). This analysis and previously published research (Hozier et al. 1982) indicate that the overall karyotype for this cell line is stable and contains normal mouse chromosomes, stable chromosome rearrangements, and stable marker chromosomes. There is at least one normal mouse homolog present for almost every mouse chromosome, and the chromosome-11 homologs, on which the *tk* gene is located (Kozak and Ruddle 1977), are distinguishable from each other by the size of the centromeres (Hozier et al. 1982; Sawyer et al. 1985). Banded karyotype analyses of small-colony mutants approximately 3 weeks after induction show an association between the small-colony phenotype and readily observable (at the 230–300-band resolution level) chromosomal abnormalities (primarily translocations involving the larger centromeric chromosome 11, carrying the functional *tk* gene) in 30 of 51 induced mutants studied (Hozier et al. 1982; Moore et al. 1985b). Using an early clonal analysis of mutants (~2 weeks after induction), Hozier et al. (1985) showed that 28 of 30 small-colony mutants had the chromosome-11 rearrangements. Furthermore, in this early clonal analysis, a significant proportion of the small-colony mutants (34%) had dicentric chromosomes involving chromosome 11 (Hozier et al. 1985). All large-colony mutants studied (17 of 17 evaluated 3 weeks after induction and 8 of 8 evaluated 2 weeks after induction) showed normal karyotypes at the 230–300-band resolution level, including the chromosome-11 homologs. Similar results have been observed by Blazak et al. (1986). They found that 16 of 34 small-colony mutants had chromosome-11 abnormalities, and 4 had other chromosome abnormalities. Of eight large colonies analyzed, one had structural chromosome abnormalities not involving chromosome 11, and the other seven were normal. More recent studies (Applegate and Hozier, this volume) indicate the utility of performing karyotypic analysis in conjunction with molecular analysis.

Relative Capabilities of Specific Agents to Induce Small- and Large-colony TK Mutants

To investigate the hypothesis that chromosomal mutations (i.e., small-colony mutants) result from the clastogenic activity of the test agent, whereas large-colony mutants reflect single-gene mutations, a number of compounds have been evaluated. Agents such as 2-acetylaminofluorene (2-AAF), benzo[a]pyrene (B[a]P), dimethylnitrosamine (DMN), methyl methanesulfonate (MMS), and *N*-methyl-*N'*-nitro-*N*-nitrosoguanidine (MNNG), which are both clastogenic in mammalian cells and cause gene mutation in a variety of other systems, induce a significant number of both large- and small-colony TK mutants (Moore et al. 1985b). Hycanthone (an agent that produces primarily multilocus deletions in *Neurospora* [Ong and de Serres 1980]) was found to induce predominantly small-colony TK mutants in L5178Y/TK$^{+/-}$–3.7.2C cells (Moore-Brown and Clive 1979; Moore et al. 1985a). The clastogens, γ-irradiation (Moore et al. 1986), caffeine (Clive 1983), and acyclovir (Clive et al. 1983) induce almost exclusively small colonies.

We recently evaluated several topoisomerase-active drugs known to be potent clastogens that give very low mutant frequencies at the *hgprt* locus. Without exception, these compounds were found to be clastogenic to L5178Y/$TK^{+/-}$–3.7.2C cells and to induce almost exclusively small colonies. 4′-(9-Acridinylamino)methanesulfon-*m*-anisidide (*m*-AMSA), an antitumor

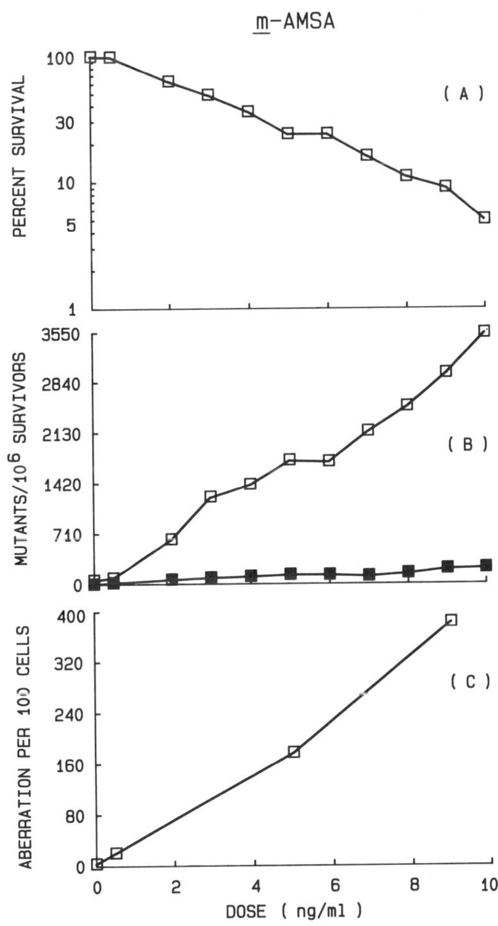

Figure 1

Cytotoxicity (*A*), mutagenicity (*B*), and clastogenicity (*C*) following *m*-AMSA treatment of $TK^{+/-}$–3.7.2C mouse lymphoma cells. Cytotoxicity is calculated according to the method of Clive and Spector (1975). The mutagenicity is divided into small-colony (□) and large-colony (■) TK mutant frequencies. Clastogenicity is expressed as total aberrations (excluding gaps) per 100 metaphases analyzed.

agent, was found by Wilson et al. (1984) to give relatively low mutant frequencies at the *hgprt* locus in V79 cells. In mouse lymphoma cells, we found *m*-AMSA to be one of the most potent mutagens and clastogens tested (Fig. 1). Almost all of the mutants were small colonies. Other topoisomerase-active drugs that give low HGPRT mutant frequencies yet induce high TK mutant frequencies, due almost exclusively to small colonies, include teniposide (DeMarini et al. 1987), ellipticine (Moore et al. 1987a), and adriamycin (Moore et al. 1987c).

Structurally similar compounds known not to induce mutations in *Salmonella* yet known or suspected to be clastogens for mammalian cells have been evaluated. These compounds, acrylamide (Moore et al. 1987b), acrylic acid (M.M. Moore, unpubl.), ethyl acrylate and methyl acrylate (Amtower et al. 1986), are clastogenic and induce almost all small-colony TK mutants.

We also evaluated two free-radical producing agents known to be non-mutagenic at the *hgprt* locus yet mutagenic at the bacterial guanine phosphorobosyl transferase (*gpt*) locus that has been integrated into the Chinese hamster ovary (CHO) genomes (Hsie et al. 1986; Stankowski and Tindall, this volume). As anticipated, hydrogen peroxide (Fig. 2) and potassium superoxide (Fig. 3) were clastogenic and induced primarily small-colony TK mutants.

Very few compounds have been defined that exclusively or even predominantly induce point mutations. One source of information is the *Neurospora* system; however, several of the compounds that appear to induce point mutations in *Neurospora* are known to be clastogenic to mammalian cells. ICR-170 was reported by de Serres and Brockman (1968) to induce exclusively point mutations. Both ICR-170 (Moore et al. 1985a) and the structurally similar ICR-191 (Fig. 4) induce primarily large-colony TK mutants. Ethyl methanesulfonate (EMS) has also been found to induce more large-colony TK mutants than small-colony mutants (Clive et al. 1979; Moore et al. 1985a). It should be noted that all three compounds do induce small-colony TK mutants, particularly at higher doses, and all three compounds are clastogenic to mouse lymphoma cells (C.L. Doerr, unpubl.).

Mutagenesis in $TK^{+/0}$ versus $TK^{+/-}$ Cells

Evans et al. (1986 and this volume) have recently reported differential mutagenicities between a $TK^{+/0}$ (R83) and a $TK^{+/-}$ (R16) mouse lymphoma cell line. The TK mutant frequency of the $TK^{+/0}$ line and the $TK^{+/-}$ cell line were approximately the same for EMS treatment. The $TK^{-/0}$ mutant frequency was significantly lower than the $TK^{-/-}$ mutant frequency following X-irradiation and treatment by other compounds that induce primarily small-colony TK mutants in L5178Y/$TK^{+/-}$–3.7.2C (Evans et al., this volume). In

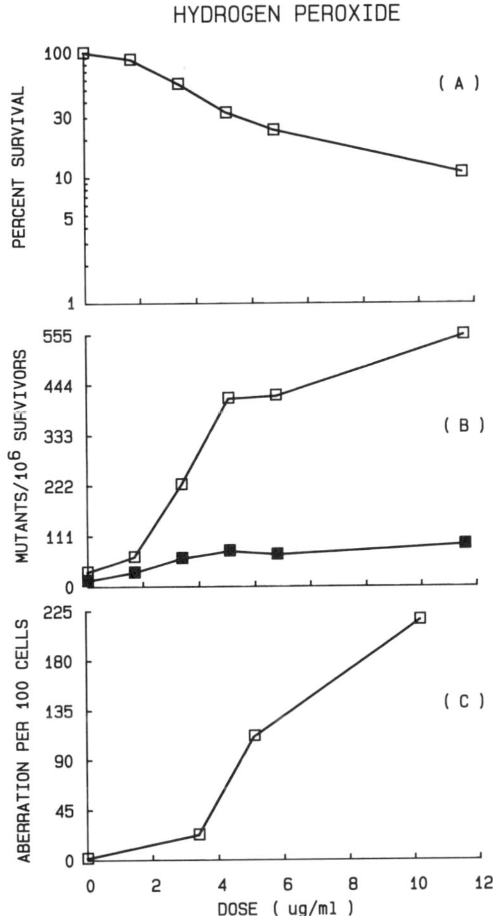

Figure 2

Cytotoxicity (*A*), mutagenicity (*B*), and clastogenicity (*C*) following hydrogen peroxide treatment of $TK^{+/-}$-3.7.2C mouse lymphoma cells. Cytotoxicity is calculated according to the method of Clive and Spector (1975). The mutagenicity is divided into small-colony (□) and large-colony (■) TK mutant frequencies. Clastogenicity is expressed as total aberrations (excluding gaps) per 100 metaphases analyzed.

collaboration with Dr. Evans, we have compared the TK mutagenicity of our $TK^{+/-}$-3.7.2C cell line with that of her $TK^{+/0}$ R83 cell line following *m*-AMSA treatment in our laboratory. Figure 5 shows that the total $TK^{-/0}$ mutant frequency is about the same as either the $TK^{-/-}$ large-colony mutant frequency or the $HGPRT^{-/0}$ mutant frequency. These results indicate the inability of the hemizygous $TK^{+/0}$ cell line to quantitate the chromosomal mutations represented by the small-colony $TK^{-/-}$ mutants.

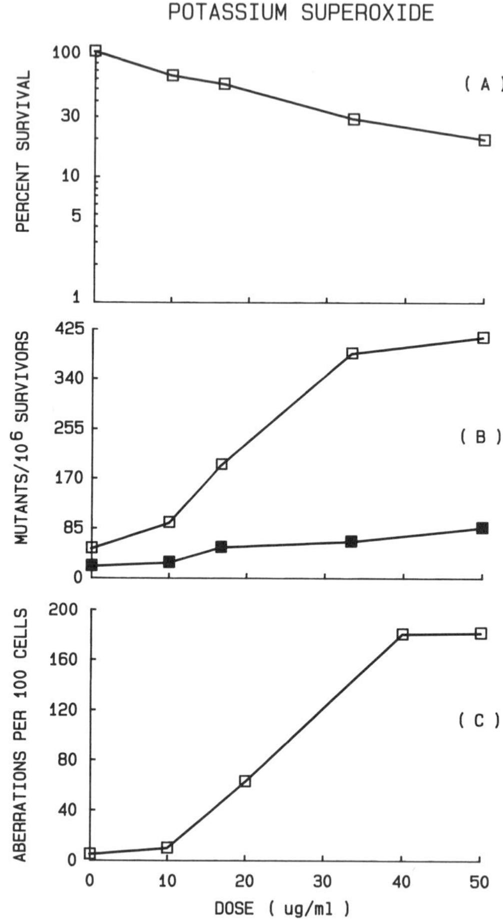

Figure 3

Cytotoxicity (*A*), mutagenicity (*B*), and clastogenicity (*C*) following potassium superoxide treatment of $TK^{+/-}$–3.7.2C mouse lymphoma cells. Cytotoxicity is calculated according to the method of Clive and Spector (1975). Mutagenicity is divided into small-colony (□) and large-colony (■) TK mutant frequencies. Clastogenicity is expressed as total aberrations (excluding gaps) per 100 metaphases analyzed.

Comparison of the *tk* and *hgprt* Loci

The observation that in mouse lymphoma cells the HGPRT mutant frequency induced by various chemicals was generally lower than the TK mutant frequency (Clive et al. 1979) contributed to the hypothesis that the two loci might differ in their abilities to detect both single-gene and chromosomal mutation (Brown and Clive 1978). Table 1 provides a representative sample

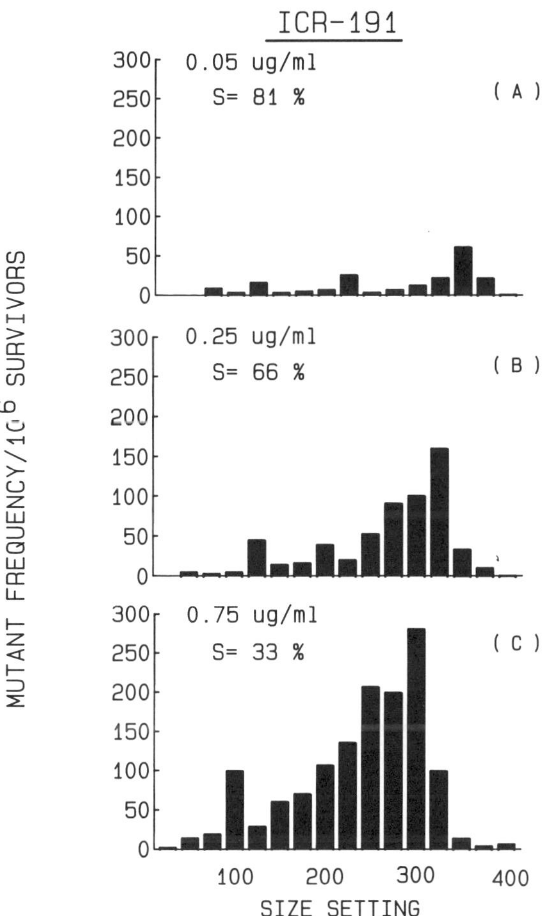

Figure 4
Evaluation of the $TK^{-/-}$ mutant colony size distribution for ICR-191-treated L5178Y/$TK^{+/-}$ -3.7.2C cells. The change in the magnitude of the mutant frequencies quantitated at adjacent size settings is plotted relative to the colony counter size settings. Colony size increases with increasing size setting.

of comparative TK and HGPRT mutagenicity data. In general, compounds that give a high frequency of large $TK^{-/-}$ mutants also induce a high HGPRT mutant frequency. One of these compounds, diethylnitrosamine (DEN), has been shown to induce predominantly point mutations in *Neurospora* (de Serres, this volume). Compounds inducing low-large $TK^{-/-}$ mutant frequencies also give relatively low HGPRT mutant frequencies in L5178Y/ $TK^{+/-}$ -3.7.2 cells.

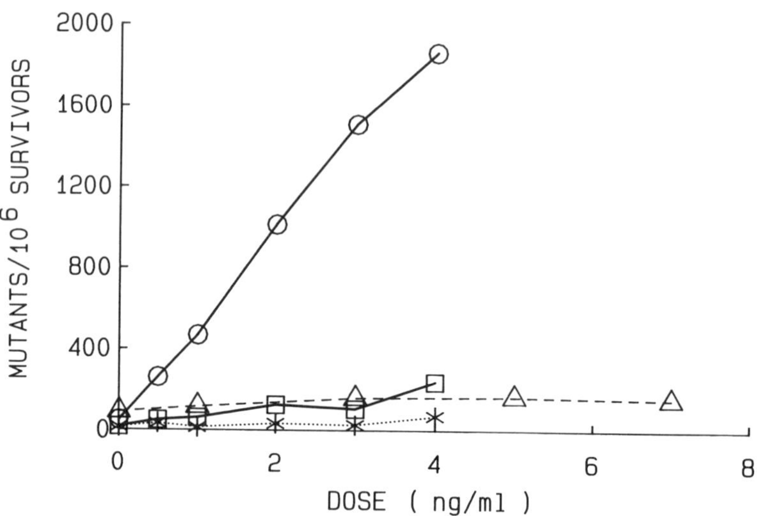

Figure 5
Comparative mutagenicity following *m*-AMSA treatment of $TK^{+/-}$–3.7.2C or $TK^{+/0}$ R83 cells. (○) Small-colony $TK^{-/-}$–3.7.2C mutant frequency; (□) large-colony $TK^{-/-}$–3.7.2C mutant frequency; (*) $HGPRT^{-/0}$ mutant frequency ($TK^{+/-}$–3.7.2C cells); and (△) $TK^{-/0}$ R83 mutant frequency.

Expression Time

The *tk* locus of L5178Y/$TK^{+/-}$–3.7.2C cells requires 2–3 days for expression, whereas the *hgprt* locus requires 6 days (Moore and Clive 1982). An evaluation of the TK mutant frequency at various times posttreatment indicated that for compounds inducing significant small-colony TK mutant frequencies, the TK mutant frequency declined after 2–3 days expression (Clive et al. 1979; Moore-Brown and Clive 1979; Moore and Clive 1982). When the small- and large-colony frequencies were evaluated separately, the lower mutant frequency could be accounted for by a lower small-colony mutant frequency (Moore and Clive 1979; Moore et al. 1985a). This decline in small-colony mutant frequency was seen to result from a subpopulation of slow-growing mutants that decline in frequency (because of their slow growth relative to the rest of the population) during the time allowed for expression. This raised the possibility that the difference in the TK and HGPRT mutant frequency might be due simply to the difference in expression time. Experiments reported by Moore and Clive (1982) and O'Neill et al. (1982) partially address the issue of long expression times and the underestimation of mutant frequencies.

Table 1
Mutagenicity Observed at the *tk* and *hgprt* Loci of $TK^{+/-}$ − 3.7.2 Mouse Lymphoma Cells

Compound	Concentration (μg/ml)	Survival (% control)	Mutant frequency ($\times 10^{-6}$)	
			tk	*hgprt*
m-AMSA	0	100	67	15
	0.004	15	2102	71
2-AAF	0	100	52	16[a]
	50	12	624	17[a]
B[a]P	0	100	57	18[a]
	3	31	601	—
	3.50	22	—	28[a]
	3.75	14	—	51[a]
DEN[a]	0	100	78[b]	20
	2500	32	475[b]	399
DMN[a]	0	100	55[b]	8
	74	24	519[b]	99
EMS	0	100	36	7
	650	14	1408	640
Hycanthone[a]	0	100	54	3
	12.5	23	231	5
MMS[a]	0	100	41	7
	10	20	393	24

[a]Data from Clive et al. (1979).
[b]BrdU selection; 72-hr expression.

Moore and Clive (1982) reported a new technique that allowed for the sequestering, expression, and selection (SES) of mutants in Linbro wells. Briefly, in this SES technique, 25 or 50 EMS-treated cells were placed in a large number of Linbro wells. The selective agent was added to the wells at day-3 posttreatment for TK mutants and at day-6 posttreatment for HGPRT mutants. The wells were then scored for mutant growth, and mutant frequencies were calculated. Mutant colonies containing hundreds of cells were counted equally with the slowest-growing mutant colonies containing approximately 20 cells. Using this SES technique, the quantitated mutant frequency is based on the number of mutant cells (either expressed or unexpressed) present in the population at the time the cells were plated into the Linbro wells—not at the time the selective agent was added (as is the case in conventional soft agar cloning). For the *tk* locus, SES analysis yielded an approximately eightfold higher frequency at day-1 posttreatment than was obtained by either the SES technique or conventional cloning at day-3 posttreatment. For the *hgprt* locus, there was some indication that the day-1 mutant frequency might be higher than the day-6 mutant frequency. When the day-1 posttreatment frequencies (obtained with the SES technique) were compared for the two loci, the *tk* locus appeared to be more sensitive (eight to ten times) to the same EMS insult. It should be noted that this analysis does not address potential differences in mutabilities of the two loci, but only potential differences in numbers of viable mutants recovered. For additional details, see Moore and Clive (1982).

O'Neill et al. (1982) performed a series of experiments using CHO cells that were growth-arrested at about day-2 posttreatment. Cells were held without division until they were plated for thioguanine (TG) selection on day-9 posttreatment. This method prevents the decline in mutant frequency resulting from any slow-growing HGPRT mutants. This procedure would not, however, protect against any significant decline in frequency prior to the growth arrest. Using this technique, O'Neill et al. found that the EMS-induced mutant frequency was about the same as that seen in the normal cloning procedure. Both sets of experiments indicate that the HGPRT mutant frequency does not decline significantly during the expression period, and that any quantitative difference between the TK and HGPRT mutant frequency is not likely accounted for by the difference in the expression time.

DISCUSSION

Our evaluation of the research performed to date indicates that both small and large TFT-resistant colonies are in fact true TK-deficient mutants. The difference between the two classes appears to result from the slow growth rate of the small-colony mutants. In *Neurospora*, de Serres (this volume) observed that the ability to recover mutations affecting multiple loci was dependent on the presence of a second allele to protect against heterozygous effects. It appears that the situation is similar in mammalian cells. A number of studies (for a summary, see Adair; Evans et al.; Hsie; Little et al.; Stankowski and Tindall; all this volume) indicate a difference in the ability of newly induced mutant loci in hemizygous regions of the genome (i.e., adenine phosphoribosyl transferase [$APRT^{+/0}$], $HGPRT^{+/0}$, and $TK^{+/0}$) and mutant loci in heterozygous or homozygous regions of the genome (i.e., $TK^{+/-}$, GPT^{+}) to survive both single-gene mutations and mutations that affect multiple loci. Specifically, mutants resulting from point mutations ought to be recovered equally well by all nonessential genes (i.e., *tk* and *hgprt*). However, mutational events that affect the function of multiple linked genes will be recovered as viable mutants only if no essential gene(s) function is affected. In hemizygous regions of the genome, there will be no allele(s) to replace the missing function of newly mutated essential gene(s). In heterozygous or homozygous regions of the genome, a newly induced mutant will survive only if any affected essential gene(s) have allele(s) that can cover for their lost function. It should be noted that many of these mutants would be expected to have only partial function (compared to wild type) of an essential gene and might be slow growing. Mutagenic mechanisms that may affect the target gene and other loci include, but may not be restricted to, multilocus deletion, chromosome rearrangement, mitotic recombination, gene conversion, and nondisjunction.

On the basis of these principles, TK and HGPRT mutants, when analyzed at the DNA level, would show some similarities and some differences. Both types of mutants would be expected to show point mutations (affecting one or a few base pairs) and large-scale events (affecting all or part of the target gene). The type of events observed would be mutagen specific. X-rays or γ-rays would induce primarily large-scale events; EMS would induce both types of damage. Deletions (including entire single-gene deletions) affecting only the expression of the target gene would be recovered at both the *tk* and *hgprt* loci. The nonrevertible mutants first observed by Chu (1971) likely resulted from deletions in or of the *hgprt* locus, as seen with molecular analyses by O'Neill et al. (this volume) and Little et al. (this volume). The two loci will differ, however, in their ability to recover multilocus deletions and other events affecting the expression of multiple loci. These chromosomal mutations would more likely be recovered by selecting for TK mutants rather than HGPRT mutants. The molecular analysis of these mutlilocus events requires techniques that probe not only the target gene but also the adjacent loci. It may be possible to recover some HGPRT mutants that result from chromosome rearrangements (Cox and Mason 1978). In fact, de Serres (1958) found X-ray-induced chromosomal rearrangements associated with single-gene mutation in the *ad-3* region of *Neurospora.* However, as predicted by Chu (1971), the presence of linked essential genes and/or essential DNA sequences would limit the recovery of induced mutations affecting both *hgprt* and linked loci. On the basis of the comparative mutagenicity studies presented by Evans et al.; Little et al.; Stankowski and Tindall (all this volume), there does appear to be a quantitative difference in the ability of the *tk* and *hgprt* loci to permit the recovery of mutations affecting the expression of multiple loci. This difference is true both in rodent (Evans et al.; Stankowski and Tindall; both this volume) and human cells (Little et al., this volume).

Because of the significance of both single-gene mutations and chromosomal aberrations in human health, it is necessary that chemical and physical agents be evaluated for their ability to induce both types of mutation. From all analyses, it appears that the quantitation of induced small- and large-colony TK mutants allows this evaluation. Most importantly, the chromosomal mutations quantitated in this manner are particularly significant because they represent only viable alterations. Only those alterations compatible with cell viability are a significant risk for human carcinogenicity or mutagenicity.

ACKNOWLEDGMENTS

The authors gratefully acknowledge Shirley Milton for editing and typing the manuscript and Theresa Jakos for preparing the figures.

REFERENCES

Amtower, A.L., K.H. Brock, C.L. Doerr, K.L. Dearfield, and M.M. Moore. 1986. Genotoxicity of three acrylate compounds in L5178Y mouse lymphoma cells. *Environ. Mutagen.* (suppl. 6) **8:** 4.

Blazak, W.F., B.E. Stewart, I. Galperin, K.L. Allen, C.J. Rudd, A.D. Mitchell, and W.J. Caspary. 1986. Chromosome analysis of trifluorothymidine-resistant L5178Y mouse lymphoma cell colonies. *Environ. Mutagen.* **8:** 229.

Brown, M.M.M. and D. Clive. 1978. The utilization of trifluorothymidine as a selective agent for $TK^{-/-}$ mutants in L5178Y mouse lymphoma cells. *Mutat. Res.* **53:** 116.

Chu, E.H.Y. 1971. Mammalian cell genetics. III. Characterization of X-ray-induced forward mutations in Chinese hamster cell cultures. *Mutat. Res.* **11:** 23.

Chu, E.H.Y. and H.V. Malling. 1968. Mammalian cell genetics. II. Chemical induction of specific locus mutations in Chinese hamster cells *in vitro. Proc. Natl. Acad. Sci.* **61:** 1306.

Clive, D. 1983. Viable chromosomal mutations affecting the TK locus in L5178Y/ $TK^{+/-}$ mouse lymphoma cells: The other half of the assay. *Ann. N.Y. Acad. Sci.* **407:** 253.

Clive, D. and J.F.S. Spector. 1975. Laboratory procedure for assessing specific locus mutations at the TK locus in cultured L5178Y mouse lymphoma cells. *Mutat. Res.* **31:** 17.

Clive, D., W.G. Flamm, M.R. Machesko, and N.J. Bernheim. 1972. A mutational assay system using the thymidine kinase locus in mouse lymphoma cells. *Mutat. Res.* **16:** 77.

Clive, D., K.L. Johnson, J.F.S. Spector, A.G. Batson, and M.M.M. Brown. 1979. Validation and characterization of the L5178Y/$TK^{+/-}$ mouse lymphoma mutagen assay system. *Mutat. Res.* **59:** 61.

Clive, D., N.T. Turner, J.C. Hozier, A.G. Batson, and W.E. Tucker, Jr. 1983. The genetic toxicity of acyclovir. *Fundam. Appl. Toxicol.* **3:** 587.

Cox, R. and W.K. Mason. 1978. Do radiation-induced thioguanine-resistant mutants of cultured mammalian cells arise by HGPRT gene mutation or X-chromosome rearrangement? *Nature* **276:** 629.

DeMarini, D.M., K.H. Brock, C.L. Doerr, and M.M. Moore. 1987. Mutagenicity and clastogenicity of teniposide (VM-26) in L5178Y/$TK^{+/-}$-3.7.2C mouse lymphoma cells. *Mutat. Res.* **187:** 141.

de Serres, F.J. 1958. Studies with purple adenine mutants in *Neurospora crassa.* III. Reversion of X-ray-induced mutants. *Genetics* **43:** 187.

de Serres, F.J. and H.E. Brockman. 1968. Homology tests on presumed multilocus deletions in the *ad-3* region of *Neurospora crassa* induced by the acridine mustard ICR-170. *Genetics* **58:** 79.

de Serres, F.J. and H.V. Malling. 1971. Measurement of recessive lethal damage over the entire genome and at two specific loci in the *ad-3* region of a two-component heterokaryon of *Neurospora crassa. Chem. Mutagens* **2:** 311.

de Serres, F.J. and R.S. Osterbind. 1962. Estimation of the relative frequencies of X-ray-induced viable and recessive lethal mutations in the *ad-3* region of *Neurospora crassa. Genetics* **47:** 793.

Evans, H.H., J. Mencl, N.-F. Horng, M. Picanati, C. Sanchez, and J. Hozier. 1986. Locus specificity in the mutability of L5178Y mouse lymphoma cells: The role of multilocus lesions. *Proc. Natl. Acad. Sci.* **83:** 4379.

Hozier, J., J. Sawyer, D. Clive, and M. Moore. 1982. Cytogenetic distinction between the TK^+ and TK^- chromosomes in the L5178Y $TK^{+/-}$ 3.7.2C mouse-lymphoma cell line. *Mutat. Res.* **150:** 451.

———. 1985. Chromosome 11 aberrations in small colony L5178Y $TK^{-/-}$ mutants early in their clonal history. *Mutat. Res.* **147:** 237.

Hsie, A.W., L. Recio, D.S. Katz, C.Q. Lee, M. Wagner, and R.L. Schenley. 1986. Evidence for reactive oxygen species inducing mutations in mammalian cells. *Proc. Natl. Acad. Sci.* **83:** 9616.

Kozak, C.A. and F.H. Ruddle. 1977. Assignment of the genes for thymidine kinase and galactokinase to *Mus musculus* chromosome 11 and the preferential segregation of this chromosome in Chinese hamster/mouse somatic cell hybrids. *Somatic Cell Genet.* **3:** 121.

Malling, H.V. and F.J. de Serres. 1973. Genetic alterations at the molecular level in X-ray induced *ad-3B* mutants of *Neurospora crassa. Radiat. Res.* **53:** 77.

Meyer, M., K. Brock, K. Lawrence, B. Casto, and M.M. Moore. 1986. Evaluation of the effect of agar on the results obtained in the L5178Y mouse lymphoma assay. *Environ. Mutagen.* **8:** 727.

Moore, M.M. and D. Clive. 1982. The quantitation of $TK^{-/-}$ and $HGPRT^-$ mutants of L5178Y/$TK^{+/-}$ mouse lymhoma cells at varying times posttreatment. *Environ. Mutagen.* **4:** 499.

Moore, M.M., A. Amtower, G.H.S. Strauss, and C. Doerr. 1986. Genotoxicity of gamma irradiation in L5178Y mouse lyphoma cells. *Mutat. Res.* **174:** 149.

Moore, M.M., K.H. Brock, C.L. Doerr, and D.M. DeMarini. 1987a. Mutagenesis and of L5178Y/$TK^{+/-}$–3.7.2C mouse lymphoma cells by the clastogen ellipticine. *Environ. Mutagen.* **9:** 161.

———. 1987c. Mutagenicity and clastogenicity of adriamycin in L5178Y/$TK^{+/-}$–3.7.2C mouse lymphoma cells. *Mutat. Res.* (in press).

Moore, M.M., A. Amtower, C. Doerr, K.H. Brock, and K.L. Dearfield. 1987b. Mutagenicity and clastogenicity of acrylamide in L5178Y mouse lymphoma cells. *Environ. Mutagen.* **9:** 261.

Moore, M.M., D. Clive, B.E. Howard, A.G. Batson, and N.T. Turner. 1985a. In situ analysis of trifluorothymidine-resistant (TFT^r) mutants of L5178Y/$TK^{+/-}$ mouse lymphoma cells. *Mutat. Res.* **151:** 147.

Moore, M.M., D. Clive, J.C. Hozier, B.E. Howard, A.G. Batson, N.T. Turner, and J. Sawyer. 1985b. Analysis of trifluorothymidine-resistant (TFT^r) mutants of L5178Y/$TK^{+/-}$ mouse lymphoma cells. *Mutat. Res.* **151:** 161.

Moore-Brown, M.M. and D. Clive. 1979. The L5178Y/$TK^{+/-}$ mutagen assay system, in situ results. *Banbury Rep.* **2:** 71.

Moore-Brown, M.M., D. Clive, B.E. Howard, A.G. Batson, and K.O. Johnson. 1981. The utilization of trifluorothymidine (TFT) to select for thymidine kinase-deficient ($TK^{-/-}$) mutants from L5178Y/$TK^{+/-}$ mouse lyphoma cells. *Mutat. Res.* **85:** 363.

Muller, H.J. 1927. The artificial transmutation of the gene. *Science* **66:** 84.

O'Neill, J.P., R. Machanoff, and A.W. Hsie. 1982. Phenotypic expression time of mutagen-induced 6-thioguanine resistance in Chinese hamster ovary cells (CHO-HGPRT system): Expression in division-arrested cell cultures. *Environ. Mutagen.* **4:** 421.

Ong, T.-M. and F.J. de Serres. 1980. Genetic analysis of *ad-3* mutants induced by hycanthone, lucanthone and their indazole analogs in *Neurospora crassa. J. Environ. Pathol. Toxicol.* **4:** 1.

Russell, W.L. 1952. X-ray induced mutations in mice. *Cold Spring Harbor Symp. Quant. Biol.* **16:** 327.

Sawyer, J., M.M. Moore, D. Clive, and J. Hozier. 1985. Cytogenetic characterization of the L5178Y $TK^{+/-}$3.7.2C mouse lymphoma cell line. *Mutat. Res.* **147:** 243.

Webber, B.B. and F.J. de Serres. 1965. Induction kinetics and genetic analysis of X-ray-induced mutations in the *ad-3* region of *Neurospora crassa. Proc. Natl. Acad. Sci.* **53:** 430.

Wilson, W.R., N.M. Harris, and L.R. Ferguson. 1984. Comparison of the mutagenic and clastogenic activity of amsacrine and other DNA-intercalating drugs in cultured V79 Chinese hamster cells. *Cancer Res.* **44:** 4420.

Genetic and Molecular Characterization of Genomic Regions Surrounding Specific Loci of the Mouse

LIANE B. RUSSELL AND EUGENE M. RINCHIK
Biology Division
Oak Ridge National Laboratory
Oak Ridge, Tennessee 37831

OVERVIEW

Mutations detected by the mouse specific-locus test (SLT) include multilocus deletions as well as intragenic lesions. Genetic analyses have characterized sets of presumed overlapping deletions and have mapped previously unrecognized genes to the regions surrounding each of several specific loci.

Molecular entry to one of these regions, dilute short ear (*d se*), was achieved by utilizing a viral integration at, or near, a marker locus. Presumed deletions were shown to be, in fact, deleted for DNA sequences, and the physical map was oriented relative to the earlier functional map. Presently, a random-clone approach is being used for initiating molecular characterization of regions, which, in aggregate, span a minimum of 9 cM. Mapping to subregions already identified by functional units will facilitate the generation of comprehensive molecular maps and the identification of numerous structure-function correlations for the regions.

Results of the genetic and molecular analyses of multilocus deletions have enhanced the value of the SLT by adding qualitative to quantitative capabilities. Studies of the heterozygous effects of deletions (which are the predominant lesions induced by many mutagens) provide information important to assessment of genetic risk. Long deletions are, further, providing tools for targeted mutagenesis studies that will generate information on the number of loci within segments of defined length that are capable of mutating to detectable alleles, as well as providing new mutations important for strategies of refining molecular and functional maps.

INTRODUCTION

For cellular mutagenesis assays, the implications of the possible induction of multilocus deletions have only recently been addressed (Evans et al.; Moore et al.; this volume; D. DeMarini, pers. comm.). In contrast, a germ-cell mutagenesis test developed almost four decades ago (Russell 1952) took cognizance of the likelihood of deletion induction and incorporated a simple indicator of this event. In the decades that followed, genetic analyses of large numbers of recovered mutations provided inductive evidence for sets of deletions overlapping at specific loci (Russell and Russell 1960; Russell 1971; Bernstine et al. 1978; Russell et al. 1979, 1982). More recently, it has been possible to obtain molecular proof that segments of DNA are indeed deleted in many of these presumed deletion mutations (Rinchik et al. 1985, 1986).

Particularly valuable among mutagenesis assays are those that allow mutations to be recovered and fixed in breeding stocks, rendering them indefinitely available for analysis that, with time, can become more and more detailed as increasingly powerful techniques are developed. The mouse SLT has the capability of detecting, and subsequently recovering, mutations of a variety of types—with respect to both the phenotypic effect and the structural nature of the lesion. Mutations are first observed in heterozygosity with a viable recessive marker; as a result, not only viables but also lethals and sublethals can be detected, as can marker-locus phenotypes ranging from null to near-wild type. Any one from this array of mutations can subsequently be propagated over appropriate balancer chromosomes.

By subjecting large numbers of mutations that involve the same marker locus to complementation studies, we have been able to deduce the existence of several "functional units" (including both known genes and previously nonrecognized presumed genes) in the regions immediately surrounding some of the marked loci and to derive complementation and deletion maps. Each set of mutations is subclassifiable into complementation groups, defined on the basis of which of the functions are missing (Russell 1971; Russell et al. 1982; Rinchik et al. 1986; Russell and Montgomery 1987). That such deficiencies, as has been assumed, indeed result from deletions of genetic material has recently been demonstrated by molecular techniques (Rinchik et al. 1986).

The existence of deletion mutants greatly facilitates the selection of molecular sequences mapping to specific regions of the mouse genome; and the availability of *functional* maps will allow correlations between structure and function to be established (Rinchik 1987). A few early approaches to this exciting goal will be summarized.

In turn, past and ongoing genetic and molecular analyses are serving to enhance the value of the mouse SLT for mutagenesis and genetic-risk assessment by adding qualitative capabilities to its quantitative ones. Even with what we have learned already, the test can, on the basis of phenotype alone, identify multilocus deletions and intragenic mutations at some of the loci (Russell 1986).

Characterized deletions can be used to answer questions pertinent to genetic risk. The most obvious question is whether, and to what extent, deletions (which are recovered in recessive-mutation assays) have detectable *heterozygous* effects. Furthermore, large deletions can provide tools for deriving estimates of the number of genes with detectable phenotypic effects that are found within a certain recombinational or physical length in different parts of the genome, a parameter of some importance in estimating total genetic damage caused by environmental agents. Some of these approaches will be discussed.

RESULTS AND DISCUSSION

Structural Analysis of Functionally Mapped Regions

In developing the specific-locus test for the detection of heritable mutations in the mouse, Russell (1952) incorporated two very closely linked markers (*d* and *se*; recombination distance, 0.16 cM) into a set of seven marked specific loci at which mutations are readily detectable. With induced mutation rates of the order of 1×10^{-4} per locus, it was highly unlikely that any double mutant, *d se*, was the result of two independent mutational events; therefore, such a mutant could at least provisionally be assumed to be a multilocus deletion. The number of *d se* relative to *d* or *se* mutants was found to be appreciable following parental exposures to certain types of radiation and to vary with treatment conditions (Russell 1971).

It is, perhaps, not surprising that the first in-depth genetic analysis with radiation-induced and spontaneous specific-locus mutations was carried out for the *d-se* region of chromosome 9 (Russell and Russell 1960; Russell 1971). A similar study followed for the *c* (albino) region of chromosome 7 (Gluecksohn-Waelsch 1979; Russell et al. 1979, 1982; Russell 1979; Russell and Raymer 1979), and analyses are in progress for the *p* (pink-eye dilution) region of chromosome 7 and the *a* (nonagouti) region of chromosome 2 (L.B. Russell and C.S. Montgomery, unpubl.). Each of these studies involves the intercrossing of large numbers of mutants (e.g., over 100 in the *d-se* region, almost 40 in the *c* region and 43 to date in the *p* region), followed by the phenotypic (postnatal and prenatal) analysis of many hundreds of genotypically different types of progeny (e.g., close to 1000 types of *d*-, *se*-, and *d-se*-mutant combinations).

The product of each analysis is a complementation map and the deduced presence of functional units surrounding the marked locus or loci. In the *d-se* region, for example, the ongoing expansion of the genetic analysis (Russell 1971; Rinchik et al. 1986; Russell and Montgomery 1987) to date has identified 12 functional units and 27 complementation groups; in the *c* region, 8 functional units and 13 complementation groups have been identified (Russell et al. 1982).

Molecular entry to the *d-se* region was made possible by the discovery that the original *d* mutation that is used as a marker in the SLT was caused by the integration of murine leukemia retroviral DNA (*Emv-3*) into or near the *d* locus (Jenkins et al. 1981; Copeland et al. 1983a). A unique-sequence clone derived from chromosome-9 DNA flanking the viral integration site of *d*/*d* mice (Copeland et al. 1983a) was used to probe genomic DNA from a large number of *d-se*-region mutants that had previously been characterized with regard to their position in the complementation map (Rinchik et al. 1985, 1986). Many of these presumed deletions were shown to be, in fact, deleted for chromosome-9 DNA, known to be located near the proviral integration site. This probe also detected a deletion breakpoint-fusion fragment in one of the prenatally lethal *d* mutants. Cloning of this fusion fragment provided another unique-sequence clone, from the opposite "end" of the deletion, that was subsequently mapped between two of the functional units of the complementation map. With the addition of this second distant probe, the physical map was oriented relative to the functional map (Fig. 1). The probes also made it possible to discriminate between mutations that, on the basis of prior genetic (functional) mapping, had been assigned to the same complementation group, thus expanding by at least five the currently distinguishable classes of mutations in the region (Rinchik et al. 1986). Obviously, the use of the limited number of probes represents only a beginning in the correlations that will be possible between the genomic composition of the *d-se* region and the several functions that have defined important developmental processes, such as prenatal and neonatal survival (eight separate units), cartilage formation (one unit), neurological functioning (two units), and melanocyte morphology (two units).

Another murine leukemia retroviral DNA integration, *Emv-15* (Copeland et al. 1983b), closely but not inseparably associated with the lethal yellow (A^y) mutation (L.D. Siracusa et al. 1987), may provide a similar method of access to the *a*-locus region of chromosome 2, for which a complementation map is now in process of being derived. While the *a* region has yielded fewer radiation-induced mutations for analysis than have other marked regions (Russell and Russell 1959), it has given evidence of genetic complexity. For example, A^y was found by us to be genetically separable from *a*. Analysis of rare recombinants, following introduction of closely flanking markers (Russell et al. 1963; L.D. Siracusa et al. 1987), demonstrates these loci to be organized as follows: centromere-A^y-(0.12 cM)-a^x (where a^x is a radiation-induced lethal allele of *a*). A^y-a and A^y-a^t recombination frequencies are similar. The physical distance involved could be of the order of 100 kb of

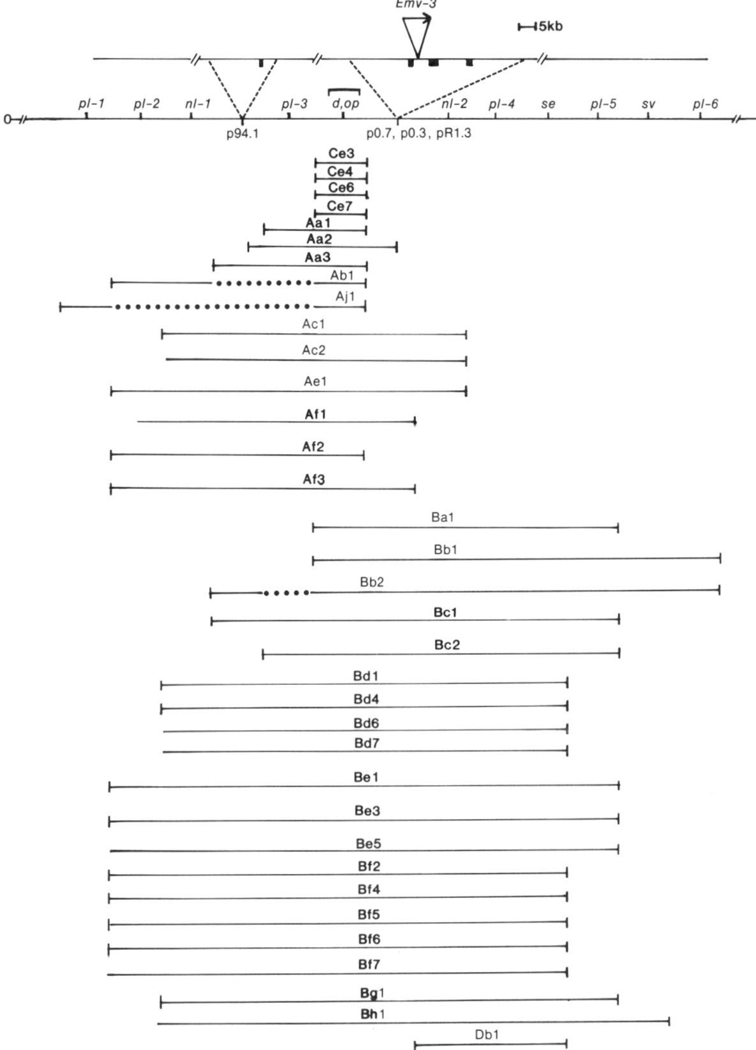

Figure 1

Correlation of the *d-se* region physical and complementation maps. The placement of cloned probes, with their associated regions of DNA, onto the complementation map is indicated. Solid boxes (top line) indicate probe sequences. Horizontal bars represent deletions, aligned to demonstrate their extent with respect to both functional units and DNA markers. (Reprinted, with permission, from Rinchik et al. 1986.)

DNA; however, it is not yet known whether *Emv-15* lies within or outside the A^y-*a* segment. Nevertheless, at present this viral sequence provides the most promising means of molecular entry to the *a* region, and a unique-sequence clone from chromosome-2 DNA flanking the viral integration site has been derived (Siracusa et al. 1986).

Virus-assisted studies represent a specific-gene approach to the molecular characterization of deletions (as would studies utilizing, for example, a cDNA clone for a gene located in the region of interest; Rinchik 1987). An alternative strategy is to access these large regions from many points (Table 1). Our current work therefore utilizes, in addition to specific-gene strategies, a random-clone approach in which DNA sequences are recovered that map to segments corresponding to the longest deletions. These random clones are then submapped to component regions marked by the functional units of the complementation maps.

The random-clone approach is now being used for the *c* and *p* regions of chromosome 7. These regions are defined by the longest characterized deletion in each location: 6–11 cM (which may correspond to a physical length of >10,000 kb) surrounding *c*, and 3 cM or more surrounding *p*.

Random clones are derived from libraries that are enriched for chromosome-7 sequences by fluorescence-activated chromosome sorting, using a long X-7 translocation chromosome, or by metaphase-chromosome microdissection and microcloning, using the same chromosome. Appropriate somatic cell hybrids have also been used for enrichment purposes (Albritton 1986). Genomic subtractive hybridization (GSH) can be used to provide enrichment for portions of the regions. GSH requires homozygously (or hemizygously) deleted segments. Such homozygosity can be achieved (1) for some of the smaller deletions whose homozygotes die late enough in development to be identifiable, and (2) for segments that are common to complementing dele-

Table 1
Molecular Characterization of Deletions

Specific-gene approach
Virus or transgene insertions (e.g., *Emv-3, Emv-15)*
Genes with known products (e.g., tyrosinase, LDHA)
Random-clone approach[a]
Homozygous deletions
Late-dying homozygotes
Overlapping regions of complementing deletions
Hemizygous deletions (somatic cell hybrids; sex chromosomes)
Interspecies hybrids

Development of detailed molecular maps by means of walking or jumping (via deletion breakpoint-fusion fragments) subsequent to obtaining access to regions.

[a]Derivation from enriched DNA libraries, using (A) fluorescence-activated chromosome sorting (special chromosomes); (B) metaphase-chromosome microdissection, microcloning (special chromosomes); (C) somatic-cell hybrids; (D) genomic subtractive hybridization (GSH).

tions and are defined by their region of overlap. This approach is currently being developed in our laboratory.

One strategy for the mapping of random clones is the use of homozygous or hemizygous deletions. To obtain the latter, a mouse-hamster somatic-cell-hybrid line was constructed that retains a deletion-carrying mouse chromosome but not its homolog (Albritton 1986). The genesis of this line was as follows. One of the longest *c*-region deletions, c^{26DVT}, was introduced by genetic recombination into a T(X;7)5Rl X-autosome translocation chromosome (Russell 1983) that carries the wild-type *hprt* gene on its X-chromosomal portion. Males of this genotype (i.e., lacking a second X chromosome) furnished the cells that were fused to $hprt^-$ Chinese hamster cells. By selection for *hprt*, the *c*-region-deleted T(X;7)5Rl X^7 chromosome was retained, and the cell lines were screened for loss of the nontranslocated chromosome 7 (which would have "covered" the deletion). This desired cell line was, in fact, constructed by one of our graduate students, L. Albritton, but it turned out to be of limited value due to the relatively high rate of random loss of nonselected portions of the X^7 chromosome. Such losses introduce ambiguity into the mapping of probes to a specific, functionally mapped deletion.

A more promising strategy for the random-clone approach is the use of interspecies hybrids. From an appropriate cross, offspring are derived that carry a deleted *Mus musculus* chromosome opposite a homologous *Mus spretus* chromosome. Because restriction-fragment-length-polymorphisms (RFLPs) between *M. spretus* and *M. musculus* DNAs are readily demonstrable (Benoit et al. 1985), the derivation of each member of a pair of homologous chromosomes is easily identified in the F_1. Thus, for any particular DNA probe, Southern analysis of F_1 and parental DNAs will indicate whether the sequence is located within the deletion segment (*M. musculus*-specific band absent) or outside (*M. musculus* band present). Clones mapping to the longest deletion in a region can thus be selected from enriched genomic libraries, and they can then be submapped to any one of the segments marked by the functional units by repeating the procedure with a mapping panel of *M. musculus*/*M. spretus* hybrids carrying different deletions. This approach is being exploited at our laboratory for extensive mapping in four genomic regions.

Employing all three of the random-clone mapping methods listed in Table 1 in her pilot studies, Albritton (1986) was able to assign clone 12A (Disteche and Adler 1984) to the segment marked by the juvenile-lethal functional unit in the *c* region (Russell et al. 1982). In the *p* region, another student, J. Mendel, used the interspecies cross to study the $p^{46DFiOD}$ mutation, 1 of about 40 *p*-lethals that is deleted for *p*, ruby-2 (*ru-2*), and lactate dehydrogenase A (*Ldh-1*). To determine whether a particular 14-kb *Eco*RI *M. musculus* fragment (one of many fragments hybridizing to a lactate dehydrogenase-A

[LDHA] cDNA clone) contains the expressed gene for LDHA, offspring of the *M. musculus* ($p^{46DFiOD}$)/*M. spretus* cross were classified according to LDH activity in various tissues. In all F_1 animals examined, the 14-kb band cosegregated with normal LDH activity, whereas lack of this band cosegregated with reduced activity (Mendel et al. 1986).

Within each of the genomic regions discussed, the mapping of random clones to subregions (identified by the functional units) can now proceed by the various methods outlined. In turn, this will allow the generation of comprehensive molecular maps for these regions by utilizing a combination of cosmid walking, large-fragment DNA electrophoresis, and chromosomal "jumping" via deletion breakpoint-fusion fragment cloning (of the type that has already proved useful for analysis of the *d-se* region). Eventual studies on the expression of selected clones, both in adults and at various developmental stages, for which molecular map positions have been determined should allow detailed structure-function correlations for regions which, in aggregate, represent about 1% of the mouse genome.

Enhancement of the Qualitative Capabilities of the SLT

For at least three marked regions (*d-se*, *c*, and *p*), we now have molecular proof of the earlier genetic hypothesis concerning the deletional nature of certain types of mutations. The maps for these three regions strongly suggest that the combination of complementing overlapping deletions results in the total absence of the marked locus. (The *c* locus will be taken as an example in the subsequent discussion.) Therefore, the wild-type gene at this locus cannot be an essential one, and any new *c*-locus mutation found to be nonviable in the homozygous state may, without further testing, be assumed to be a multilocus deletion involving nearby functional units. Another type of phenotypic finding may justify the opposite conclusion: Since the combination of complementing overlapping deletions produces the null phenotype at the marked locus (e.g., total absence of pigment), a mutation that does *not* produce the null phenotype is presumably not a deletion of the marked locus.

For four of the seven loci marked in the SLT, the enhanced confidence in complementation maps that has resulted from recent molecular findings permits a qualitative grouping of mutations on the basis of simple phenotypic diagnoses (Fig. 2). Mutations can be classified as multilocus deletions, presumed intragenic mutations, and a third category that could belong to either of these groups or represent single-gene deletions, which we assume will be subclassifiable by molecular means after the structural mapping (outlined in the previous section) has been accomplished. Even with the present classification, achieved by simple phenotypic means, marked qualitative differ-

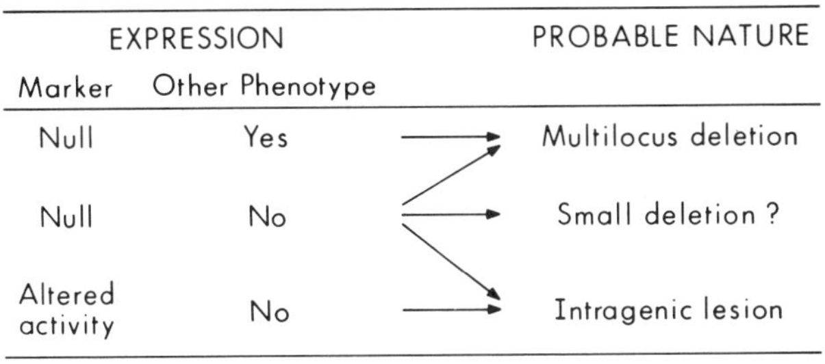

EXPRESSION			PROBABLE NATURE
Marker	Other Phenotype		
Null	Yes	→	Multilocus deletion
Null	No	→ (also ↗ Multilocus deletion, ↘ Intragenic lesion)	Small deletion ?
Altered activity	No	→	Intragenic lesion

Figure 2
Classification of mutations on the basis of phenotypic criteria. For basis of this classification, see text.

ences between spontaneous and induced mutations, and between mutations induced by different mutagens or in different germ-cell types, are readily demonstrable (Table 2). For example, a comparison between mutations induced in postgonial stages and oocytes by neutrons and those induced in spermatogonial stem cells by ethylnitrosourea (ENU) shows 71% and 0%, respectively, to be multilocus deletions, whereas 0% and 26%, respectively, are intragenic mutations. These new qualitative capabilities greatly enhance the value of a test whose quantitative capabilities (the screening of >2000 loci/person-hour) are already unrivaled among mutation-detection systems for mammalian germ cells.

Table 2
Effect of Mutagenic Treatment on the Qualitative Spectrum of Specific-locus Mutations at *d, se,* and *c*

			Distributions of mutations[a]		
Germ-cell stage	Mutagen	No. of mutations	intragenic mutations (%)	viable nulls[b] (%)	multilocus deletions (%)
Spontaneous	—	45[c]	15.5	80.0	4.4
Spermatogonia	low LET radiations	132	11.4	65.2	23.5
Spermatogonia	neutrons	54	3.7	61.1	35.2
Postgonial stages and oocytes	low LET radiations	57	1.8	42.1	56.1
Postgonial stages and oocytes	neutrons	21	0	28.6	71.4
Spermatogonia	ENU	95	26.3	73.7	0

[a]For criteria, see Fig. 2.
[b]Could be intragenic mutations, single-gene deletions, or multilocus deletions.
[c]Includes mosaic mutants from experimental groups.

Heterozygous Effects of Deletions

Since radiations and many chemical mutagens are now known to be effective inducers of deletions, determinations of whether and to what extent deletions have *heterozygous* health effects are important to genetic risk assessment. That multilocus deletions are overall more deleterious than intragenic mutations may be inferred by extrapolating from the known early lethality of all autosomal monosomies (Gropp 1982) to the effects of sectional hemizygosities. To date, no deletions longer than 10 cM have been recovered in the heterozygous state, the most extensive one, to our knowledge, being $c^{12FR60H(b)}$, which is at least 6 but not more than 11 cM in length. Surprisingly, certain deletions of over 2 cM survive as homozygotes until birth (e.g., c^{112K}/c^{112K}), and some smaller ones, (e.g., c^{20FATw}/c^{20FATw}) sometimes live up to 4 months (Russell et al. 1982).

Many of the multilocus deletions in the mouse have heterozygous effects so obvious as to be detectable even in noninbred backgrounds. Thus, of 33 different *d se* deletions in heterozygosity with wild-type, 17 had markedly reduced survival, and 13 had clearly reduced weaning weight (Russell et al. 1966). Similar observations have been made for several deletions in the *c* region (Russell et al. 1979).

Other multilocus deletions fail to show discernible heterozygous effects even in carefully controlled experiments. This is the case for d^{9R250M}, which we have shown to span at least five functional units: *ash*, *pl-2*, *nl-1*, *pl-3,* and *d*, *op* (Russell 1971; Rinchik et al. 1986; Russell and Montgomery 1987). This mutation, a molecularly corroborated deletion, was made congenic with an inbred strain carrying the spontaneous d^v mutation (the original, virus-induced, dilute). The two strains were compared with respect to the following vital characteristics: littersize-adjusted weaning weight, life-time productivity (various parameters) of females, lifespan of females and males, and short-term male reproductive performance. With the exception of one very slight, and doubtful, difference, all measures failed to distinguish between the deletion and nondeletion stocks (L.B. Russell and E.L. Phipps, unpubl.).

The preliminary results indicate that, as might be predicted, location (and thus content) of a deletion, more so than length, determines the severity of heterozygous effect. Since, however, the lengths of deletions are to date known only within relatively wide limits, critical comparisons will have to await the outcome of further genetic and molecular mapping studies.

The heterozygous effects of deletions must be considered when one compares mutation rates at different specific loci. With an agent that produces primarily intragenic mutations, differences may be attributable to relative size of the loci, frequency of mutable bases or sequences, etc. Where the spectrum obtained with a deletion inducer (such as high-linear energy transfer [LET] radiation) differs from that obtained with an intragenic-mutation inducer, such as ENU, the most mutable loci in the former array (e.g., *s*, piebald spotting) may merely be those surrounded by genetic material that is unimportant to *heterozygous* well-being, such as the material located in the region between *ash* and *d*.

Deletions as Tools for Estimating the Number of Mutable Loci

Long deletions are prime tools for a type of experiment that can yield information of major importance to risk assessment, as well as furnish material that will refine our capability for relating DNA structure to function. The experiment is targeted mutagenesis, and its objective is to recover all detectable intragenic mutations mapping to chromosomal segments that correspond to long deletions in specific genomic regions. The protocols instituted by us utilize ENU, which can generate mutations at the rate of 1×10^{-3} per locus, and which, we believe, induces primarily very small genetic lesions in chromosomes of spermatogonial stem cells (see earlier section). The experiments are designed to detect recessive-lethal as well as recessive-visible mutations hemizygously expressed opposite a deletion.

The procedure involves mutagenizing males (spermatogonial stem cells) that are homozygous for a marker(s) in the region of interest, and mating to wild-type females (G_0 generation). G_1 animals heterozygous for the mutagenized (marked) chromosome are then mated to animals heterozygous for a long deletion in the corresponding intact segment. Induced lethals are recognized in G_2 by absence of one marked class of segregants, whereas visibles are recognized by being expressed in all members of this marked class and nowhere else. It should be noted that, even though lethals are detectable by the absence of a class, they are propagatable as stocks derived from G_2 carriers, or from the appropriate G_1 parent. Intercrossing between independently derived mutants will establish whether the same or different loci are involved.

As progress is made in the subregion mapping of molecular clones (by procedures described above), single-gene mutations recovered in targeted mutagenesis experiments will be mapped to these subregions, thus increasing the level of detail at which correspondence between cloned segments of DNA and specific functions can be established.

For the area of genetic-risk assessment, the deletion-facilitated targeted mutagenesis experiments can provide much-needed information on the number of loci within segments of defined recombinational or physical length that are capable of mutating to detectable alleles. Detectability levels can be refined with time. Eventually, we should also know whether this number is different or similar in various chromosomal regions, and this will affect limits on total-genome estimates for such loci.

ACKNOWLEDGMENTS

Research sponsored by the Office of Health and Environmental Research, U.S. Department of Energy, under contract DE-AC05-840R21400 with the Martin Marietta Energy Systems, Inc.

REFERENCES

Albritton, L.M. 1986. "The albino locus in the laboratory mouse *Mus musculus:* Analysis by somatic cell genetics and interspecific meiotic recombination." Ph.D. thesis, University of Tennessee, Knoxville.

Benoit, R., P. Barton, A. Minty, P. Danbas, A. Weydert, F. Bonhomme, J. Catalan, D. Chazottes, J.-L. Guenet, and M. Buckingham. 1985. Investigation of genetic linkage between myosin and actin genes using an interspecific mouse backcross. *Nature* **314:** 181.

Bernstine, E.G., L.B. Russell, and C.S. Cain. 1978. Effect of gene dosage on expression of mitochondrial malic enzyme activity in the mouse. *Nature* **271:** 748.

Copeland, N.G., K.W. Hutchison, and N.A. Jenkins. 1983a. Excision of the DBA ecotropic provirus in dilute coat-color revertants of mice occurs by homologous recombination involving the viral LTRs. *Cell* **33:** 379.

Copeland, N.G., N.A. Jenkins, and B.K. Lee. 1983b. Association of the lethal yellow (A^y) coat-color mutation with an ecotropic murine leukemia virus genome. *Proc. Natl. Acad. Sci.* **80:** 247.

Disteche, C.M. and D. Adler. 1984. Localization of cloned mouse chromosome 7-specific DNA to lethal albino deletions. *Somatic Cell Mol. Genet.* **10:** 211.

Gluecksohn-Waelsch, S. 1979. Genetic control of morphogenetic and biochemical differentiation: Lethal albino deletions in the mouse. *Cell* **16:** 225.

Gropp, A. 1982. Value of an animal model for trisomy. *Virchows Arch. Pathol. Anat.* **395:** 117.

Jenkins, N.A., N.G. Copeland, B.A. Taylor, and B.K. Lee. 1981. Dilute (*d*) coat-color mutation of DBA/2J mice is associated with the site of integration of an ecotropic MuLV genome. *Nature* **293:** 370.

Mendel, J.E., E.G. Bernstine, L.B. Russell, P.A. Lalley, and D.E. Housman. 1986. Genomic cloning of mouse lactate dehydrogenase A: Use of deletion mutants to distinguish an expressed gene from its pseudogene counterparts. In *Proceedings of the 14th Molecular and Biochemistry Workshop*, Jackson Laboratory, Bar Harbor, Maine.

Rinchik, E.M. 1987. Molecular analysis of heritable mouse mutations. *Environ. Health Perspect.* (in press).

Rinchik, E.M., L.B. Russell, N.G. Copeland, and N.A. Jenkins. 1985. The dilute-short ear (*d-se*) complex of the mouse: Lessons from a fancy mutation. *Trends Genet.* **1:** 170.

———. 1986. Molecular genetic analysis of the dilute-short ear (*d-se*) region of the mouse. *Genetics* **112:** 321.

Russell, L.B. 1971. Definition of functional units in a small chromosomal segment of the mouse and its use in interpreting the nature of radiation-induced mutations. *Mutat. Res.* **11:** 107.

———. 1979. Analysis of albino-locus region of the mouse. II. Fractional mutants. *Genetics* **91:** 141.

———. 1983. X-autosome translocations in the mouse: Their characterization and use as tools to investigate gene inactivation and gene action. *Prog. Top. Cytogenet.* **3A:** 205.

———. 1986. Information from specific-locus mutants on the nature of induced and spontaneous mutations in the mouse. In *Genetic toxicology of environmental chemicals*, Part B: *Genetic effects and applied mutagenesis* (ed. C. Ramel et al.), p. 437. A.R. Liss, New York.

Russell, L.B. and C.S. Montgomery. 1987. Structural and functional analyses of the dilute-short-ear region of the mouse. *Genetics* (in press).

Russell, L.B. and G.D. Raymer. 1979. Analysis of the albino-locus region of the mouse. III. Time of death of prenatal lethals. *Genetics* **92:** 205.

Russell, L.B. and W.L. Russell. 1960. Genetic analyses of induced deletions and of spontaneous nondisjunction involving chromosome 2 of the mouse. *J. Cell. Comp. Physiol.* (suppl. 1) **56:** 169.

Russell, L.B., M.N. Cupp McDaniel, and F.N. Woodiel. 1963. Crossing-over within the a-"locus" of the mouse. *Genetics* **48:** 907.

Russell, L.B., C.S. Montgomery, and G.D. Raymer. 1982. Analysis of the albino-locus region of the mouse. IV. Characterization of 34 deficiencies. *Genetics* **100:** 427.

Russell, L.B., W.L. Russell, and E.M. Kelly. 1979. Analysis of the albino-locus region of the mouse. I. Origin and viability. *Genetics* **91:** 127.

Russell, L.B., M.S. Steele, and H.M. Thompson, Jr. 1966. Deficiencies in the mouse. *Biology Division Semiannual Report* ORNL-3999, p. 90. Oak Ridge National Laboratory, Oak Ridge, Tennessee.

Russell, W.L. 1952. X-ray-induced mutations in mice. *Cold Spring Harbor Symp. Quant. Biol.* **16:** 327.

Russell, W.L. and L.B. Russell. 1959. The genetic and phenotypic characteristics of radiation-induced mutations in mice. *Radiat. Res.* (suppl.) **1:** 296.

Siracusa, L.D., L.B. Russell, E.M. Eicher, N.G. Copeland, and N.A. Jenkins. 1986. Molecular genetic approach to the study of the agouti locus of mouse chromosome 2. *Mouse News Letter* **76:** 83.

Siracusa, L.D., L.B. Russell, E.M. Eicher, D.J. Corrow, N.G. Copeland, and N.A. Jenkins. 1987. Genetic organization of the agouti region of the mouse. *Genetics* (in press).

Comments

O'Neill: I am having some difficulty reconciling the concept of an essential gene on chromosome 11 near the *tk* locus as an explanation for the small mutant colonies with the existence of Helen Evans' cell strain monosomic for chromosome 11. You propose that strain L5178Y 3.7.2C has two copies of this essential gene, and that loss of one copy leads to a small colony. By that rationale, a cell monosomic for chromosome 11 has only one copy of this essential gene and should only form small colonies. The only way I can see to reconcile the small colony mutants with the proposed essential gene is to postulate that strain L5178Y 3.7.2C has only one copy of that essential gene and that it is on the chromosome 11 which contains the inactive *tk* locus. The active *tk* locus must be on the other chromosome 11 with no active essential gene. The monosomic chromosome 11 strain of Evans must then contain both the *tk* and the essential gene loci on that single chromsome 11.

Evans: The LYR and LYS strains don't give small and large TFT^r colonies similar to those seen in the 3.7.2C strain.

O'Neill: No. I'm saying your monosomic line to me should only give rise to small colonies.

Chasin: Does that line in fact give small colonies?

Evans: No. We don't get a distribution of small and large colonies, as seen in strain 3.7.2C. I don't know that the LYR and 3.7.2C strains can be compared with regard to colony size. We get tiny TFT^r mutants, and a high proportion of these are induced in the LYS strain. However, the heterozygous and hemizygous R lines are similar to each other in the relative distribution of tiny and normal-sized TFT^r colonies that are induced by various mutagens.

Moore: Actually, I have done some colony sizing with Helen's lines. We didn't get too far with it, because her cells grow differently than ours. Under our test conditions, Helen's R16 and S line form mostly "medium" size and very tiny colonies. That is, the "large" colonies don't get as large as our large colonies making it difficult to see a bimodal distribution. Of course our counter doesn't count colonies smaller than 0.3 mm so you cannot evaluate them. The R83 line doesn't produce very many mutants. If you size the mutant colonies, however, the large colonies are about the same size as our large colonies. Almost all of the R83 TFT^r mutant colonies are large.

O'Neill: I am not trying to be negative as regards your small colony theory. I am saying there is something testable in here, because if you're going to say 3.7.2C has two essential genes, and you say that, when you knock out one copy of that essential gene, you get a small colony, then somewhere in this monosomic line there has to be that essential gene. If one essential gene gives you small colonies with L5178Y 3.7.2C, only one essential gene should give you small colonies in the monosomic chromosome 11 strain also.

Calos: There could be second site mutations somewhere else. The strains aren't necessarily isogenic.

O'Neill: Certainly. I'm just saying if you are going to make that comparison of something on chromosome 11, it can't be that simple. That's all I'm saying.

Clive: I think you can take a look at what isn't happening in these cells and say that there definitely is something different in the single chromosome 11 present in the R83 from what is present in the 3.7.2C. If that extra chromosome 11 is really completely nonessential, as it would appear to be from R83, we would have lost it in all the generations that the cells have been growing. These cells just don't carry extra DNA very well. They tend to lose it, unless there is something essential holding that chromosome there. So, I agree, there probably is something different about the single chromosome 11 in R83. Weren't there an extra couple of bands in it, John?

Hozier: It's unusual in appearance. The R83 chromosome 11 doesn't look exactly like 3.7.2C chromosome 11s.

Evans: I think it would be more valuable to compare the heterozygous and hemizygous LYR strains. We have compared the distribution of small colonies and large colonies induced, although the sizes are different than in strain 3.7.2C in that the small ones in our case are all less than 0.3 mm (which are not included in the case of strain 3.7.2C). Even though the clastogen-induced mutant frequency is much lower for R83 than R16, we find that the relative distribution of tiny and normal-sized TFT^r colonies, induced by clastogens is similar for the two strains. Therefore, it seems that in the case of the LYR strains at least, the TFT^r colony size can't be a simple matter of "essential gene" dosage (if it is assumed that strain LYR16 has two active copies and strain LYR83 has only one copy of the essential gene).

Hozier: There is residual gene present on the 3.7.2C 11A by hybridization techniques.

Clive: So it's intragenic by even Fred's (de Serres) definition. The 11B *tk* was revertable, so that was probably a base-pair event.

de Serres: One of the problems that I encountered is that when you do something like this, when you get a *tk* mutation, you may do things in the immediately adjacent regions that are probably best labeled "cryptic mutations." One of the things that I have been trying to do with slides that I didn't get to in my lecture was to look at the heterozygous effects of multilocus deletions. What I am finding in the very large multilocus deletion mutants is that you find other genetic damage in the immediately adjacent genetic regions that are expressed in terms of heterozygous effects when you make certain combinations of *ad-3* mutants. So, don't think just in terms of what has happened at the locus of interest, but there may be other things going on in the immediately adjacent genetic regions.

If you look at Lee's (Russell) complementation maps of dilute short-ear, you will see she has "skipping" mutations. Look at the complementation map of the *ad-3* region, where I tried to show that individual map as a series of mutants overlapping the deletions. If you look at the testers, a lot of those testers are point mutations at the *ad-3* locus or point mutations at the *ad-3B* locus, with very closely linked sites of demonstrable, recessive lethal damage. Now, what I am finding out in our current research is that when you look at multilocus deletion mutants in the *ad-3* region that appear to have definite boundaries, and if you go beyond these boundaries into the immediately adjacent genetic regions, you find what I am calling cryptic mutations, which are only expressed when you do the appropriate experiments. It may be that kind of thing that is going on in these situations where you appear to have skipping mutations such as those found by Lee in the dilute, short-ear region in the mouse.

Chu: That's a beautiful demonstration of genetic complementation in the mouse in vivo by Lee Russell, but in somatic cells it may be possible to take a further experimental approach. Distal to *tk*, there is a galactokinase gene. I remember reading Don Clive's 1983 paper in the New York Academy of Sciences in which you stated that between *galk* and *tk* there is some essential gene. I want to know, Don, whether you have studied further genetic complementation in this chromosome region with two selectable loci, both in the mouse and the humans.

Clive: We have looked at that. John (Hozier) has done some of the cytogenetics on it. They do appear to be closely mapped, in the two terminal bands of chromosome 11. Is that correct?

Hozier: Yes, probably. The strongest evidence that they are very closely linked is that you can cotransfect with TK and galactokinase.

Chu: There is no crossing-over between them?

Hozier: Yes. So, if that gene is cloned, and I have heard that it may have been by now, it is an excellent, additional marker on the distal end of chromosome 11 on which to do these kinds of studies.

Clive: We have worked out selective techniques for each of the two markers.

Chu: Yes. It can work both ways.

Clive: When we tried to select for double mutants, we just didn't get any when, by the single mutant frequencies, we should have recovered ten to fifty.

Liber: With respect to recoverability of mutations at the *hgprt* locus, I just want to point out that Cox and Masson, a number of years ago, published a paper in which they stated that they found a large number of deletions and translocations—in fact, up to 40% of neutron-induced mutations—in human fibroblasts. So, I sometimes wonder if excess mutations at autosomal loci can be attributed to multilocus deletions, and whether or not we have to also think about mechanisms that may require homologous DNA in order to operate. I think that sort of mechanism would account for all of the data which have been presented here.

De Marini: In Cox and Masson's paper in 1978, I think 40% of their mutants showed clear chromosome aberrations at the *hprt* locus, based on the level of cytogenetic analysis.

Albertini: This may be a very trivial point, but I am just putting together the fact that I think Don (Clive) said there are 10% or 20% of small colonies spontaneously in the background.

Clive: We worked with EMS, hycanthone and methyl iodide and took the mutants that resulted and tested them for sensitivity to the particular mutagen that induced them, as well as to the other two. In almost all cases, the mutants seemed to be less resistant to the mutagen that induced them than the wild-type was. From this we concluded that these mutants were not being enhanced by selection by the mutagens.

Davidson: I have first a comment and then a question. I think the results with small and large colonies are basically uninterpretable and will remain uninterpretable because of the thousands of types of insults that one can give to a cell that will affect the growth rate. What Ernie (Chu) was saying, I think, is completely correct. You have to start looking at a real second marker, such as *galk*, before any of this is going to make sense.

On the other hand, it strikes me that if there is any real causal relationship between the development of small colonies and TK deficiency, you could make a prediction. Given the fact that you have 1% background small colonies with which to start, if you took that population of cells that form without selection, the small colonies, you should find in there a very greatly increased percentage of $TK^{-/-}$ cells, if they are close enough, so that you are going to be inactivating one by the other. If you just took the small-colony population and looked for $TK^{-/-}$ cells among them without any mutagen treatment, you should find a tremendous enrichment. Are the $TK^{-/-}$ cells in there?

de Serres: I don't think that logically follows. If that's a heterozygous effect of a multilocus deletion, then you are going to have the same probability of that ocurring at many gene loci throughout the whole genome. Just because some TK mutations are small colonies, that doesn't necessarily mean that all small colonies are going to be TK mutations.

Davidson: I'm not saying all of them will be.

de Serres: But that's what's implied in your argument.

Davidson: No. You would predict that if there is a real locus that affects colony size, and that this is a pre-existing mutation, and also, you are suggesting inactivation of one by a clastogenic event, then you should expect an increased frequency of $TK^{-/-}$ cells among the preexisting small-colony mutants.

de Serres: If that's the only basis for the small-colony size. What I'm saying is that it's due to a heterozygous effect of a multilocus deletion, and it's a different form of expression than Lee (Russell) sees, dilute short ear, of heterozygous effects of multilocus deletions in that region. In *Neurospora*, the multilocus deletions show reduced growth rate. If I have multilocus deletions at the *ad-3A* locus and the *ad-3B* locus, they grow more slowly. If I put two of these multilocus deletions together, they grow still more slowly. So, I get the equivalent of small colonies in *Neurospora* as a result of the heterozygous effects of multilocus deletions.

Moore: Let me make two comments. One, I absolutely agree with the statement that we will only really be able to sort out what is happening when we can do some other analysis that includes linked loci. Specifically, we will know what is happening when we can do the molecular analysis at both the *tk* locus and the linked genes on both sides of the *tk* locus.

The other thing that I want to emphasize is that not all TFT^r small colonies are going to be chromosomal mutations for chromosome 11. As you say, there are lots of other reasons why colonies can be slow growing. It may be a point mutation in some other gene, that's possible—maybe a chromosomal event someplace else in the genome. That is why we do not intend to say that all small-colony mutants are going to be chromosome 11 abnormalities. We would not expect that.

Davidson: But have you or anybody else ever checked to see what the frequency of $TK^{-/-}$ cells is among mutants selected for small-colony phenotype?

Clive: We have looked at small colonies on viable count plates and have never found a TK-deficient one yet. Bill Blazak has looked at small colonies occurring on viable count plates and done cytogenetics on them and has found chromosome aberrations, but he has found them in chromosomes other than 11. So, apparently, any chromosome aberration can produce this phenotype. It's about the right frequency. If you look for spontaneous chromosome aberrations in the 3.7.2C cells, you see around 1–5%, and that's about the frequency you find in these small colonies on unselected plates.

Glickman: The question that I have for Fred de Serres has to do with the utility of the *Neurospora* system and the detection of the gross versus the smaller events. Particularly, what I wanted to ask you is this: You have also done work in base analogs, but in base analogs did you find multilocus deletions?

de Serres: Only with 2-aminopurine. We have already published an abstract on that, and I am getting a paper out right now. We found about a sixfold increase over spontaneous with 2-aminopurine in the wild-type strain. There seemed to be some multilocus deletions, a very small percentage; I think it was about 12% as I remember. Then, when you go into the repair-deficient strain, although the number of mutants that you get in the repair-deficient strains doesn't increase quantitatively, quantitatively the spectrum of multilocus deletions is markedly different than in the wild-type strains. I think the multilocus deletions go up to 68%. We really don't have any idea what is going on with 2-aminopurine, but we have done two other purine analogs, 6-HAP (6-hydroxyaminopurine) and AHA (2-amino-N6-hydroxyadenine). HAP is by orders of magnitude more mutagenic, and AHA *ad-3* is the one that gives us almost 1% mutants. Both of those latter two purine analogs give only point mutations, no multilocus deletions, and the story is the same in the wild-type strain as it is in the repair-deficient strain. You get no change quantitatively in the repair-deficient strain from what you get in the wild-type strain and the mutants are predominantly, if not exclusively, point mutations in the two strains. However, something different is happening with 2-aminopurine. I frankly don't understand it, and I don't know whether Herman (Brockman) does, but I wouldn't generalize about purine analogs in general on the basis of our 2-AP data.

Glickman: I would like to get some understanding of the mechanisms involved. What are these multilocus deletions? In the case of ionizing radiation we have a clear story. Reducing the dose from 1000 rads to 10 rads yields a threefold drop in these events. That fits perfectly with a model, but what needs to be asked is, "What other mechanisms are available to generate multilocus deletions?" You do have one of the better systems to demonstrate their occurrence.

de Serres: The thing that is going to drive that machine is the molecular analysis, which is what we are trying to learn how to do. We are going to do DNA cloning and sequencing. We can describe until we get to be 100 years old, but it's not until you can do the molecular analysis that you are going to get the answers. That's the only way we will be able to resolve that problem.

Evans: I have to mention strain LYS, which is unable to repair DNA double-strand breaks. After exposure to X-irradiation, 60% of the LYS TFT^r colonies are tiny, whereas, with strains LYR16 and LYR83, only 18% of the TFT^r colonies are tiny. Another interesting thing is that even UV and EMS induce a high proportion of tiny colonies in strain LYS. So, I would suggest that multilocus deletions may be formed in strains unable to rejoin double-strand breaks even after treatment with agents that normally produce primarily base-substitutions in repair-proficient cells.

Brockman: We have a similar situation with nitrous acid in *Neurospora*, where in the repair-proficient strain nitrous acid caused no multilocus deletions, even down to a very low percent survival of about 5%. However, in the strain that is homokaryotic for a defect in nucleotide excision repair, the incidence of multilocus deletions was as high as 40%.

De Marini: That's why I think it's important to look qualitatively at the mutants. Veronica (Maher) has published a great deal of work looking at the mutability of XP cells and other repair-deficient cells. You can find higher mutant frequencies in those repair-deficient strains, but when one actually can do the qualitative analysis of those mutants, you may find that you are getting a different spectrum. It's not just that you're getting more, but the quality of them may be different.

Chasin: In that vein, wasn't it said that the AS52 cells also had a repair deficiency?

Stankowski: We don't think they do. The differences we see in survival vary at most by threefold. For most of the things that we have tested, it's less than 10%. AS52 isn't more sensitive to everything or any particular type of agent. We see variations. CHO-K1-BH4 might be more resistant to one agent, and AS52 is more resistant to another. We think the differences we see are not due to repair deficiencies, like you see where you get thousandfold differences in drug resistance, transport mutants that don't take up drugs, or strand break-repairing deficient mutants.

Tindall: We believe the difference in survival that we observe with mutagen treatment simply reflects clonal differences between the AS52 and the CHO-K1-BH4 cell lines. There are no notable differences between these cell lines except the *gpt* integration.

Hutchinson: I would like to point out the extraordinary interest in an excision-deficient strain that shows a high rate of deletions with nitrous acid. Nitrous acid is a cross-linking agent for DNA. This does reinforce the idea that damage to two strands is somehow critical.

Brockman: We find the same thing with mitomycin C and with 1,2:7,8-diepoxyoctane.

Jensen: I would like to comment a little more about the AS52. I'm sure that you people have heard this complaint every time that you've talked about this: the sensitivity of this might well be noncharacteristic; that is, this is *gpt* put in with integrative DNA, and therefore, it is going to fall out easily. In that regard, I'm really worried about the clastogenicity of ethanol and methanol.

Stankowski: They're very weak only in the presence of S9. There are data in the literature that have demonstrated both those agents are converted to the corresponding aldehydes by S9. Those aldehydes are indeed clastogenic, and that's why we think we're picking them up.

Jensen: What I really wanted to ask was have you tried other integrations of *gpt*, isolated it in altogether different cell lines, and seen whether they have similarities?

Tindall: That's a very good question. Franklin Hutchinson's lab is currently in the process of isolating new cell lines with single copy *gpt* integrations. I'm sure they will be interested in this question. Also, Gerry Adair mentioned earlier that he has data with *aprt* cDNA inserts in hamster cells that are consistent with our observations. Finally, we have assayed a number of point mutagens and toxic nonmutagens in the AS52 cell line. These agents appear not to affect the stability of the *gpt* insert in AS52 cells.

Hsie: Actually, the fact that EMS, ICR, and UV did not cause a differential mutagenic response between AS52 cells and the parental K_1-BH_4 cells suggests that there is nothing unique or unusual about the mutagenic response of the AS52 cells.

Stankowski: We have only seen major differences in reactivity between *hprt* and *gpt* for clastogens.

Chu: There is a recent paper published in *Somatic Cell Molecular Genetics* on the systems. I think by Goring or someone. I don't remember the data, but they used the herpes simplex I *tk* gene being transfected into it, also. I don't know whether this would be another system for comparison for mutagens.

Tindall: Yes, there are a number of similar systems that use various selectable markers, e.g., *gpt*, *tk*, and others. A limitation of many of these other systems, however, is that the observed spontaneous mutant frequencies are often extremely high. Such high spontaneous mutant frequencies affect the confidence in which one can evaluate induced mutational processes.

I isolated the AS52 cell line by evaluating the spontaneous and UV-induced mutant frequencies of several different *gpt*-transfected clones. The AS52 cell line was selected for further analysis because it demonstrated a reasonable dose-response curve following UV-irradiation as well as low (map10^{-5}) spontaneous mutant frequency. Indeed, when we performed the molecular analysis, we found that the AS52 cell line has a single copy *gpt* integration. In short, we first selected for a cell line that had the characteristics that we thought would be useful in a system using the *gpt* gene as a mutational target. Further experiments with this cell line indicated that it is quite comparable with the CHO/HPRT system with the following two exceptions. First, the small size of the *gpt* gene (456 base pairs) and its linkage to an ampicillin resistance gene at the site of integration makes this cell line potentially useful for the rapid isolation and sequence characterization of mutations induced at a chromosomally integrated target gene. Second, this cell line appears to be capable of recovering mutations induced by clastogens and radiomimetic agents that are not detected or only weakly detected at the *hprt* locus. We believe the site of *gpt* integration allows the recovery of mutants that are not recovered at the hemizygous X-linked *hprt* locus. Either the *gpt* gene has integrated into an autosome such that extensive multilocus deletions affecting both *gpt* and adjacent essential sequences will be viable due to the presence of complementary essential sequences on the homologous autosome or the *gpt* gene has simply integrated into a region that will allow extremely large deletions. The telomeric region of a chromosome might be a good example of the latter. We are currently in the process of defining the site of the *gpt* integration in AS52 cells. We will be continuing to use the AS52 cell line to study mutagenesis, hopefully, to more clearly define deletions that could not have been recovered in the CHO/HPRT system.

Chu: May I recall another paper I read? I think a group of people in San Francisco in the Laboratory of Radiation Biology, Nernam or something like that, had done some transfection, so that he selects for transformation depending on the SV40 T antigens. He actually located two sites in order to select one for survival, the one he transfected in at the same time as the neomycin resistance. The conclusion from that paper was that in order to have a survival to be maintained in a malignant kind of situation, it always retained one copy of that integrated gene. If you do the molecular analysis, that band is always there.

This reminds me of the Sumiono quote in his book, *Gene Duplication*, in 1970, where he used the term "forbidden mutation." That is, you must have one gene duplication, then if you have that one essential gene you will not be allowed to change. So, in Nernam's case, it was a very nice comparable situation, shuttling into two loci, and one is always present.

When we are studying the situation for mutagenesis, I think we have to be careful about the position, the number of copies, and all these kinds of questions.

Clive: Not only where you put the gene, but what you take away from it, and how you modify the environment are other questions. We have been able to increase the mutability of the *tk* locus by at least twentyfold just by preselecting for a TK-deficient mutant. Apparently, the *tk* gene was rearranged elsewhere in the genome in the process. The resulting *tk* was still heterozygote, but had a twentyfold higher mutability, both spontaneous and induced.

Tindall: I agree. I think position effects are potentially very important.

Also, I think chromatin structure may have an important role in mutagenesis. Unfortunately, at this point in time, we can't speak in sophisticated terms about either of these. There are data, however, that indicate that differential DNA repair activity may depend upon gene location. Most certainly, this observation will impact the study of mutagenesis in mammalian cells.

Chasin: I was intrigued with Martha's results showing that AMSA yields such a high frequency of apparent deletions of the essential gene along with TK, as indicated by the appearance of small colonies. These deletions must span at least two loci. If AMSA is working through topo-2 inhibition and inducing cleavages at those sites, and if it is true that topo-2 sits at the bottom of loops, and the loops are on the order of 100 kb, let's say, (the higher end of many estimates) then AMSA may be a good candidate for making these relatively small deletions.

To get to my question then, since the X-linked *hprt* locus may be able to support deletions of only a limited size, have you tried AMSA on the *hprt* locus as well as on *tk*?

Moore: Yes. The *hprt* locus gives a very low frequency.

De Marini: Comparable to what you see at $tk^{+/0}$ in Helen's LYR83 hemizygous cells that are monosomic for chromosome 11.

Moore: AMSA is a particularly interesting compound because it does produce such a high mutant frequency. When we analyzed gross aberrations at the highest dose (9% survival), we had an average of 4 visible aberrations per cell. Therefore, a lot of those small colonies following AMSA treatment are probably not associated with chromosome 11. There is a whole lot else going on across the genome.

De Marini: AMSA is a perfect candidate after X-rays to begin to do molecular analysis. Maybe we will see something that might tell us something about the distance between the bases of those loops. We don't know. However, there might be some kind of regular pattern. There are all kinds of things you can envision. For example, the junctions of rearrangements may identify the cut site for DNA topoisomerase. They should be very interesting mutants to analyze.

Hutchinson: What level of total colony survival did you have in those experiments, in which 1% $TK^{-/-}$ small colonies were formed?

Moore: Are you familiar with the way "survival" is calculated in the mouse lymphoma assay? Survival becomes a complex issue. We do not plate immediately after treatment for a survival figure. We do a combination of relative growth in suspension, during the expression period and the plating efficiency. The plating efficiency, in the very worst case would be 20–30–40%, roughly speaking.

Hutchinson: So that means that your mutation rate can't be off by more than an order of magnitude. That's what I was really getting at. Mutation rates can be very tricky.

Hutchinson: Your survival wasn't 10^{-3}?

De Marini: No.

Clive: There is one finer chemical scalpel, if you will, for dissecting out these mutations other than AMSA and that is methotrexate. Methotrexate is peculiar in that it induces no HGPRT-deficient mutants. This places a lower bound on the size of the deletion, or whatever it is that methotrexate induces. It produces a large proportion of large-colony mutants at the *tk* locus and this can put an upper limit on the size of the mutations. So far, John has analyzed three of these, and they are all showing complete loss of the *tk* gene, but we have no idea how much more DNA is gone.

Glickman: Methotrexate also induces recombination, and there are some problems with that.

O'Neill: I have one other comment on something Martha (Moore) was saying earlier concerning small colonies. I think one of the problems with interpreting studies in V79 or CHO cells is the way people tend to do monolayer cultures, where you stain plates after 8 or 9 days. Most do not see small colonies. When I look back on earlier studies of expression, I had a lot of data with MMS that I didn't like, because there was this background growth on the plates employing an in situ expression, which I suspected were problems with cross-feeding. Now I wonder about slow growing mutants. However, nobody is going to let them go long enough to see a small colony, if they really are slow-growing mutants.

Moore: Are you talking about your experiments where you plate out on the large dishes and perform the expression and selection without subculture?

O'Neill: Yes. I think any experiments. The problem is, if you have a mutant that grows slowly, you might not see it in a monolayer culture, because you stain them so soon (8–9 days).

Moore: That might be. However, with mouse lymphoma cells, we do allow our 6TG plates to go for the same length of time. We don't see any small colonies, or we see very, very few small colonies.

O'Neill: But if they really are slower growing, they are not going to be there because of the long expression period.

Moore: That's right.

O'Neill: What I am saying is, even if you growth arrest CHO cells during expression, you are probably not going to see any small colonies, because even if they were there, you didn't let the cells grow long enough to give you a small colony.

Moore: I see what you're saying.

O'Neill: I think there's a lot of people who only do one assay and don't understand the other assays. In fact, you couldn't see them even if they occurred, which is not to say that they did or did not occur, but simply that you couldn't have seen it anyway.

Moore: So, what we need to do, Pat, is we need to use your growth arrest technique, and we need to plate in soft agar.

O'Neill: That's probably a reasonable approach.

Differential Recovery of Mutants In Vivo

hgprt Mutation In Vivo in Human T Lymphocytes: Quantitative Considerations

RICHARD J. ALBERTINI, JANICE A. NICKLAS, LINDA M. SULLIVAN, TIMOTHY C. HUNTER, AND J. PATRICK O'NEILL
Genetics Laboratory
University of Vermont
Burlington, Vermont 05401

OVERVIEW

Somatic gene mutation occurring in vivo in human T lymphocytes may be assessed in one of two ways, i.e., by autoradiography or by clonal assay. Studies of T-cell receptor (TCR) gene-rearrangement patterns in in-vivo-derived-mutant clones isolated from the latter show that there is not necessarily a one-to-one correlation between in vivo somatic gene mutations and the measured mutant frequencies. Specifically, in some instances, the number of T-cell mutants over-represents the number of responsible in vivo mutations. This and other complications involved in quantitatively relating measured mutant frequencies to in vivo mutation are considered.

INTRODUCTION

Somatic gene mutation occurs in vivo in humans. Although its direct measurement is elusive, it is generally assumed by analogy with in vitro systems that in vivo mutations may be quantitated in individuals by determining the frequencies of resultant mutants. 6-Thioguanine-resistant (TG^r) T lymphocytes in peripheral blood represent one class of in-vivo-derived mutant somatic cells in humans. They arise from in vivo mutation of the gene for hypoxanthine-guanine phosphoribosyl transferase (HGPRT, *hgprt* gene) and can be assessed by one of two methods, i.e., a short-term autoradiographic assay and a more definitive clonal assay (Albertini 1985). Both are being exploited for human mutagenicity monitoring, and mutant colonies isolated in clonal assays are allowing studies of fundamental mutational mechanisms.

Gene mutation is probably an initial event in certain somatic genotoxic diseases including cancer. Furthermore, somatic gene mutation may reflect germinal gene mutation, and its assessment may be useful for monitoring human populations for risk of heritable genetic damage. Therefore, methods that allow quantitation of in vivo somatic gene mutation are valuable for public health purposes. However, in developing these methods, care must be taken that the measured indicators of gene damage truly have a genetic basis, i.e., are mutant cells, and that the frequencies of these mutants truly reflect the frequencies of the underlying mutational events, which are the genetic parameters of interest.

It has now been amply demonstrated that the TG^r T cells of human peripheral blood are HGPRT mutants because they can be developed in vitro into populations of sufficient size for analysis. These unusual lymphocytes are T lymphocytes by surface-marker analysis, maintain their TG^r phenotype in vitro in the absence of selection, and are deficient in HGPRT activity (Albertini et al. 1982; Albertini 1985). Furthermore, we and other investigators have shown that many of the in-vivo-derived mutants show *hgprt* gene structural alteration, as defined by Southern blots (Albertini et al. 1985; Turner et al. 1985; Bradley et al. 1987; Nicklas et al. 1987). Although we know that the TG^r T cells are mutants, it remains to be shown the fidelity with which their measured in vivo frequencies quantitatively reflect the underlying mutational events.

One characteristic of the human in vivo T-cell population is its heterogeneity. It differs in this regard from the mammalian populations used in most in vitro mutagenicity studies. Actually, there are many levels of T-cell heterogeneity. At the extreme, each T cell acquires a uniqueness as it differentiates in the thymus in terms of its ability to recognize specific antigen. This process, which is achieved by rearrangements of TCR genes, occurs early in life and might be completed in humans shortly after adolescence (Hedrik et al. 1984; Royer et al. 1984; Saito et al. 1984; Toyonaga et al. 1984; Born et al. 1985). There are at least three TCR genes, i.e., the α, β, and γ genes, located on the long arms of chromosomes 14 and 7 and the short arm of chromosome 7, respectively (Caccia et al. 1985; Isobe et al. 1985; Murre et al. 1985). These genes rearrange their variable, joining, diversity, and constant regions in vivo in each cell to confer the uniqueness of the T-cell receptors found in specifically reactive mature T lymphocytes. Each gene rearranges independently in a cell to yield a great potential diversity of DNA restriction fragment patterns. Once rearranged, the TCR gene pattern persists throughout the life of the T cell and in all of its clonal descendants. Presumably, these cells persist for many years, resulting in the patterns persisting throughout the life of the individual.

We are finding these TCR gene-rearrangement patterns to be quite useful for fundamental studies of in vivo gene mutation in human T lymphocytes. Because they are fixed in most T cells at a time when the thymus is active, they provide one marker of the in vivo independence of HGPRT mutant clones isolated in clonal assays of adults. Information regarding this in vivo independence of separate mutant isolates is crucial for assessing the number of mutational events reflected in an observed number of mutants.

This paper describes the methods used for these studies. It presents results with spontaneous in vivo HGPRT mutants derived from normal adults and provides evidence suggesting that, in some instances, simple measurements of HGPRT T-cell-mutant frequencies do not quantitatively reflect the underlying in vivo mutational events.

RESULTS

We have now completed 115 clonal assays of spontaneous HGPRT T-cell-mutant frequencies in normal adults. The method has been recently described in detail and involves direct cloning of T lymphocytes in vitro in microtitre wells in limiting dilutions (1–2 cells/well; nonselective conditions and 1×10^4, 2×10^4, or 5×10^4 cells/well; selection in 10 μM TG) to define cloning efficiencies in the absence and presence of TG, respectively (O'Neill et al. 1987). T-cell growth factor and irradiated feeder cells are required for cloning and cell growth. Cloning efficiencies are calculated from the Poisson relationship $P_O = e^{-x}$, where x is the average number of clonable cells/well. A mutant frequency for an individual sample is the ratio of cloning efficiency with TG to cloning efficiency without TG. The mean mutant frequency for the 115 clonal assays is 6.5×10^{-6}, with a standard deviation of 4.8×10^{-6} (median value of 5.3×10^{-6}, with 10th and 90th percentile values of 2.2×10^{-6} and 12.0×10^{-6}, respectively).

Elsewhere, we report results of Southern blot analyses of 50 wild-type and 164 HGPRT mutant clones isolated in single or replicate clonal assays of blood samples from 8 normal adult males (included in the above 115 assays) (J.A. Nicklas et al., in prep.). Of the 164 mutant colonies, 16 (10%) showed gross structural alterations in HGPRT detectable on Southern blots hybridized with a full-length or modified HGPRT cDNA probe (Brennand et al. 1983; J.A. Nicklas et al., in prep.). Blots from 47 of the wild-type and 154 of the mutant colonies were rehybridized with a TCR β-gene probe which is the insert from Jurkat 2 (Yanagi et al. 1984), obtained from T.W. Mak, containing nucleotides 100–870 cloned into the *Pst*I site of pBR322. Some blots were rehybridized also with a TCR γ-gene probe, which is the 700-bp *Hin*dIII-*Eco*RI insert of pH 60, a genomic probe containing TCR γ J1 sequences (Lefranc et al. 1986). It is these studies of TCR gene rearrangement patterns, also reported elsewhere (Nicklas et al. 1987), and our attempts to use them in quantitating in vivo mutational events that are considered here.

Forty-four of the 47 wild-type colonies studied for TCR gene rearrangements showed unique patterns, suggesting that they arose from independent in vivo progenitors. TCR gene rearrangement patterns can be used also as indicators of the in vivo independence of HGPRT mutants and, therefore, to make quantitative judgements regarding the underlying, in vivo, mutational events. Figure 1 shows a Southern blot of *Hin*dIII-cleaved DNAs from one wild-type and 11 HGPRT mutant T-cell colonies isolated from a single individual. None of the mutants showed *hgprt* alterations detectable at this level of study (Fig. 1A). A priori, it is possible that all 11 mutants are the progeny of a single cell that suffered *hgprt* mutation, that all 11 mutants resulted from 11 different mutational events, or that some combination of these possibilities occurred. As shown in Figure 1B, however, the 11 mutant colonies showed 7 different TCR β-gene rearrangements. Unique patterns are seen with mutants 2, 4, 6, and 7. Mutants 1, 3, and 5 appear to have identical rearrangements, as do the pairs 8-10 and 9-11. Thus, with one restriction endonuclease analysis of 1 of the 3 TCR genes, it is clear that 7 of

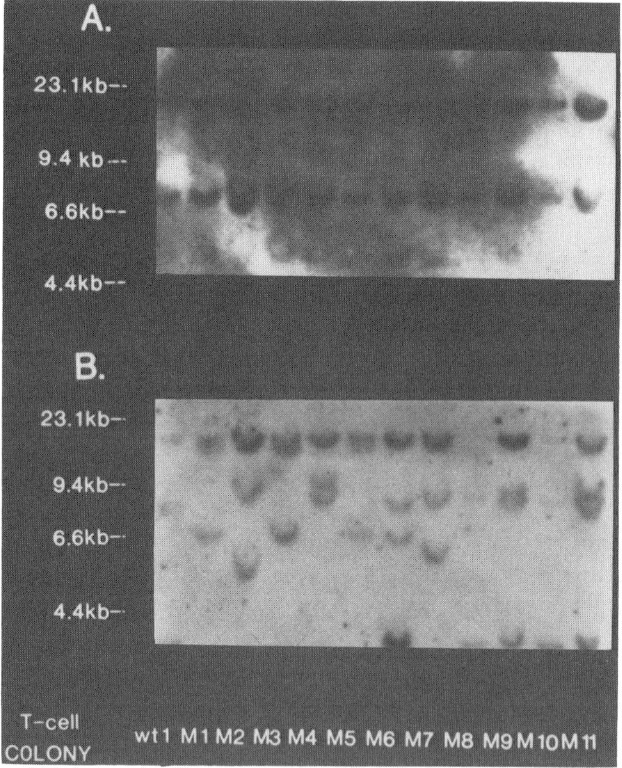

Figure 1

Southern blot analyses of DNA from T cell colonies. DNAs from 1 unselected wild-type and 11 TG-selected mutant colonies were cleaved with *Hin*dIII and hybridized with an *hgprt* gene cDNA probe (pHPT 30) (*A*) or with TCR gene cDNA probe (Jurkat 2) (*B*).

the 11 mutant colonies were derived from independent, in vivo, mutational events. The use of *Bam*HI showed that mutants 8-10 and 9-11 had different TCR β-gene rearrangements (data not shown). Therefore, this collection of 11 HGPRT mutants resulted from at least 9 in vivo mutational events.

The number of shared TCR gene patterns among the HGPRT mutants isolated from a single individual and, therefore, the number of underlying in vivo mutations, differ. Mutant frequency determinations were made twice on a normal male 6 weeks apart, with values of 9.3×10^{-6} and 7.4×10^{-6}. A total of 6 wild-type and 31 HGPRT mutant colonies, isolated from both assays, were studied. Two of the 31 mutants showed structural changes within *hgprt* (data not shown). Five of the mutant colonies isolated from both experiments showed the same TCR β-gene pattern. Four of the five are shown in Figure 2 (arrows) in a TCR β *Bam*HI Southern blot of 1 wild-type and 20 HGPRT mutant colonies. No HGPRT gross structural alterations were noted in these five mutants. Three of the five were studied with three restriction endonucleases; the other two were studied with only two. Nonetheless, the 31 mutants from this individual can be conservatively attributed to 27 underlying mutational events. Two wild-type colonies from this individual also shared a TCR β-gene pattern that differed from those of the five mutants.

The above finding suggest that most isolated HGPRT mutant T cells result from independent in vivo mutational events, even though there is some distortion due to in vivo clonal amplification of mutants. However, these findings were obtained in normal adults with "normal" mutant frequencies.

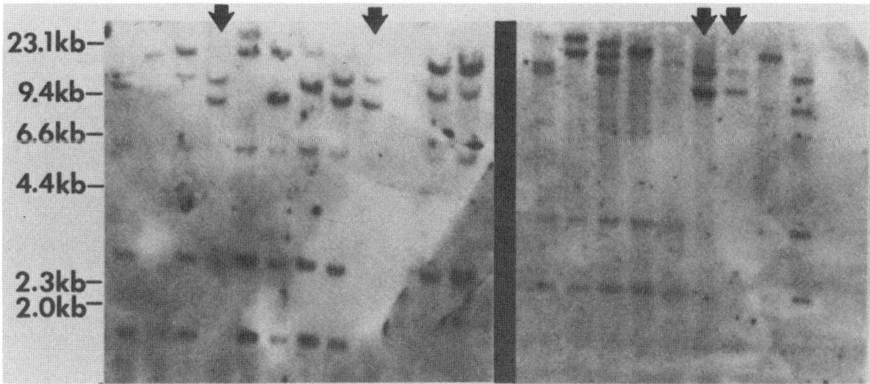

Figure 2

Southern blot analysis of DNAs T-cell from colonies. DNA for one unselected wild-type (first lane to right of black vertical separation) and 20 TG-selected mutant colonies were cleaved with *Bam*HI and probed with a TCR β-gene cDNA probe (Jurkat 2). Arrows indicate four mutants with identical TCR β-gene rearrangement patterns.

Hints that this disparity between in vivo mutational events and resultant derived mutants may be much greater in mutagenized individuals or in others with "high" mutant frequencies comes from recent findings in a normal female oncology nurse who administered cancer chemotherapeutic agents. Mutant frequency values were elevated in this woman, reaching $\sim 500 \times 10^{-6}$ in the most recent clonal assay. Study of *Hin*dIII-digested DNA obtained from 7 wild-type and 28 HGPRT mutant colonies isolated from this assay showed that 27 of the 28 mutant colonies had the same TCR β-gene rearrangement pattern (data not shown). Southern blot analysis of HGPRT showed no changes using the usual restriction endonucleases, perhaps because hybridization fragment losses can be missed in the face of two X chromosomes. Molecular, antigenic, and cytogenetic studies of hundreds of mutants and wild-type colonies isolated from several clonal assays of this individual are currently under way, but our early indication is that the 100-fold increase in measured mutant frequency may not be due to an increase in in vivo *hgprt* mutation at all, but its rather due to clonal amplification of mutants. Rather than needing minor correction, such a finding indicates that in some individuals, measured HGPRT mutant frequencies may bear little relationship to underlying *hgprt* mutational events. Thus, although the mutants are real, their increased frequency in this woman is not a quantitative reflection of an increase in somatic gene mutation in vivo.

DISCUSSION

Selection in vivo against TG^r T lymphocytes in humans has long been suspected. The first evidence for this was the low frequency of mutant T cells found in women who were heterozygous for the Lesch-Nyhan mutation. Rather than 50% as expected on the basis of random X-chromosomal inactivation, the frequency of HGPRT mutant T cells in heterozygotes is usually less than 10%, even when the relative proportion of mutant and wild-type skin fibroblasts are as expected in the same individual (Strauss et al. 1980; Dempsey et al. 1983). The HGPRT mutant T cells arising de novo from somatic mutation also seem to be selected against in vivo. Autoradiographically determined TG^r T-cell mutant frequencies rise within 2 weeks after cyclophosphamide infusion in multiple sclerosis patients, but fall to pretreatment levels within a few months (Ammenheuser et al. 1986). Finally, the slope of the increase in HGPRT mutant T-cell frequencies with age in humans is quite shallow, indicating that there is not a great accumulation with time (Trainor et al. 1984; Vijayalaxmi and Evans 1984). This too probably represents in vivo selection against the mutants. This in vivo negative selection is advantageous when using the mutant T-cell system for human mutagenicity monitoring, because rises in mutant frequencies should be temporally associated with their responsible environmental influences.

Other complications in relating measured T-cell mutant frequencies with in vivo mutation must be considered. One is the question of domains or restricted circulations of T lymphocytes. It might be asked how representative are the circulating T lymphocytes in humans to the T-lymphocyte population as a whole. Also, it is generally understood that the lymphocyte population itself is heterogeneous according to class, i.e., B lymphocytes, T lymphocytes, and within the latter, helper and suppressor cells. Even the latter subclasses are themselves divisible into still further subclasses. At present, we have no evidence for or against differential susceptibility to mutation among the various classes and subclasses of T cells. However, there are physiological shifts in the relative proportions of these different subclasses of lymphocytes within an individual at different times. It can be imagined that a relative increase in a "mutation-susceptible" subclass of lymphocytes that would go unrecognized might be reflected as an increase in measured T-cell mutant frequency. Such an increase would not necessarily have the implication of an increase in in vivo somatic gene mutation as regards human health risks.

Some of the above concerns are, at present, theoretical. We have attempted to address them, in part, in the clonal assay by determining that the same subpopulations of T lymphocytes are represented in nonselection wells, defining cloning efficiencies, and in selection wells, defining mutant fraction (Albertini 1985). Therefore, mutant frequency, which is a ratio of cloning efficiencies under the two conditions, is at least a ratio of two like entities.

The phenomenon addressed in this paper is another dimension of the complex quantitative relationship between in vivo mutational events and their resultant measured T-cell mutants. It should have been expected, from the biology of the circulating T lymphocytes, that these cells would periodically undergo clonal amplification in the peripheral blood, lymph nodes, etc., in response to antigenic or other stimuli. Furthermore, clonal amplification will probably be a stochastic process and irregularly involve the lymphocyte population as a whole. If this amplification involves a cell that has already undergone *hgprt* mutation, there will be a great excess of mutants vis-à-vis mutations when the former are expressed as a mutant frequency. Perhaps more likely, clonal amplification itself may generate spontaneous HGPRT mutant T lymphocytes if sufficient cell generations are involved. In this instance, it might be anticipated that any in vivo mitogenic influence will result in an increase in HGPRT or other somatic mutants, the mutational process simply being a concomitant of cell division.

What effect might this amplification phenomenon have on the use of the mutant T-lymphocyte system for human mutagenicity monitoring? First of all, the phenomenon must be recognized. Also, the distorting effect of clonal amplification will be in one direction only, i.e., that of increasing measured mutant frequencies. We show above that the vast majority of humans tested have HGPRT T-cell mutant frequencies in the low range, i.e., 6.5×10^{-6} with a standard deviation (S.D.) of 4.8×10^{-6}. We have data on 82 autoradiographic assays for TG^r T-cell mutant frequencies and the results are quite similar, i.e., mean value of 8.7×10^{-6} with S.D. of 6.1×10^{-6}. On the other hand, the chemotherapy nurse whose results are presented above, clearly showed values well above this range. Therefore, the scheme for human mutagenicity monitoring for somatic gene mutation using T-cells might involve simple quantitation of HGPRT mutant frequencies either by autoradiography or by clonal assay as the initial step. Clonal assay and study of the mutants might be reserved for blood samples showing high mutant frequencies. Elevated mutant frequency values with either assay may or may not represent an increase in in vivo, somatic, gene mutation, but low-mutant-frequency values will be reassuring that such an increase has not occurred. Since both autoradiography and clonal assay for HGPRT mutants are now performed on cryopreserved mononuclear cell samples, an aliquot from the same blood sample can be reserved for molecular studies should an elevated mutant frequency be found.

Finally, and perhaps most importantly, it must be recognized that the HGPRT mutant T-cell assay might have a value for fundamental studies, not only for somatic gene mutation, but of normal and pathological lymphocyte kinetics in vivo in humans. T-cell mutation may be useful as a probe for processes other than those damaging the marker gene under study. It is clear that, at present, we do not know the full import of the results that our experiments are providing.

ACKNOWLEDGMENTS

This research was supported by National Cancer Institute grant 2-RO1-CA-30688-04A1, Department of the Environment contract DE-AC02-83ER60147-A-003. We thank Inge Gobel for preparing this manuscript.

REFERENCES

Ammenheuser, M.M., J.B. Ward, Jr., and J.M. Killian. 1986. A longitudinal study of the frequency of 6-thioguanine-resistant lymphocytes from multiple sclerosis patients receiving cyclophosphamide. *Environ. Mutagen.* (suppl.6) **8:** 3.

Albertini, R.J. 1985. Somatic gene mutations in vivo as indicated by the 6-thioguanine-resistant T-lymphocytes in human blood. *Mutat. Res.* **150:** 411.

Albertini, R.J., K.S. Castle, and W.R. Borcherding. 1982. T-cell cloning to detect the mutant 6-thioguanine-resistant lymphocytes present in human peripheral blood. *Proc. Natl. Acad. Sci.* **79:** 6617.

Albertini, R.J., J.P. O'Neill, J.A. Nicklas, N.H. Heintz, and P.C. Kelleher. 1985. Alterations of the *hprt* gene in human *in vivo* derived 6-thioguanine-resistant T-lymphocytes. *Nature* **316:** 369.

Born, W., J. Yague, E. Palmer, J. Kappler, and P. Marrack. 1985. Rearrangement of T-cell receptor β-chain genes during T-cell development. *Proc. Natl. Acad. Sci.* **82:** 2925.

Bradley, W.E.C., J.L. Gareau, A.M. Seifert, and K. Messing. 1987. Molecular characterization of 15 rearrangements among 90 human in vivo somatic mutants shows that deletions predominate. *Mol. Cell. Biol.* **7:** 956.

Brennand, J., D.S. Konecki, and C.T. Caskey. 1983. Expression of human and Chinese hamster hypoxanthine-guanine phosphoribosyl transferase cDNA recombinants in cultured Lesch-Nyhan and Chinese hamster fibroblasts. *J. Biol. Chem.* **258:** 9593.

Caccia, N., G.A.P. Bruns, H.R. Kirsch, G.F. Hollis, V. Bertness, and T.W. Mak. 1985. T-cell receptor α chain genes are located on chromosome 14 at 14 q 11-14 q 12 in humans. *J. Exp. Med.* **161:** 1255.

Dempsey, J.L., A.A. Morley, R.S. Seshadri, B.T. Emmerson, R. Gordon, and C.I. Bhagat. 1983. Detection of the carrier state for an X-linked disorder, the Lesch-Nyhan syndrome, by the use of lymphocyte cloning. *Hum. Genet.* **64:** 288.

Hedrick, S.M., D.I. Cohen, E.A. Nielsen, and M.M. Davis. 1984. Isolation of cDNA clones encoding T-cell-specific membrane-associated proteins. *Nature* **308:** 149.

Isobe, M., J. Erikson, B.S. Emanuel, P.C. Nowell, and C.M. Croce. 1985. Location of gene for β subunit of human T-cell receptor at band 7q35, a region prone to rearrangements in T-cells. *Science* **228:** 580.

Lefranc, M.-P., A. Forster, and T.H. Rabbitts. 1986. Rearrangement of two distinct T-cell γ-chain variable-region genes in human DNA. *Nature* **319:** 420.

Murre, C., R.A. Waldmann, C.C. Morton, K.F. Bongiovanni, T.A. Waldmann, and T.B. Shows. 1985. Human γ-chain genes are rearranged in leukaemic T-cells and map to the short arm of chromosome 7. *Nature* **316:** 549.

Nicklas, J.A., T.C. Hunter, L.M. Sullivan, J.K. Berman, J.P. O'Neill, and R.J. Albertini. 1987. Molecular analyses of *in vivo hprt* mutations in human T-lymphocytes. I. Study of low frequency "spontaneous" mutants by Southern blots. *Mutagenesis* (in press).

O'Neill, J.P., M.J. McGinniss, J.K. Berman, L.M. Sullivan, J.A. Nicklas, and R.J. Albertini. 1987. Refinement of a T-lymphocyte cloning assay to quantify the *in vivo* thioguanine resistant mutant frequency in humans. *Mutagenesis* **2:** 87.

Royer, H.D., O. Acuto, M. Fabbi, R. Tizard, K. Ramachandran, J.E. Smart, and E.L. Reinherz. 1984. Genes encoding the Ti β subunit of the antigen/MHC receptor undergo rearrangement during intrathymic ontogeny prior to surface T3-Ti expression. *Cell* **39:** 261.

Strauss, G.H., E.F. Allen, and R.J. Albertini. 1980. An enumerative assay of purine analogue resistant lymphocytes in women heterozygous for the Lesch-Nyhan mutation. *Biochem. Genet.* **18:** 529.

Saito, H., D.M. Kranz, Y. Takagaki, A.C. Hayday, H.N. Eisen, and S. Tonegawa.

1984. A third rearranged and expressed gene in a clone of cytotoxic T-lymphocytes. *Nature* **312:** 36.

Toyonaga, B., Y. Yanagi, N. Suciu-Foca, M. Minden, and T.W. Mak. 1984. Rearrangements of T-cell receptor gene YT35 in human DNA from thymic leukaemia T-cell lines and functional T-cell clones. *Nature* **311:** 385.

Trainor, K.J., D.J. Wigmore, A. Chrysostomou, J.L. Dempsey, R. Seshadri, and A.A. Morley. 1984. Mutation frequency in human lymphocytes increases with age. *Mech. Ageing Dev.* **27:** 83.

Turner, D.R., A.A. Morley, H. Haliandros, R. Kutlaca, and B.J. Sanderson. 1985. *In vivo* somatic mutations in human lymphocytes frequently result from major gene alterations. *Nature* **315:** 343.

Vijayalaxmi and H.J. Evans. 1984. Measurement of spontaneous and X-irradiation-induced 6-thioguanine-resistant human blood lymphocytes using a T-cell cloning technique. *Mutat. Res.* **125:** 87.

Yanagi, Y., Y. Yoshikai, K. Leggett, S.P. Clark, I. Aleksander, and T.W. Mak. 1984. A human T-cell specific cDNA clone encodes a protein having extensive homology to immunoglobulin chains. *Nature* **308:** 145.

In Vivo Somatic Mutations in the Glycophorin A Locus of Human Erythroid Cells

RONALD H. JENSEN, WILLIAM L. BIGBEE, AND RICHARD G. LANGLOIS
Biomedical Sciences Division
Lawrence Livermore National Laboratory
Livermore, California 94550

OVERVIEW

An assay system for detecting in vivo human somatic cell mutations was developed using monoclonal antibodies specific for the M and N forms of the polymorphic erythrocyte surface antigen glycophorin A (GPA). Pairs of antibodies, each conjugated with a different fluorophor, are used to label erythrocytes in heterozygous MN blood samples to identify rare variant cells that fail to express one of the two *gpa* alleles. Flow cytometry and cell sorting are used to enumerate the frequency of variant cells from each sample. Variants occur at a frequency of about ten per million normal erythrocytes in samples from normal donors. In these samples, we detected NΦ and MΦ hemizygous variant cells and NN and MM homozygous variant cells, suggesting that these variants are caused by either gene loss mutagenic events or missegregation or recombination events. Preliminary analysis on samples from cancer patients being treated with mutagenic drugs indicate that significant increases in frequency occur during chemotherapy. Samples from individuals that survived the atomic bomb in Hiroshima show that radiation exposure induces increases in variant cell frequency that persist for many years. Variant cell frequencies for some survivors differ dramatically from the dose-response, and these deviations are the basis for a stochastic model describing the effects of radiation on the long-lived stem-cell pool in individuals. This assay may provide a cumulative biodosimeter for measuring exposure of humans to radiation over their entire life.

INTRODUCTION

Exposure of mammalian cells to chemical pollutants or high energy radiation can result in cytotoxicity, mutagenesis, carcinogenesis, and teratogenesis. In general, the interaction of toxic chemicals or radiation with somatic cells of whole mammals may cause developmental abnormalities or cancer. Extrapolation of the molecular events being discovered by experiments in vitro to the pathological effects detected in vivo is a difficult process. One way to investigate the relationship between the molecular events and the whole-animal effects is to analyze somatic cells taken from the whole organism and attempt to determine effects of the whole body exposure to genotoxic agents. The accumulation of evidence from epidemiological studies of human populations and from carcinogenicity testing in rats and mice continues to indicate a high risk of carcinogenesis for individuals exposed to mutagens. Thus, mutagenesis in the somatic cells of whole organisms is a desirable method for investigating the effects of genotoxic exposure and allows a shorter extrapolation of the insights gathered to the effects on the organism.

We developed a system for analyzing human peripheral blood and detecting rare variant erythrocytes that appear to be the progeny of mutated erythroid precursor cells (Langlois et al. 1986). This system uses fluorescently labeled monoclonal antibodies to label cells for the presence of the cell-surface antigen, GPA, and flow cytometric analysis to enumerate the frequency of variant cells. GPA is the major cell surface glycoprotein of human erythrocytes and is expressed at about one-half million copies per cell (Anstee 1980). It occurs in two allelic forms that are very similar to each other, differing only in 2 amino acids of the 131 amino acids and 16 tetrasaccharides that compose the molecules. These two allelic forms are the basis for the MN blood group and are codominantly expressed, so that individuals who are heritably heterozygous express both forms on their erythrocytes. In blood samples from an *mn* heterozygote, variant cells would possess only one of the two allelic forms of *gpa*. Expression of one of the two allelic forms of *gpa* would guarantee that the variant cells are capable of expressing this cell-surface antigen in a native state, so that phenocopies would not be included as false positives in the enumeration of variant cells.

RESULTS

To perform flow cytometric analysis, blood samples were simultaneously labeled with a GPA (public)-specific monoclonal antibody (10F7) and a GPA (M)-specific monoclonal antibody (6A7). Antibody 10F7 binds specifically to GPA, but equally well to each allelic form (Langlois et al. 1985). Thus, all normal erythrocytes in human blood will bind equal amounts of this antibody, regardless of the blood type. Antibody 6A7 binds only the *m*-allelic form of *gpa* (Langlois et al. 1985), and therefore binds MN blood erythrocytes to only half the extent of binding to MM homozygous erythrocytes and not at all to NN erythrocytes. Figure 1a shows that a mixture of cells from different blood types can be separated extremely well using this antibody-labeling technique. In the analysis shown, separation along the Y-axis is due to the amount of GPA(M) on each cell. As expected MM cells are separated from MN cells by a factor of two in fluorescence intensity, whereas NN cells are about 1/100 as

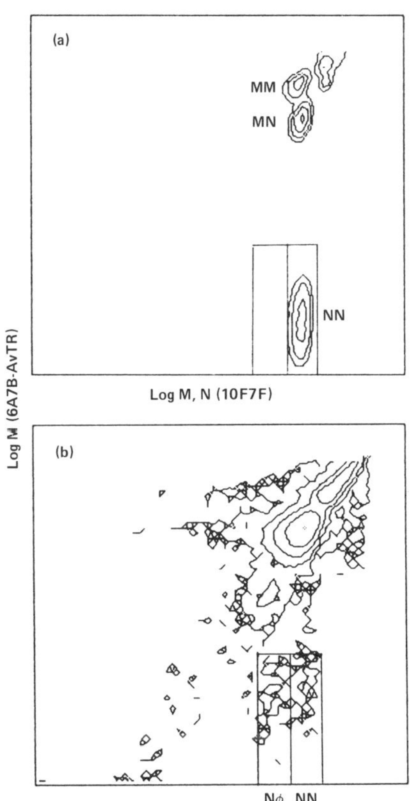

Figure 1
Bivariate flow distributions of erythrocytes stained with GPA-specific monoclonal antibodies. Logarithmic amplification was used on both axes, and the height of each peak is indicated by contours. GPA (public) antibody, 10F7, was conjugated directly with fluorescein, which is detected by excitation at 488 nm and emission at 514 nm. GPA (M) antibody, 6A7, was conjugated with biotin, and secondary labeling was performed with avidin-Texas Red. Texas Red is detected by excitation at 590 nm and emission at 630 nm. (*a*) Control cell mixture containing equal numbers of MM, NN, and MN cells; (*b*) flow distribution from analysis of 10^6 erythrocytes from a normal MN donor. Contours differ by factors of ten in events per channel with the lowest contour corresponding to one event per channel.

bright as MN cells. Boxes shown in Figure 1 define the position of hemizygous variant cells (NΦ) or homozygous variant cells (NN). In our experiments, we perform this same labeling and flow analysis on about 10^6 erythrocytes from each heterozygous MN sample and sort the variant cells from the windows defined. Microscopic analysis of the sorted cells is used to confirm their cellular nature and the fluorescence properties measured by the flow cytometer.

Blood samples from 54 different people were analyzed by this method to determine the frequency of hemizygous NΦ variant cells. A histogram of the results of these analyses (Fig. 2) shows that variant cells occur at very low frequencies (1×10^{-6}–33×10^{-6}) with a rather large variation among individuals [S.D.] $= 6.91 \times 10^{-6}$). Independently, we found that technical difficulties in measuring these very rare events lead to rather low precision (coefficient of variation =50%), so that real biological variations among

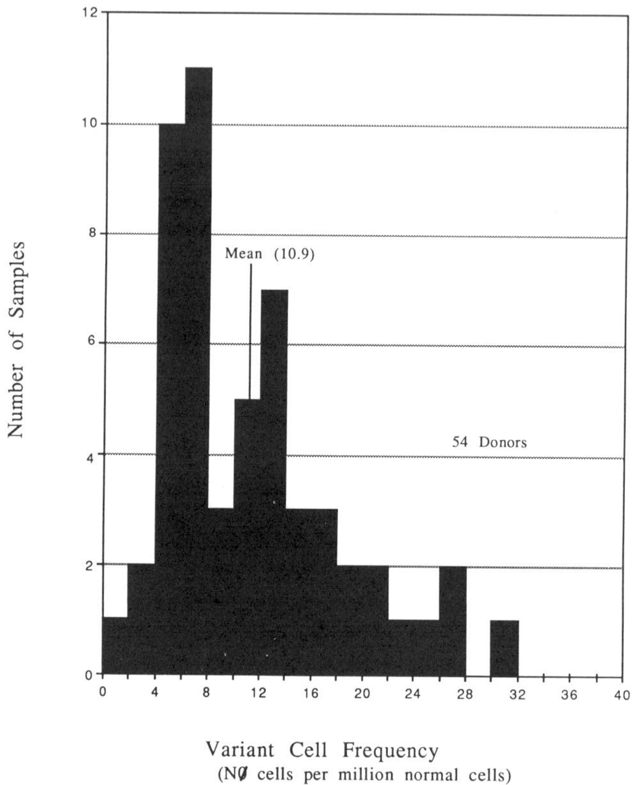

Figure 2

Hemizygous variant cell frequencies of normal donors. Blood samples from 54 donors were analyzed for NΦ variant cell frequency.

individuals are difficult to discern. Nevertheless, by analyzing variant cell frequencies versus age of donor we find a trend toward a higher frequency of NΦ variant cells among older indivdiuals (Fig. 3). Efforts to improve precision in the assay should allow more meaningful analysis of age effects.

Previous indications that cancer patients being treated with mutagenic drugs show increasing frequency of hemizygous variant cells (Langlois et al. 1986) are being sustained by more complete studies. One study entails analysis of samples from different patients, each suffering a different form of neoplasia and being treated with variable mutagenic adjuvants. For patients who were started on treatment more than 6 weeks prior to analysis, we found a significant increase in the frequency of hemizygous variant cells (frequencies in the range from 5×10^{-6} to 131×10^{-6}). Longitudinal studies on samples from patients being treated with mutagenic chemotherapy for breast cancer

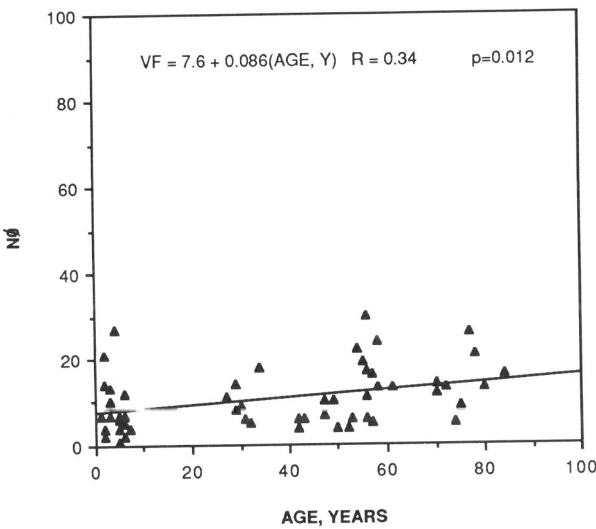

Figure 3

Hemizygous variant cell frequencies as a function of age of normal donors. NΦ variant cell frequencies from the same samples as shown in Fig. 2. Variant cell frequency (per million normal cells) is plotted on the abscissa. The solid line is a linear regression fit to the data, with the equation of regression and statistical fit parameters shown at the top.

by one of three standard adjuvant therapies, cyclophosphamide, adriamycin, and 5-fluorouracil (CAF), cyclophosphamide, methotrexate, and 5-fluorouracil (CMF), and vinblastine, methotrexate, and 5-fluorouracil (VMF) are in progress. The preliminary results on ten patients indicate a two- to fourfold increase in variant cells frequency after 6–8 weeks of therapy with CAF or CMF, each of which contains mutagens, and no increase after therapy with VMF, which contains no mutagen (W.L. Bigbee et al. in prep.).

In concept we would predict that exposure to high-energy radiation also should result in somatic cell mutations, and as a result we investigated blood samples from survivors of the atomic bomb over Hiroshima, Japan, who were exposed to radiation 40 years ago (Langlois et al. 1987). Two different antibody combinations were used to enumerate four different variant cell phenotypes from these blood samples (hemizygous NΦ and MΦ variant cells and homozygous NN and MM variant cells). We found that variant cell frequencies of three of these cell types were significantly higher for a cohort of survivors exposed to radiation (estimated doses of 0.14 to 8.84 gray [Gy]) than for a matched control cohort of survivors (individuals with estimated doses of less than 0.01 Gy). One cell type (homozygous NN cells) showed unpredictably high frequencies in the control cohort. We believe that the assay for this variant cell type is sensitive to glycosylation effects on the GPA, which may be caused by genetic or physiological factors in vivo or degradation of glycoproteins during cell storage and fixation. A different monoclonal antibody is being substituted into our assay to prevent this artifact.

To reduce the effect of measurement errors on each individual being analyzed, we calculated a mean hemizygous variant cell frequency for each donor including means from all replicate samples taken from the same donor. These values include both NΦ and MΦ variant cell frequencies, since our analyses show that these two cell types have similar spontaneous and induced frequencies for most people. Figure 4a shows the hemizygous frequencies of the Hiroshima survivors that were analyzed. There is a significant relationship between the estimated dose that each survivor received (T65DR) and the hemizygous variant cell frequency. T65DR is an internationally accepted dose estimate [Milton and Shohoji 1968] that will be superseded by a new procedure in 1987. Clearly, some individuals differ substantially from the calculated linear fit shown in Figure 4a, but averaging the dose response over quartile groups of donors (as shown in Fig. 4b) results in a remarkably good linear fit.

The largest deviations from this general increase in frequency of hemizygous variant cells with increased estimated radiation dose are for individuals that received more than 4 Gy. Some of these people display frequencies that are much higher than expected (up to 10^{-3}), whereas samples from people exposed to very high doses (7 to 8 Gy) show frequencies similar to unexposed

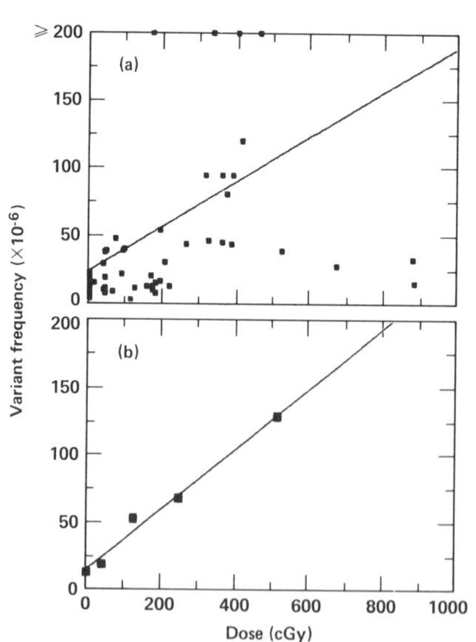

Figure 4

Hemizygous variant cell frequency for survivors of atomic bomb at Hiroshima. Variant frequency for each donor is the mean of all NΦ and MΦ measurements performed on that donor. (*a*) Variant cell frequencies for all donors, 43 exposed to 0.1 Gy or more, and 20 control ($<$ 0.01 Gy exposure). Frequencies $>200 \times 10^{-6}$ are plotted on the upper boundary. The values for these donors is 336, 330, 244, and 666×10^{-6} (from low to high dose). The line represents a linear regression to give the equation $VF = 23 \times 10^{-6} + 0.17 \times 10^{-6}/cGy$; $R = 0.33$ and $P = 0.004$. (*b*) Variant cell frequencies of pooled cohorts of survivors. Exposed donors were grouped into four equal-sized subgroups on the basis of estimated exposure. Control donors made a fifth subgroup at zero dose. Each point represents a mean hemizygous variant cell frequency and mean dose for one group. The line corresponds to a linear regression yielding the fit parameters $VF = 15 \times 10^{-6} + 0.22 \times 10^{-6}/cGy$; $R = 0.99$, $P < 0.001$.

Table 1
Analyses on Selected Exceptional Donors

Donor number	Estimated dose (T65DR)[a]	Measured variant cell frequency (× 10^{-6})[b]		Expected variant cell frequency[c] (× 10^{-6})
		NΦ	MΦ	
72	1.73	478	42	53
		882	60	
77	3.88	16	175	100
		13	175	
60	6.72	34	44	163
		13	22	
30	8.84	17	n.d.	209
		14	15	

[a]Estimated dose in Gy (1 Gy = 100 rads).
[b]For each donor two blood samples were taken at 1–8 months apart. Variant cell frequencies are shown separately for each sample. n.d. = not done.
[c]Variant cell frequencies calculated from regression in Fig. 4.

individuals. Independent measurement of the two hemizygous variant cell types (Table 1) indicate that the general correlation of frequencies between these two cell types does not hold up.

DISCUSSION

The development of this GPA-based assay for measuring the frequency of the rare variant erythrocytes in human peripheral blood has provided a new approach for measuring genetic damage to somatic cells in people exposed to potentially genotoxic agents. Because erythrocytes do not carry DNA, we cannot confirm the mutagenic nature of the events that cause GPA-variant cells. However, we have observed elevated frequencies in cancer patients being treated with mutagenic chemotherapy and elevated frequencies in individuals exposed to high-energy radiation. The magnitude of these increases is similar to the magnitude of the increases observed for the in vivo clonogenic hypoxanthine phosphoribosyl transferase (HPRT) assay (Albertini et al. 1982; Morley et al. 1983; Albertini 1985; Messing and Bradley 1985) and also for somatic cell mutations of cultured cells in the *hprt*, thymidine kinase (*tk*), dehydrofolate reductase (*dhfr*), and human leukocyte antigen (*hla*) loci (Sankaranarayanan 1982; Brown and Thacker 1984, Sanderson et al. 1984; Vijayalaxmi and Evans 1984; Stankowski and Hsie 1986), thus supporting a mutational origin for GPA-variation erythrocytes.

One important advantage of the GPA-based assay is the ability to measure several mutational effects simultaneously. The occurrence of NΦ and MΦ variant cells in samples from *mn* heterozygotes must result from independent genetic lesions in erythroid precursor cells, since the loss of expression is for an autosomal locus on chromosome 4. In addition, our previous measurements on heritably acquired mutant alleles at the GPA locus indicate independence of expression of the allelic forms at this locus. Although most of our results are on the frequency of NΦ cells, many coanalyses on the loss of expression of *m* allele and *n* allele show that these events occur at a similar frequency. In addition, we can measure the occurrence of homozygous variant cells, MM and NN. Although we have reason to believe our present measurement on NN cells has some phenocopy type artifacts, the frequency of MM-variant cells circumvents this difficulty. The occurrence of such cells must be the result of mitotic recombination or chromosomal missegregation, events that are independent of the mutational events that cause hemizygous GPA-variant cells (Eves and Farber 1983; Wasmuth and Hall 1984).

An important exception to the observation of similar frequencies of the NΦ and MΦ gene-loss variant cells was found in the analysis of blood samples from atomic bomb survivors who were exposed to high doses of radiation (see Table 1). We can explain these results as well as the low frequencies obtained on samples from very-high-dose survivors using a stochastic model for occurrence and expression of gene-loss mutations in a stem-cell pool whose size is altered by radiation-induced cytotoxicity. The parameters for this model (Table 2) enable us to use the effects of cytotoxicity, induced mutation frequency, and expression effects of these mutations to infer variant cell frequencies. Our model is based on the assertion that the stem-cell pool in individual survivors (which is the only erythroid cell type whose lifetime is long enough to show effects 40 years after exposure) was small enough to show statistical fluctuations in the resulting variant cell frequency. If we assume that the stem-cell pool per individual is 10^6 to 10^7 cells, radiation-

Table 2
Mutation Model for Long-lived Stem Cells

Dose (Gy)	Number of surviving stem cells[a]	Frequency of induced mutations[b]	Number of induced mutations[c]	Increase in variant cell frequency[d]
0.5	10^6–10^7	10×10^{-6}	10–100	0.1–1
4.0	10^4–10^5	100×10^{-6}	1–10	10–100
8.0	10^2–10^3	200×10^{-6}	0.02–0.2	10^3–10^4

[a]Number of surviving cells calculated with $N(D) = N(0)e^{-D/D_0}$, with $D_0 = 1$ Gy.
[b]Frequency of induced mutations from regression in Fig. 4.
[c]Number of induced mutations per donor is product of columns 2 and 3.
[d]Increase in variant cell frequency due to a single mutation is 1/column 4.

induced toxicity to these cells would decrease the viable stem-cell population to about 10^4 to 10^5 with 4 Gy exposure. At a frequency of 100 GPA mutations per million (as estimated from our regression) this would induce only 1–10 GPA mutant cells per donor. This is small enough that Poisson fluctuations among donors and between the two allelic forms in each donor would be significant. In addition, the pool size is small enough that a mutated cell that is formed within that pool would replicate as the pool size expands to its natural size of 10^6 to 10^7 cells and increase the variant frequency by 10×10^{-6} to 100×10^{-6}. As can be seen in Table 2, cytotoxicity is so prevalent at very high doses (e.g., 8 Gy) that most individuals would have no *gpa* mutations recorded in their stem-cell pool. Thus, most of these individuals would show no increase in variant cell frequency, but a few would be expected to display extremely high frequencies. Our model suggests that we must screen about 100 such donors in order to find a high-frequency highly exposed donor.

Two points about our model are somewhat notable. First, to be consistent with our observations, the stem-cell pool must be in the range of 10^6 to 10^7 cells for a normal individual. This estimate is somewhat lower than generally estimated (Cronkite and Feinendegen 1976; Messner et al. 1980), but the colony-forming assays used for the previous estimates do not require the 40 year cell lifetime required for effects from the atomic bomb exposure to be detected today. This stringent requirement for stem cells means that these cells must retain a high self-renewal capability not presently indicated for colony forming cells (Boggs et al. 1982).

In addition, our model indicates that this GPA-loss assay would be a poor biodosimeter for people who have received large acute doses of radiation. The variant cell frequency would vary substantially from Poisson fluctuations, and no reliable relationship could be established for relating variant frequency to dose received. Nevertheless, the size of the long-lived stem-cell pool does appear to be large enough to measure effects reliably when individuals are acutely exposed to less than 1 Gy, an exposure level at which biodosimetry would be an important addition to our monitoring capabilitites. In addition, the persistence of the measured effect indicates that multiple low-level exposures should result in an increasing variant cell frequency, thus serving as a cumulative biodosimeter.

ACKNOWLEDGMENTS

We thank B. Nisbet and R. Barlett for technical assistance. We also thank Dr. M. Akiyama of the Radiation Effects Research Foundation, Hiroshima, Japan, for his valuable collaboration with us. Work at Lawrence Livermore National Laboratory is performed under the auspices of the U.S. Department of Energy contract W-7405-ENG-48. It was partially supported by U.S. Environmental Protection Agency grant R811819-01.

REFERENCES

Albertini, R.J. 1985. Somatic gene mutations in vivo as indicated by the 6-thioguanine-resistant T-lymphocytes in human blood. *Mutat. Res.* **150:** 411.

Albertini, R.J., K.L. Castle, and W.R. Borcherding. 1982. T-cell cloning to detect the mutant 6-thioguanine-resistant lymphocytes present in human peripheral blood. *Proc. Natl. Acad. Sci.* **79:** 6617.

Anstee, D.J. 1980. Blood group MNSs-active sialoglycoproteins of the human erythrocyte membrane. In *Immunobiology of the erythrocyte* (ed. S.G. Sandler et al.), p. 67. A.R. Liss, New York.

Boggs, D.R., S.S. Boggs, D.F. Saxe, L.A. Gress, and D.R. Canfield. 1982. Hematopoietic stem cells with high proliferative potential. *J. Clin. Invest.* **70:** 242.

Brown, R. and J. Thacker. 1984. The nature of mutants induced by ionizing radiation in cultured hamster cells. *Mutat. Res.* **129:** 269.

Cronkite, E.P. and L.E. Feinendegen. 1976. Notions about human stem-cells. *Blood Cells* **2:** 269.

Eves, E.M. and R.A. Farber. 1983. Expression of recessive APRT-mutations in mouse CAK cells resulting from chromosome loss and duplication. *Somatic Cell Mol. Genet.* **9:** 771.

Langlois, R.G., W.L. Bigbee, and R.H. Jensen. 1985. Flow cytometric characterization of normal and variant cells with monoclonal antibodies specific for glycophorin A. *J. Immunol.* **134:** 4009.

———. 1986. Measurements of the frequency of human erythrocytes with gene expression loss phenotypes at the glycophorn A locus. *Hum. Genet.* **74:** 353.

Langlois, R.G., W.L. Bigbee, S. Kyoizumi, N. Nakamura, M.A. Bean, M. Akiyama, and R.H. Jensen. 1987. Evidence for increased somatic cell mutations in the glycophorin A locus in atomic bomb survivors. *Science* **236:** 445.

Messner, H.A., A.A. Fauser, J. Lepine, and M. Martin. 1980. Properties of human pluripotent hemopoietic progenitors. *Blood Cells* **6:** 595.

Messing, K. and W.E.C. Bradley. 1985. In vivo mutant frequency rises among breast cancer patients after exposure to high doses of gamma-radiation. *Mutat. Res.* **152:** 107.

Milton, R.C. and T. Shohoji. 1968. Tentative 1965 radiation dose (T65D) estimation for atomic bomb survivors, Hiroshima-Nagasaki. *Atomic Bomb Casualty Commission Tech. Rep.* p. 1.

Morley, A.A, K.J. Trainor, R. Seshardri, and R.G. Ryall. 1983. Measurement of in vivo mutations in human lymphocytes. *Nature* **302:** 155.

Sanderson, B.J.S., J.L. Dempsey, and A.A. Morley. 1984. Mutations in human lymphocytes: Effect of X- and UV-irradiation. *Mutat. Res.* **140:** 223.

Sankaranarayanan, K. 1982. Gene mutations. In *Genetic effects of ionizing radiation in multicellular eukaryotes and the assessment of genetic radiation hazards in man.*, p. 51. Elsevier/North-Holland, Amsterdam.

Stankowski, L.F. and A.W. Hsie. 1986. Quantitative and molecular analysis of radiation-induced mutation in AS52 cells. *Radiat. Res.* **105:** 37.

Vijayalaxmi and H.J. Evans. 1984. Measurement of spontaneous and X-irradiation-induced 6-thioguanine-resistant human blood lymphocytes using a T-cell cloning technique. *Mutat. Res.* **125:** 87.

Wasmuth, J.J. and L.V. Hall. 1984. Genetic demonstration of mitotic recombination in cultured Chinese hamster cell hybrids. *Cell* **36:** 697.

Comments

Calos: Dick, how do you interpret the individual variation that you see?

Albertini: I think there are two ways. One is for individuals with normal mutant frequencies, where, I think, the numbers are alright because most of the time we have close to a 1-to-1 correlation between mutants and mutations. On the other hand if we see a high mutant frequency, we do not know if it is due to increased mutation or due to clonal amplification, for which there can be many pathological and physiological reasons. The other person that we might have seen this on was a kidney transplant recipient. Those people, as you know, are undergoing a great deal of in vivo T-cell clonal amplification. So, if a mutant frequency is high, this must be looked at to see if it represents a lot of mutants or a lot of mutations. We now are freezing all samples for quantitative assays for purely monitoring purposes. For monitoring purposes, I think that a normal mutant frequency is alright; a high one has to be looked at.

I really don't know what is turning these cells on, and it may be different from case to case. We are actively studying this now.

Glickman: At what populations have you looked? To what exposed populations have you applied the assay?

Albertini: Cancer patients receiving cytotoxic chemotherapy make up one group. Many of these agents are mutagenic. Another group, with Drs. Jerry Williams and Stan Order at Johns Hopkins, are the radioimmunoglobin-treated people. These are patients with hepatomas or Hodgkins disease, who are being treated with a polyclonal antibody to ferritin ligated to yttrium. They get pretty large total-body-radiation doses that are well calibrated.

Regarding the Hiroshima group, Dr. Hakoda in Hiroshima has looked at that group. I was surprised, as a matter of fact, that he did see mutant frequency elevations 40 years after the exposure, because there is selection in vivo against HPRT-deficient cells. We know that from the Lesch-Nyhan heterozygotes, and we have seen it in some longitudinal studies that were done by Dr. Ammenhauser in Galveston. Nonetheless, mutant frequencies are increased in Hiroshima survivors. Also, we currently have a longitudinal study of T-cell mutant frequencies in breast cancer patients receiving adjunctive chemotherapy, and a few other groups like this. Those are the exposed groups we have looked at so far. Dr. Karen Messing, in Montreal, has looked at patients who have been exposed to radiation for diagnostic purposes. Also, we have looked at some Chernobyl victims.

Hutchinson: Are there plans afoot to use the two techniques on the same population?

Jensen: What actually started us on the A-bomb population was that I had heard from Mort Mendelsohn, when he was in Hiroshima, that the people at Radiation Effects Research Foundation were doing the HGPRT assay on exposed survivors, and so I contacted the group that was collaborating with Dick Albertini. Therefore, many of the people that we have assayed are the same people that they have assayed. The next paper that the people at Hiroshima will want to put together, I'm sure, will be a correlation of those two assays plus chromosome aberration frequency on the same people.

Hutchinson: That may be complicated, however, by the factor that Albertini raised about selection.

Albertini: The 40 year interval since the exposure is really too long for our assay.

Jenson: Every one will be complicated by different kinds of things like that, I'm sure. I don't expect us to correlate. Theirs is a different tissue than our tissue, and so it may well be that we are looking at effects that are altogether different.

Albertini: I think that stem cell mutation is detected in Ron's system. I think that, in adults, we are not going to be picking up stem cell mutations in lymphocytes anymore. I think some of what Dr. Hakoda is seeing in Hiroshima, and this is just a guess because we haven't done all the right markers, might have been stem cell mutations because they occurred in people who were children at the time of the atomic bomb blast. These survivors are now adults, in their 40s or 50s. He may still be seeing stem cell mutations that are coming through. That, as I pointed out, we can tell by the T-cell-receptor gene configurations.

Auletta: If you are monitoring, and if the aim of monitoring ultimately is public health, what about the public in the public health side of the equation, which is, ultimately, what are you going to tell these people? Sooner or later, someone is going to come to you and ask, "What does this mean?" or "What did you find?" or "How many mutants or mutations do I have?" Is there any thought to that, or how you are going to handle it?

Albertini: Yes. First of all, the nurses we studied could be a case in point. They are nurses at our institution who wanted the study done, and fortunately it turned out that there was not a great mutagenic effect in chemotherapy nurses using laminar flow hoods.

I really don't know what to tell people from a point of view of their health. Health pronouncements ought not be done at our present state of knowledge. Nonetheless, health outcome is the important issue in monitoring, and determining the health implications of genotoxicity markers is a critical need in our field. We can begin making this determination with our current technology by collecting samples from known exposed persons and asking if indicators of genetic damage in these samples, in fact, do change the relative risks for subsequent diseases in the individuals tested. Frankly, this is not known at the present time. This is why I don't monitor people, at the present time, in industrial settings unless I am doing it for research and unless the people know why we are doing it and know that we really do not know what it means in terms of their health.

Auletta: I think that is the really big question that has to be addressed at some point.

Albertini: I don't think you can ever answer it in the laboratory. This needs a field study with results of monitoring being compared with disease endpoints.

Glickman: In terms of the lab studies that have been done, the correlation, as I understand it, is that you can generally separate the smokers from the nonsmokers. These are very difficult questions to answer. For example, it takes a lot of nurses being exposed to cyclophosphamides, to get a frequency of risk estimate. In general, the risks seem to be relatively small, compared to the ones about which we know what to do.

Albertini: Right. That's exactly what happened. The noise of the system—the smoking, the age effect, of which there is a slight one—all buried whatever other effect there was.

Molecular Analysis of the *tk* and *hgprt* Loci

A Study of the Specificity of Spontaneous and UV-induced Mutation at the Endogenous *aprt* Gene of Chinese Hamster Ovary Cells

BARRY W. GLICKMAN, ELLIOT A. DROBETSKY, AND ANDREW J. GROSOVSKY
Biology Department
York University
Toronto, Ontario
Canada M3J 1P3

OVERVIEW

The spectrum of spontaneous and UV-induced mutation of an endogenous mammalian cell gene has been determined at the DNA sequence level. Thirty independent spontaneous adenine phophoribosyl transferase (*aprt*$^-$) mutations and 34 UV-induced mutants were cloned and sequenced. Among spontaneous mutants, single-base substitutions predominated (27/30), and these were largely G:C→A:T transitions (22/27), the majority of which occurred at "hot-spot" sites. Among the UV-induced mutants sequenced, single-base substitutions also predominated (26/34). These included 17 G:C→A:T transitions and a single A:T→G:C transition. Transversions accounted for the remaining eight mutations. However, the G:C→T:A transversion was not among them. Six tandem double- or closely neighboring double-base substitutions, one double mutation consisting of a G:C→T:A transversion and an adjacent frameshift, and one single frameshift mutation were also recovered. The observed specificity of UV-light induced mutations at the *aprt* locus is consistent with the argument that G:C→A:T transitions result primarily from the (6-4) pyrimidine pyrimidone lesion, while transversions and frameshifts are targeted by other photolesions. Finally, there is a striking resemblance between the distribution of UV-induced and spontaneous mutants. In particular, both share common sites of multiple events and a considerable degree of overlap in regional specificity. This is interpreted as an indication that DNA context plays a significant role in mutation fixation in mammalian cells.

INTRODUCTION

Mutational specificity provides a practical approach to exploring the molecular mechanisms of mutagenesis. The description of mutation at the molecular

Banbury Report 28: Mammalian Cell Mutagenesis

level can be expected to permit new insights into the nature of DNA lesions, the influence of neighboring sequences, and of DNA repair. The approach used here involves the cloning and sequencing of mutants at the *aprt* locus of Chinese hamster ovary (CHO) cells. This locus is well suited for mutational studies. It codes for an enzyme in the nucleotide salvage pathway and is thus not essential for cell survival in ordinary growth medium. Simple, single-step selections for both forward and reverse mutations are well characterized, hemizygous strains to simplify cloning are available, and an apparent absence of pseudogenes and the small (2.6 kb) size of the functional CHO *aprt* gene all contribute to making the system attractive. A potential exclusion of some mutational events in the hemizygous system has not hampered the recovery of adequate numbers of mutants, and we have now expanded this technology to include studies in heterozygous strains. Here, we report and compare the DNA sequence alterations of 30 spontaneous and 34 UV-induced APRT$^-$ mutants.

Conditions for the Selection of *aprt*$^-$ Mutations

The CHO strain D422 used in this study is hemizgyous at the *aprt* locus (Bradley and Letovanec 1982). The conditions used for the selection and phenotypic characterization of spontaneous APRT$^-$ mutants have been detailed recently (Grosovsky et al. 1986). Cells were maintained in α-minimal essential medium supplemented with 2.5% fetal calf serum and 2.5% heat-inactivated horse serum. Each spontaneous APRT$^-$ mutant was obtained by establishing independent cultures at an initial inoculum of 500 cells that were grown to approximately 2×10^6 cells per culture. The cells were then reseeded at a density of 5×10^5 cells per 100-mm petri dish in medium containing 0.4 mM of the toxic adenine analog 8-azaadenine (Sigma). Only one mutant clone was analyzed from each independent culture.

UV-induced mutants were collected by irradiating independent cultures containing 5×10^5 cells with a dose of 5 J/m^2, using a 254-nm UV source. After exposure, the medium was changed, and a 5-day phenotypic expression period was allowed. Selections were performed as above by seeding two replicates of 5×10^5 per 100-mm petri dish in medium containing 0.4 mM 8-azaadenine. Only one *aprt*$^-$ clone was picked from each culture.

Cloning Strategy

The rapid cloning and sequencing technology depends on the in vivo recombinational rescue of a defective λ-vector carrying the mutant gene as first described by B. Seed. In short, genomic DNA is cut with *Hin*dIII and *Bgl*II and size-fractionated on a 0.7% agarose gel to enrich for a 4.3-kb fragment. DNA in the range of 3.5–5.0 kb is recovered and ligated into an A_{am} B_{am} λ-L47 derivative (λ-PDJ11). After in vitro packaging, this primary library is then amplified on a host carrying the plasmid pSDL12 (Levinson et al. 1984), a Pivx derivative carrying the *supF* gene, capable of suppressing the amber mutations of the λ vector. Cloned into this plasmid is a 0.7-kb fragment of CHO DNA that is downstream of the *aprt* gene on the 4.3-kb *Hin*dIII/*Bgl*II fragment. Homologous recombination occurs between the library phage containing the 4.3-kb *aprt* fragment and the plasmid, resulting in the incorporation of the plasmid, and its *supF* gene into the phage genome. These recombinants can be selected by titering the amplified library on a nonsuppressor host.

Homologous recombination occurs in a manner that allows the plasmid sequences and the entire *aprt* gene to be recovered on a single 5.7-kb *Bam*HI fragment. This fragment is circularized and used to transform an appropriate F′ host strain (XS127). Single-stranded plasmid DNA can be obtained for dideoxy sequencing by infecting the transformed bacteria with M13 phage, since the pSDL12 plasmid carries an M13 origin of replication. A series of oligonucleotide primers that anneal at chosen points along the *aprt* gene, but not to the M13 sequence, is used for sequence determinations.

RESULTS

Spontaneous Mutations

The wild-type *aprt* allele from strain D422 was independently cloned and sequenced three times (data not shown) and was largely comparable to that recently reported by Nalbantoglu et al. (1986a). The entire coding sequence of each mutant allele was determined along with most, and in many cases all, of the intron and flanking sequences, in order to be certain that the observed alterations were unique. As an additional control, two of the mutants (nos. 59 and 60) were independently cloned and sequenced three times each. In both cases, each independent clone carried the same unique sequence alteration. Single-base substitutions occurred in 27 of the 30 mutated sequences (Table 1). Of these, 22 (~82%) are G:C→A:T transitions. The A:T→G:C transition has not been observed. The five remaining single-base substitutions are transversions. One tandem-base substitution (no. 20) was also observed. With two exceptions, each base substitution yields an amino acid substitution. The exceptions (nos. 63 and 104) occurred within intron 3 near the 3′ splice site (Fig. 1). In mutant 104, a G:C→T:A transversion has destroyed the AG dinucleotide at the splice junction, whereas in mutant 63, an A:T→T:A

Table 1
Sequence Analysis of Spontaneous *aprt*$^{-}$ Mutants

Mutant sequence	Alteration[a]	Position	Predicted amino acid change	Target sequence
2	TS C→T	241	Ser→Phe	CGCCT C CTTCC
5	TS C→T	1324	Pro→Leu	TGGCC C CTCCC
13	TS G→A	1351	Gly→Asp	CCTGG G CTGTG
16	TS C→T	241	Ser→Phe	CGCCT C CTTCC
17	TV G→T	1422	Gly→Cys	AGTAT G GCAAG
20	TV C→G	242		
	TV T→A	243	Phe→Ile	GCCTC CT TCCGA
21	TS C→T	58	Pro→Ser	ACTTC C CCATC
27	TS G→A	210	Asp→Lys	CCAGG G ATATC
29	TS C→T	241	Ser→Phe	CGCCT C CTTCC
39	TS G→A	1309	Gly→Glu	CAGGG G ATTCT
43	TV G→T	1591	Glu→STOP	AACTA G AAATC
48	TV C→A	1780	Cys→Trp	ATGTG C GCTGC
55	TS C→T	222	Leu→Phe	CGCCC C TCCTG
57	TS C→T	241	Ser→Phe	CGCCT C CTTCC

59[b]	TS C→T	253	Ser→Phe	AGCTT C CATCC
60[b]	TS C→T	1786	Ala→Val	CGCTG C CTGTG
61	TS C→T	241	Ser→Phe	CGCCT C CTTCC
63	TV T→A	1573	Splice recognition site	[c]
65	TS C→T	1384	Pro→Leu	GCTGC C AGGCC
66	TS C→T	253	Ser→Phe	AGCTT C CATCC
68	TS C→T	1383	Pro→Ser	AGCTG C CAGGC
69	Deletion	1401–1418		[c]
70	TS C→T	1303	Ser→Phe	AGACT C CAGGG
75	TS C→T	222	Leu→Phe	CGCCC C TCCTG
84	TS C→T	241	Ser→Phe	CGCCTC CTTCC
86	Frameshift	1600		[c]
93	TS C→T	241	Ser→Phe	CGCCT C CTTCC
94	TS C→T	222	Leu→Phe	CGCCC C TCCTG
100	TS C→T	59	Pro→Leu	CTTCC C CATCC
104	TV G→T	1581	Splice recognition site	[c]

[a]TV indicates transversion, and TS indicates transition.
[b]These mutants were independently cloned three times and sequenced in their entirety, and the same sequence alterations were found, thereby confirming the cloning methodology.
[c]See Fig. 1 for target.

A. DELETION

B. FRAMESHIFT

C. SPLICE JUNCTION MUTANTS

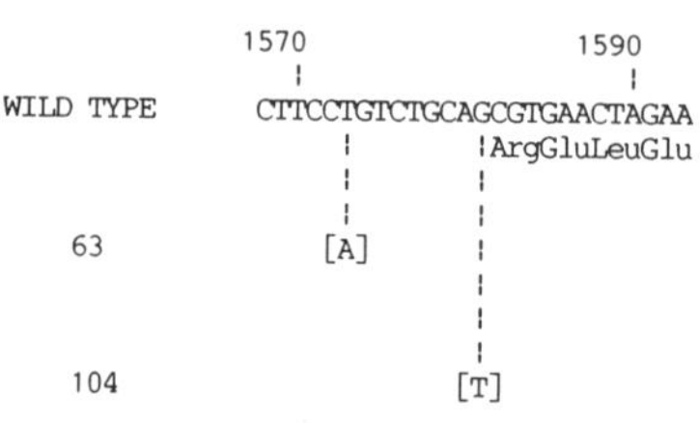

Figure 1
Sequence alterations of specific spontaneous mutants. Target sequences for (*A*) APRT deletion mutant 69, (*B*) frameshift mutant 86, and (*C*) splice junction mutants 63 and 104. (*A*) Mutant 69: the 17 deleted bases are bracketed and the two-base-pair repeat is underlined. 5 bases of surrounding sequence on either side are also shown. (*B*) Mutant 86: The frameshift results from the loss of one A in the run of three As set off from the surrounding sequence. The target nucleotide in 86 is also indicated by spacing from the surrounding sequence. (*C*) Splice junction mutants 63 and 104: These have both occurred within the consensus 3′ splice acceptor sequence of intron 3. The wild-type sequence and the altered nucleotide in each case is shown; the beginning of exon 4 is indicated by the accompanying amino acid sequence.

transversion has resulted in the introduction of an AG dinucleotide 7 bp upstream of the normal splice junction. One deletion and one frameshift event have also been detected. In mutant 86, a frameshift of the coding sequence resulted from the deletion of a single A in a run of three As, whereas a deletion of 17 bp was observed in mutant 69 (Fig. 1). The changes observed in the splice region can be reasonably expected to account for the $APRT^-$ phenotype.

There are several positions of multiple occurrence or hot spots. The G:C→A:T transition at position 241 was independently observed seven times, accounting for nearly 25% of all spontaneous *aprt* mutations. A G:C→A:T transition was also recovered at positions 222 and 253 two and three times, respectively. The occurrence of hot spots can be expected to be governed by both DNA sequence and protein function determinants; the results likely reflect both parameters. In two cases, the same codon has been mutated in two independent mutant clones (68 and 65; 21 and 100) by base substitution at the first or second codon position, respectively. In addition, we have observed two cases of independent mutations recovered in codons only one amino acid position apart (mutants 48 and 60; 39 and 70).

Although at least one mutant was isolated for each exon, the sites of multiple occurrence at a single nucleotide are clustered in exon 2. Exon 3 is evolutionarily conserved (Dush et al. 1985), suggesting that most sequence alterations within this region would result in a selectable phenotype. In fact, a relatively large number of mutant sites were identified within this exon. Two of the three closely neighboring amino acid sequence alterations fall within this exon, but there were no sites of multiple occurrence at a single nucleotide. Several mutational spectra will be required before a more detailed analysis of the available sites for mutation can be undertaken.

UV-induced Mutation

The UV dose of 5 J/m^2 was chosen because cytotoxicity was minimal (survival of 81%), whereas induced mutagenesis was approximately 45-fold over background. This increase was sufficient to ensure that the spectrum obtained reflects that of UV-induced mutation. The DNA from 34 independent mutants was analyzed by Southern blotting, and no indications of major deletions or rearrangements were observed. The *aprt* genes were subsequently cloned, and the entire coding sequence of each mutant was determined. Among the collection (Table 2) were 26 single-base substitutions, these included 17 G:C→A:T and 1 A:T→G:C transition. The remaining eight base substitutions included four G:C→C:G, two A:T→T:A, and two A:T→C:G transversions. No examples of G:C→T:A transversions were recovered. Four tandem double and three nontandem (closely neighboring) double mutations accounted for about 20% of the UV-induced mutations (Table 2B, C).

All of the single-base substitutions and most of the double mutations occurred at dipyrimidine sites. One tandem and one nontandem double mutation each had one base substitution that occurred at a dipyrimidine site and an additional alteration opposite an isolated pyrimidine. This observation is consistent with the targeting of mutation by UV photoproducts specific for dipyrimidines, which includes both the cyclobutane dimer and the (6-4) pyrimidine pyrimidone lesion.

The location of UV-induced mutation within the dipyrimidine target was analyzed. Each transition occurred at the 3′ side or within a run of pyrimidines. In contrast, of the eight transversions, half occurred unambiguously at the 5′ side of a potential dipyrimidine lesion. However, three of these did occur at a single site. Of the remaining four transversions, two occurred within runs of pyrimidines, and therefore are ambiguous, and two occurred on the 3′ side of a CT target.

Multiple Occurrences or Hot Spots among the UV-induced Mutations

Two sites within the *aprt* gene are characterized by multiple occurrences. Three G:C→C:G transversions were recovered at nucleotide 1351, whereas five G:C→A:T transitions and one tandem double mutation occurred at position 241. The latter site lies within a 9-bp pyrimidine tract and was also identified as a hot spot in our spontaneous spectrum.

Table 2
UV-induced Mutations in the *aprt* Gene

(a) *Single Base Substitutions*

Mutants	Position	Change	Amino acid	Target sequence (5′→3′)
UV-1	1642	G:C→A:T	Asp→Asn	TAGAT G ATCTC
UV-2	241	G:C→A:T	Ser→Phe	CGCCT C CTTCC
UV-4	241	G:C→A:T	Ser→Phe	CGCCT C CTTCC
UV-7	230	G:C→C:G	Lys→Asn	CTGAA G GACCC
UV-11	222	G:C→A:T	Leu→Phe	CGCCC C TCCTG
UV-13	241	G:C→A:T	Ser→Phe	CGCCT C CTTCC
UV-16	1770	G:C→A:T	Gly→Glu	CCCAG G AACCA
UV-18	1769	G:C→A:T	Splicing	CCCA G GAACC
UV-26	58	G:C→A:T	Pro→Ser	ACTTC C CCATC
UV-27	1597	G:C→A:T	Amber	AAATC C AGAAA
UV-29	253	G:C→A:T	Ser→Phe	AGCTT C CATCC
UV-32	1769	G:C→A:T	Splicing	CCCCA G GAACC
UV-33	1794	A:T→T:A	Leu→Gln	TGAGC T GCTGG
UV-34	1351	G:C→C:G	Gly→Ala	CCTGG G CTGTG
UV-35	241	G:C→A:T	Ser→Phe	CGCCT C CTTCC
UV-36	1854	A:T→G:C	Leu→Pro	CTCAC T TAAGG
UV-39	1303	G:C→A:T	Ser→Phe	AGACT C CAGGG
UV-40	1351	G:C→C:G	Gly→Ala	CCTGG G CTGTG
UV-41	208	G:C→A:T	Splicing	TCCCA G GGATA
UV-44	1644	A:T→T:A	Asp→Glu	GATGA T CTCCT
UV-47	241	G:C→A:T	Ser→Phe	CGCCT C CTTCC
UV-50	1309	G:C→A:T	Gly→Glu	CAGGG G ATTCT
UV-51	244	A:T→C:G	Phe→Cys	CTCCT T CCGAG
UV-52	1381	A:T→C:G	Leu→Arg	GAAGC T GCCAG

UV-54	1299	G:C→A:T	Asp→Asn	GCCTA G ACTCC
UV-56	1351	G:C→C:G	Gly→Ala	CCTGG G CTGTG

(b) *Tandem Double Mutations*

Mutants	Position	Change	Amino acid	Target sequence (5′→3′)
UV-10	59	G:C→A:T	Pro→Leu	CTTCC CC ATCCC
	60	G:C→A:T		
UV-24	213	A:T→T:A	Ile→Tyr	GGGAT AT CTCGC
	214	A:T→T:A		
UV-37	1887	−T	Frameshift	ACCAT TC TTCTC
	1888	G:C→T:A		
UV-49	241	G:C→A:T	Ser→Phe	CGCCT CC TTCCG
	242	G:C→A:T		

(c) *Nearly Tandem Double Substitutions*

Mutants	Position	Change	Amino acid	Target sequence (5′→3′)
UV-17	1301	G:C→A:T	None	CTAGA CTC CAGGG
	1303	G:C→A:T	Ser→Phe	
UV-25	1376	G:C→A:T	None	AGG G AAGCTG C CAG
	1383	G:C→T:A	Pro→Thr	
UV-43	210	G:C→A:T	Asp→Lys	CCAGG GAT ATCTC
	212	A:T→T:A	None	

(d) *Others*

Mutant	Position	Change	Amino acid	Target sequence (5′→3′)
UV-6	1853–54	+T	Frameshift	CTCAC TT AAGGG

DISCUSSION

The Spontaneous Spectrum

The spectrum of spontaneous mutation in CHO cells is characterized by two major features; a predominance of G:C→A:T transitions and a prominent hot spot. These results for mammalian cells can be contrasted with a recent study (R.M. Schaaper et al., in prep.) in which the spectrum of spontaneous mutation at the *lacI* gene of *Escherichia coli* was reported. In that spectrum, base substitutions accounted for only 34% of mutations and among them, G:C→A:T transitions were not predominant. In the case of the *aprt* spectrum, the predominance of G:C→A:T transitions in part reflects the contribution of the hot spots. However, 10 of the 18 nonhot-spot mutations were also of this type.

Our previous Southern blotting analysis of APRT$^-$ mutants (Grosovsky et al. 1986) failed to reveal any spontaneous deletion or chromosomal rearrangement events, in contrast to a parallel collection of γ-ray-induced mutants. The DNA sequencing analysis presented here confirms and extends these findings. The sequencing of 30 of these spontaneous "point mutations" revealed them to be largely attributable to base substitution events; only one frameshift and a single deletion too small to be detected by Southern blotting were recovered (Table 1).

The single frameshift mutant (no. 86) is an example of a frameshift occurring within a run of identical bases. The now classic model for frameshift mutagenesis (Streisinger et al. 1967) predicts the slippage and misalignment of a base onto the complement of a nearby base within such runs. The slippage and misalignment model was later extended (Albertini et al. 1982; Glickman and Ripley 1984) to include a role for repeats in deletion and insertion events. A recent study has reported (Nalbantoglu et al. 1986b) the sequence alterations of six deletions of the CHO *aprt* gene originally detected by Southern blot analysis. These deletions were characterized by short repeated sequences (2–7 bp) at their termini. Four of the deletion termini were clustered in a 40-bp region of the gene. The single deletion (no. 69) did not fall within this region, but it was flanked by a short direct repeat of 2 bp. This was the only deletion among the 30 mutants sequenced.

The Source of G:C→A:T Transitions

G:C→A:T transitions may reflect cytosine deamination, the most commonly encountered spontaneous DNA lesion (Lindahl and Nyberg 1974). Mutational hot spots have been associated with deamination of the modified base 5-methylcytosine, which deaminates to produce thymine, a base that is not removed from DNA by uracil glycosylase. In mammalian cells, the methylation of C is thought to occur largely at CpG dinucleotides (Bird 1980). However, the G:C→A:T transitions described here did not arise at such sequences but rather in runs of pyrimidines (Table 1), although we note that the methylation of cytosine bases at sequences other than CpG has been suggested to occur in *Neurospora* (Selker and Stevens 1985). A defect in uracil-*N*-glycosylase, such as in an *E. coli* Ung^- mutant, enhances the frequency of G:C→A:T transitions (Duncan and Miller 1980; Fix and Glickman 1986). Mammalian cells, in general, have been reported to have lower levels of uracil glycosylase as compared with prokaryotes (Duncan 1982), and CHO cells are particularly poor in this glycosylase activity. By analogy with Ung^- bacteria, a low level of this enzyme could account for the observed prevalence of the G:C→A:T transitions. Alternatively, the predominance of G:C→A:T transitions may reflect the fidelity of eukaryotic polymerases (Kunkel and Alexander 1986).

The UV-induced Mutational Spectrum and the Role of DNA Lesions

Recent controversy has centered around the relative contributions to mutagenesis of the two major UV-induced dipyrimidine photolesions, the (6-4) pyrimidine pyrimidone and the cyclobutane dimer. Brash and Haseltine (1982) have suggested that the predominance of G:C→A:T transitions among nonsense mutations in the *lacI* gene is due to the mutagenic potential of the (6-4) lesion. More recently, Glickman et al. (1986) have presented data indicating that perhaps two-thirds or more of the UV-induced mutation in *E. coli* is due to this photoproduct. Similar conclusions about the predominance of (6-4) photoproduct-induced transitions have been reached following the sequencing of several hundred *lacI* mutants in both Uvr^+ and $UvrB^-$ strains of *E. coli* (R.M. Schaaper et al., in prep.). The specificity suggested for the (6-4) photolesions from these studies is G:C→A:T transitions at the 3′ side of potential dipyrimidine targets. Franklin et al. (1985) have argued for the same specificity on the basis of physicochemical models which indicate that the 5′ pyrimidine base of a (6-4) photoproduct retains its coding properties while the 3′ pyrimidone does not. The preferential insertion of adenine opposite the cytosine pyrimidone ring would then account for the predominance of G:C→A:T events. An examination of the G:C→A:T transitions in our collection is consistent with such a model. A preference for events at the 3′ side or the middle of a run of pyrimidines was observed. None occurred at the 5′ side of the potential lesion. In sharp contrast, four of eight transversions (three at the same site) occurred at the 5′ side of a dipyrimidine. Transversions at the 5′ side of dipyrimidine sequences may reflect mutations targeted by either cyclobutane dimers or some other yet to be characterized photoproduct (Gallagher and Duker 1985; Frescoe and Gasparro 1986).

An alternative explanation for the predominance of G:C→A:T transition and the lack of mutation at, for example, potential T-T dimer sites might be the preferential insertion of adenine opposite noncoding lesions, such as has been demonstrated at apurinic sites in *E. coli* (Kunkel 1984). The preferential insertion of adenine may reflect properties of DNA polymerase or the high ratios of dATP to dGTP observed following UV-irradiation of CHO cells (Das et al. 1984). Certainly, pool biases are able to affect mutation frequencies in vivo (Meuth 1984).

A Role for Cyclobutane Dimers?

The role for cyclobutane dimers in mutagenesis is perhaps most strongly associated with the production of tandem double mutations that have long been considered to be the hallmark of UV-induced mutagenesis. The situation, however, is more complicated since three nontandem double mutations were also recovered in our study. Nontandem double mutations have been previously shown to be induced in *E. coli* by what appears to be a single event (Coleman et al. 1980). The prevalence of nontandem double events following UV-irradiation in mammalian cells confirms an earlier report with the SV40-driven shuttle vector, pZ189, where 13 of 96 *supF* mutants sequenced were closely neighboring double events (Hauser et al. 1986).

The simplest explanation is that double events result from two separate lesions. Because of the low dose used, we consider this very unlikely and point out (1) that two double events (mutants UV-24 and UV-43) involve base substitutions at an isolated pyrimidine and (2) that if lesions were so frequent, the multiple mutations should be scattered throughout the whole length of the target, and this was not the case. Finally, at the dose used, approximately 0.025 dimers/target/cell was produced, making it very unlikely that two photoproducts would form in such close proximity.

A second mechanism suggested to account for tandem and nontandem mutations reflects the nature of the mechanistic processes that are associated with mutagenesis and DNA repair. For example, it has been proposed that the bypass of noncoding lesions such as cyclobutane dimers requires a reduction in the fidelity of DNA polymerase. Hence, in order to replicate past a dimer, the replication machinery may have a high risk of failing to insert subsequent bases accurately. An active reduction in proofreading stringency may not be required for bypass. The energy relay model of proofreading suggested by Hopfield (1980) has the consequence of reducing replication accuracy with a loss of processivity. The relationship between accuracy and processivity has also been noted by Kunkel and Alexander (1986).

A Role for DNA Context in Mutagenesis?

The frequency at which different sites are represented in a mutational spectrum is likely to be a function of the distribution of the damage and the competitive efficiencies of error-free and error-prone pathways of repair. The repair processes may be of considerable importance, since region-to-region and site-to-site variations in repair have been observed (Burns et al. 1986; Madhani et al. 1986). Site 241 constitutes a mutational hot spot in the spontaneous spectrum accounting for 7 of the 30 mutations sequenced (Table 1). This reveals a possible dimension to the role of DNA context in mutagenesis. Either this site is particularly prone to spontaneous damage or replication at this site is for some reason less accurate than at other sites. The fidelity of DNA polymerase has been shown to vary considerably from site to site (Roberts and Kunkel 1986). If position 241 represents a site where fidelity is reduced, then it follows that bypass at this site may occur more frequently than at other sites. This would explain why a site like position 241 occurs as a hot spot for both spontaneous and UV-induced mutation. If one permits this

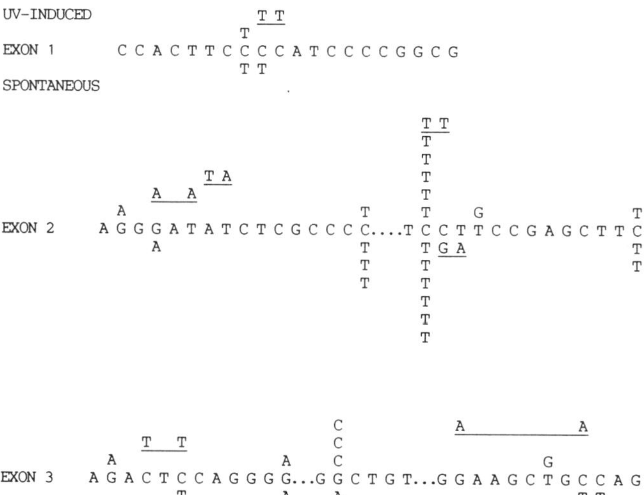

Figure 2

Clustering of multiple events in the spontaneous and induced spectra. Six regions of the *aprt* gene demonstrating the clustering of mutational events. DNA sequence is given in the 5′→3′ direction. Bases shown above the line indicate changes recovered following UV-irradiation whereas changes below the line indicate spontaneous events. The underlined bases indicate tandem and nontandem double mutations and the dots represent regions where sequence is not given.

argument, then it follows that DNA context plays an important role in mutagenesis. This reasoning suggests that factors other than damage distribution may comprise a major component of the spectrum of mutation.

The significance of DNA context in UV mutagenesis might be further clarified by an analysis of the sites of multiple mutations. If mutation is a reflection of diminished accuracy of replication at a given site, then it follows that such sites may be prone to single-base changes as a consequence of either spontaneous or UV-induced damage. Intriguingly, five of the seven double mutations overlap with sites seen in the spontaneous spectrum (Fig. 2). In four of these five cases, the base substitutions are identical. This degree of overlap would appear to lend credence to the argument for local context effects. Although the overlap might reflect a paucity of available sites at which base substitutions can be recovered, this seems unlikely considering the diversity of sites recovered in other spectra in this gene (J. de Boer and B.W. Glickman; M. Mazur and B.W. Glickman; both unpubl.) and considering also that the mutational target consists of at least the 540 bp of coding sequence. In support of the notion that DNA context contributes to the specificity of UV mutagenesis, there is also the observation that of the nine sites of UV-induced single-base substitutions shown in Figure 2, five are also represented in the spontaneous spectrum seen in Figure 1. In addition, we note that two of the double mutations are adjacent.

CONCLUSIONS

This study has demonstrated the feasibility of examining mutational specificity at endogenous cellular genes in mammalian cells. We note the similarity of the data obtained here with an endogenous mammalian gene to that observed for UV-induced *lacI* mutants in *E. coli*. This may reflect common mechanisms of UV-induced mutagenesis in prokaryotes and higher eukaryotic organisms and is encouraging considering the vast data base from bacterial test systems. Our data also confirm many observations made in shuttle vectors replicated in mammalian cells, including the preference for G:C→A:T transitions at dipyrimidines and the induction of both tandem and nontandem mutations. Shuttle vectors only monitor direct mutagen-target interactions, however, and cannot be utilized to define other potentially important mutational parameters such as chromosomal aberrations (including nondisjunction, translocations, and inversions), mitotic recombination, and gene conversion. For example, recent studies have implicated a role for somatic recombination in oncogene activation in individuals heterozgyous for the retinoblastoma gene (Cavenee et al. 1986). Such phenomena highlight the importance of developing the capability to study mutational specificity in endogenous cellular genes.

ACKNOWLEDGMENTS

The authors wish to thank Dr. P. de Jong for his assistance in the original construction of the cloning vector and Dr. Brian Seed for supplying the necessary pieces to facilitate this project. Discussions with Drs. R. Schaaper, D. Fix, D. Brash, R. Setlow, and B.A. Bridges are also gratefully acknowledged. This work was supported by grants from the Medical Research Council of Canada (MA-8677) and the National Cancer Institute of Canada.

REFERENCES

Albertini, A.M., M. Hofer, M.P. Calos, and J.H. Miller. 1982. On the formation of spontaneous deletions: The importance of short sequence homologies in the generation of large deletions. *Cell* **29:** 319.

Bird, A.P. 1980. DNA methylation and the frequency of CpG in animal DNA. *Nucleic Acids Res.* **8:** 1499.

Bradley, W.E.C. and D. Letovanec. 1982. High-frequency nonrandom mutational event at the adenine phosphoribosyl-transferase (aprt) locus of sib-selected CHO variants heterozgyous for aprt. *Somatic Cell Genet.* **8:** 51.

Brash, D.E. and W.A. Haseltine. 1982. UV-induced mutation hotspots occur at DNA damage hotspots. *Nature* **298:** 189.

Burns, P.A., F.A. Allen, and B.W. Glickman. 1986. DNA sequence analysis of mutagenicity and site specificity of ethyl methane sulfonate in Uvr$^+$ and Uvr$^-$ strains of *Escherichia coli. Genetics* **113:** 811.

Cavenee, W.K., A. Koufos, and M.F. Hanson. 1986. Recessive mutant genes predisposing to human cancer. *Mutat. Res.* **168:** 3.

Coleman, R.D., R.W. Dunst, and C.W. Hill. 1980. A double base change in alternate base pairs induced by ultraviolet irradiation in a glycine transfer RNA gene. *Mol. Gen. Genet.* **177:** 213.

Das, S.K., E.P. Benditt, and L.A. Loeb. 1984. UV irradiation alters deoxynucleotide triphosphate pools in *Escherichia coli. Mutat. Res.* **131:** 97.

Duncan, B. 1981. DNA glycosylases. In *The enzymes* (ed. P.T. Boyer), vol. 14, p. 565. Academic Press, New York.

Duncan, B.K. and J.H. Miller. 1980. Mutagenic deamination of cytosine residues in DNA. *Nature* **287:** 560.

Dush, M.K., J.M. Sikela, S.A. Khan, J.A. Tischfield, and P.J. Stambrook. 1985. Nucleotide sequence and organization of the mouse adenine phosphoribosyltransferase gene: Presence of a coding region common to animal and bacterial phosphoribosyltransferases that has a variable intron/exon arrangement. *Proc. Natl. Acad. Sci.* **82:** 2731.

Fix, D.F. and B.W. Glickman. 1986. Differential enhancement of spontaneous transition mutations in the *lacI* gene of an Ung$^-$ strain of *Escherichia coli. Mutat. Res.* **175:** 41.

Franklin, W.A., P.W. Doetsch, and W.A. Haseltine. 1985. Structural determination of the ultraviolet light-induced thymine-cytosine pyrimidine-pyrimidine (6-4) photoproduct. *Nucleic Acids Res.* **13:** 5317.

Frescoe, J.R. and F.P. Gasparro. 1986. Ultraviolet-induced 8,8-adenine dehydrodimers in oligo- and polynucleotides. *Nucleic Acids Res.* **14:** 4239.

Gallagher, P.E. and N.J. Duker. 1985. Detection of UV purine photoproducts in a defined sequence of human DNA. *Mol. Cell. Biol.* **6:** 707.

Glickman, B.W. and L.S. Ripley. 1984. Structural intermediates of deletion mutagenesis: A role for palindromic DNA. *Proc. Natl. Acad. Sci.* **81:** 512.

Glickman, B.W., R.M. Schaaper, W.A. Haseltine, R.L. Dunn, and D.E. Brash. 1986. The C-C (6-4) photoproduct is mutagenic in *Escherichia coli. Proc. Natl. Acad. Sci.* **83:** 6945.

Grosovsky, A.J., E.A. Drobetsky, P.J. de Jong, and B.W. Glickman. 1986. Southern analysis of genomic alterations in gamma-ray-induced aprt$^-$ hamster cell mutants. *Genetics* **113:** 405.

Hauser, J., M.M. Seidman, K. Sidur, and K. Dixon. 1986. Sequence specificity of point mutations induced during passage of a UV-irradiated shuttle vector plasmid in monkey cells. *Mol. Cell. Biol.* **6:** 277.

Hopfield, J.J. 1980. The energy relay: A proofreading scheme based on dynamic co-operativity and lacking all characteristic symptoms of kinetic proofreading in DNA replication and protein synthesis. *Proc. Natl. Acad. Sci.* **77:** 5248.

Kunkel, T.A. 1984. Mutation specificity of depurination. *Proc. Natl. Acad. Sci.* **81:** 1494.

Kunkel, T.A. and P.S. Alexander. 1986. The base substitution fidelity of eukaryotic DNA polymerases. *J. Biol. Chem.* **261:** 160.

Levinson, A., D. Silver, and B. Seed. 1984. Minimal size plasmids containing an M13 origin for production of single strand transducing particles. *J. Mol. Appl. Genet.* **2:** 507.

Lindahl, T. and B. Nyberg. 1974. Heat-induced deamination of cytosine residues in deoxyribonucleic acid. *Biochemistry* **13:** 3405.

Madhani, B.H., V.A. Bohr, and P.C. Hanawalt. 1986. Differential DNA repair in transcriptionally active and inactive protooncogenes: c-*abl* and c-*mos*. *Cell* **45:** 417.

Meuth, M. 1984. The genetic consequences of nucleotide precursor pool imbalance in mammalian cells. *Mutat. Res.* **126:**107.

Nalbantoglu, J., G.A. Phear, and M. Meuth. 1986a. Nucleotide sequence of hamster adenine phosphoribosyl transferase (*aprt*) gene. *Nucleic Acids Res.* **14:** 1914.

Nalbantoglu, J., D. Hartley, G.A. Phear, G. Tear, and M. Meuth. 1986b. Spontaneous deletion formation at the *aprt* locus of hamster cells: The presence of short sequence homologies and dyad symmetries at deletion termini. *EMBO J.* **5:** 1199.

Roberts, J.D. and T.A. Kunkel. 1986. Mutational specificity of animal cell DNA polymerases. *Environ. Mut.* **8:** 769.

Schaaper, R.M., B.N. Danforth, and B.W. Glickman. 1986. Mechanisms of spontaneous mutagenesis: An analysis of the spectrum of spontaneous mutation in the *Escherichia coli lacI* gene. *J. Mol. Biol.* **189:** 273.

Selker, E.V. and J.N. Stevens. 1985. DNA methylation at asymmetric sites is associated with numerous transition mutations. *Proc. Natl. Acad. Sci.* **82:** 8114.

Streisinger, G., Y. Okada, J. Emrich, J. Newton, A. Tsugita, E. Terzaghi, and M. Inouye. 1967. Frameshift mutation and the genetic code. *Cold Spring Harbor Symp. Quant. Biol.* **31:** 77.

Molecular Basis of Genome Rearrangements at the Hamster *aprt* Locus

MARK MEUTH, JOSEPHINE NALBANTOGLU, GERALDINE PHEAR, AND CAROL MILES
Imperial Cancer Research Fund
Clare Hall Laboratories
South Mimms, Hertfordshire EN6 3LD, U.K.

OVERVIEW

The mechanisms by which eukaryotic cells rearrange genes were examined by analyzing deletion and insertion mutations at the adenine phosphoribosyl transferase (*aprt*) locus of Chinese hamster ovary (CHO) cells. Sequence determination of cloned novel junctions created by the deletions revealed common properties and indicated that these rearrangements were produced by nonhomologous recombinational events between short direct repeat sequences. These events may also be influenced by DNA secondary structure as several termini clustered in a region of *aprt* rich in short direct and inverted repeat sequences. Other mechanisms also generate deletions, e.g., a gene amplification event produced both a large inverted repeat and a deletion. Two insertion mutations studied at the DNA sequence level had similar properties with respect to the target site, however, the type of DNA inserted differed significantly. Our findings show that a variety of mechanisms are responsible for genome rearrangements and also give indications of cellular functions involved in these processes.

INTRODUCTION

Genome rearrangements play an important role in the generation of mutations in both prokaryotes and eukaryotes. Deletions are the cause of several human genetic diseases (Collins and Weissman 1984; Kunkel et al. 1986). Translocations (Haluska et al. 1987), insertions (Canaani et al. 1983), and gene amplification (Alitalo et al. 1983) have been implicated in the activation of certain oncogenes. However, very little is known of the mechanisms generating these types of mutations in eukaryotes or the factors governing the formation of these rearrangements.

To approach this problem, we examined gene rearrangements occurring at the endogenous locus coding for the nonessential salvage enzyme APRT in

Banbury Report 28: Mammalian Cell Mutagenesis

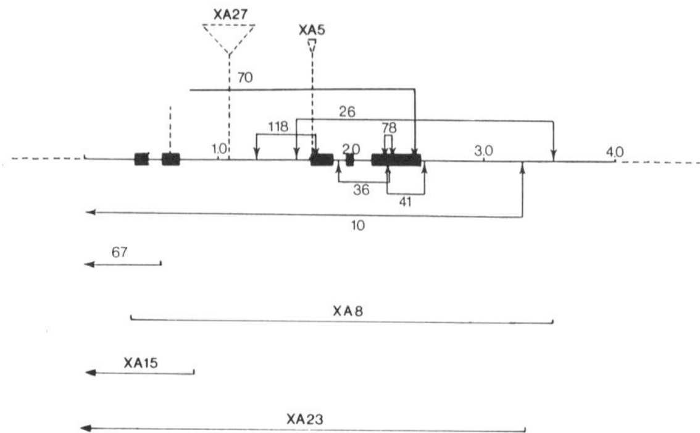

Figure 1

Map of mutations producing gene rearrangements in the 4.0-kb genomic fragment coding for *aprt*. Deletions in which termini have been precisely determined by DNA sequencing are indicated by arrows on the map. Deletions only roughly mapped by Southern blot analysis are also indicated. Insertions or other events (e.g., XA27, a probable inversion) are indicated by dashed lines. Boxed areas indicate exon sequences.

cultured CHO cells. The *aprt* locus is particularly attractive for these analyses because it is small (Lowy et al. 1980) allowing detection and mapping of deletions or insertions as short as 30 bp (Fig. 1) (Nalbantoglu et al. 1983). It is also a true endogenous structural gene having the low mutational rates inherent to such loci and making it possible to examine the role of surrounding chromosomal sequences on the generation of gene rearrangements. Although the locus is autosomal, strains with only a single copy of the gene have been identified (Nalbantoglu et al. 1983) facilitating the collection of large numbers of spontaneous or induced mutant strains. Mutant strains in such collections were first analyzed by Southern blotting to provide a rough analysis and mapping of the mutations occurring. Mutant genes in which we mapped alterations were then cloned from genomic libraries, constructed in bacteriophage λ vectors, and sequenced to determine the specific base-pair alterations involved. In this paper we summarize the properties of the gene rearrangements that have been characterized at the *aprt* locus.

RESULTS

Rearrangements Occurring in Spontaneous and Induced Mutants

Large collections of spontaneous and induced mutants were examined by Southern blotting techniques for *aprt* gene rearrangements (Table 1). Among spontaneous mutants, deletions amounted to about 7% of mutations and insertions to about 1%. Other types of mutations (involving <25 bp) were

Table 1
Types of Mutations Occurring Spontaneously or Induced by Chemical or Physical Agents at the Hamster *aprt* Locus

Mutant collection	Mutant strains in collection	Altered[a] sites	Rearrangements[b]		
			deletions	insertions	others
Spontaneous[c]	120	12	8	1	0
Induced					
ethyl methane-sulfonate	48	7	0	0	0
Benzo[a]pyrene	35	5	0	0	0
Cisplatin	30	3	0	0	0
UV light	72	2	0	0	0
γ-radiation[d]	25	1	3	1	2

[a]Mutation occurring in restriction endonuclease sites resulting in loss or gain of a site.
[b]Rearrangements as detected by altered mobility of DNA fragments on Southern blots, i.e., >30 bp in size.
[c]Data from Nalbantoglu et al. (1983).
[d]Data from Breimer et al. (1986).

mapped to restriction endonuclease sites (causing loss or gain of the site) in another 10% of the mutants.

Mutations induced by chemical agents (ethyl methanesulfonate, benzo[a]pyrene, and cisplatin) showed a noticeable shift in spectrum in the form of a disappearance of the deletions and insertions. Similarly, no deletions or insertions were observed among UV-induced mutant strains. The only mutagen that produced a significant frequency of gene rearrangement was, not surprisingly, γ-radiation (Breimer et al. 1986). But even in these mutant strains, the majority of the altered *aprt* genes showed no gross rearrangement, suggesting that large deletions (> 5 kb) were, in fact, less frequent than base-pair alterations or small deletions/insertions. The properties of the mutations producing rearrangements of the *aprt* locus are described in detail below.

Deletions

Eight large spontaneous deletions were identified and roughly mapped by Southern blot analysis (Fig. 1) (Nalbantoglu et al. 1983). Five of these had both termini within the *aprt* locus and had very similar properties (Nalbantoglu et al. 1986). All appeared to involve nonhomologous recombination events between short (2–7 bp) direct repeats that leave one copy of the repeat in the mutant gene (Fig. 2). However, there were no further significant homologies between the nucleotides surrounding these direct repeats. Certain tri- and tetranucleotides recurred at the termini (Fig. 2), suggesting a specific recom-

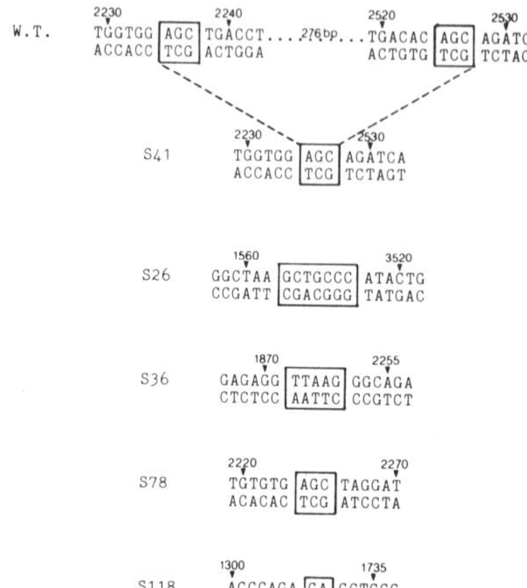

Figure 2

Deletion junctions of spontaneous mutations at the *aprt* locus. Boxed areas indicate short direct repeats involved in the formation of the deletions. For mutant S41, the surrounding wild-type cellular sequences at the two termini of the deletions are presented; however, only the junctions are shown for the rest. Numbering of nucleotides is from the sequence of the 4.0-kb fragment illustrated in Fig. 1.

binational event. No other alterations were found in the mutant genes. Deletion termini were not randomly distributed over the *aprt* gene, as four clustered within a 40-bp sequence (Fig. 3). Analysis of this region revealed some interesting structural features in the form of short direct repeats and significant stretches of dyad symmetry that could potentially form stable secondary structure (although there is still no evidence that such structures form in vivo).

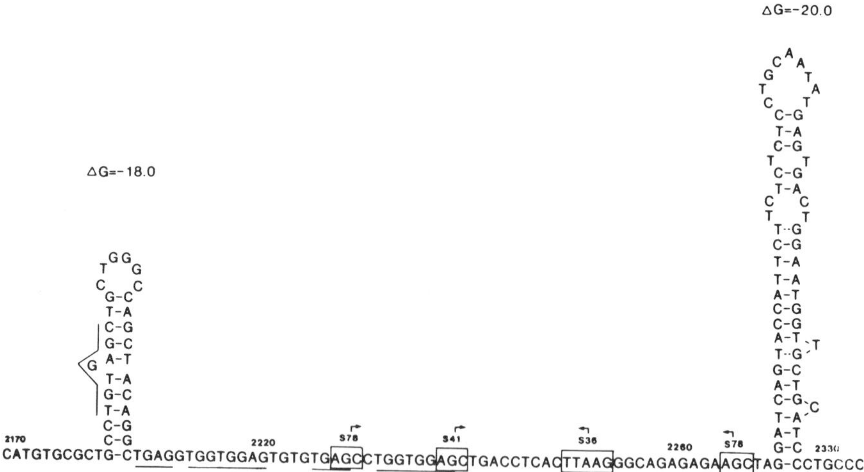

Figure 3

Potential DNA secondary structures in a region of *aprt* in which four deletion termini occur. Short direct repeats in this region are underlined. Boxed areas indicate deletion termini for the indicated mutant as is the direction of the deletion. Numbering is from the sequence of the 4.0-kb fragment illustrated in Fig. 1.

Three mutations (two spontaneous and one γ radiation induced) producing very small ($<$ 30 bp) deletions were detected as a result of their eliminating restriction endonuclease sites within the *aprt* gene (Breimer et al. 1986; Nalbantoglu et al. 1987). These deletions resembled the larger ones in that they occurred between short direct repeat sequences; however, the repeats involved in the formation of these rearrangements had no obvious similarity.

Deletion Amplification

Analyses of single-step mutants in cell culture indicate that low-level gene amplification is a randomly occurring spontaneous mutational event (Zieg et al. 1983). Consequently, it was not particularly surprising that during our analyses of single-step spontaneous mutations at the *aprt* locus, we found one (called S70) with amplified *aprt* sequences. However, since the selection we applied to our cells was negative (i.e., for the loss of enzyme activity) rather than a positive one for increased expression, we anticipated that any amplification events occurring might involve the rearrangements associated with the generation of one of the rare novel joints. Indeed, the sequence of the cloned *aprt* fragment from S70 revealed a surprisingly complex gene rearrangement. A deletion of $>$ 9 kb extending upstream of *aprt* is accompanied by an inverted duplication of flanking sequences 672 bp downstream from the novel joint. This unit of at least 18 kb is amplified three to four times, resulting in as much as an eightfold increase in copy number for some of the sequences (Nalbantoglu and Meuth 1986).

Insertions

Gene rearrangements produced by the insertion of mobile genetic elements account for a significant proportion of mutation in prokaryotes and many eukaryotes (Shapiro 1983). DNA sequences with structures similar to mobile elements are present in mammalian genomes, but attempts to prove that they are mobile have been unsuccessful, with few exceptions (e.g., Canaani et al. 1983). Insertion mutations are rare at the *aprt* locus, but two (one spontaneous, called S88; one γ-radiation induced, XA5) have now been identified and characterized at the DNA sequence level (Nalbantoglu et al. 1983 and in prep.; Breimer et al. 1986). The inserted fragments are small (285 and 58 bp); they are accompanied by a deletion of 12–13 bp at the target site, rather than duplication of flanking sequences, and no terminal repeats are present. The target sites have some similarity but do not appear to be governed by homology with the inserted fragment, and one has a strong resemblance (ten of twelve bp match) to the Jeffreys' minisatellite core sequence (Fig. 4) (Jeffreys et al. 1985). Curiously, two other small deletions were detected in spontaneous mutants at the same site where XA5 occurred (Nalbantoglu et

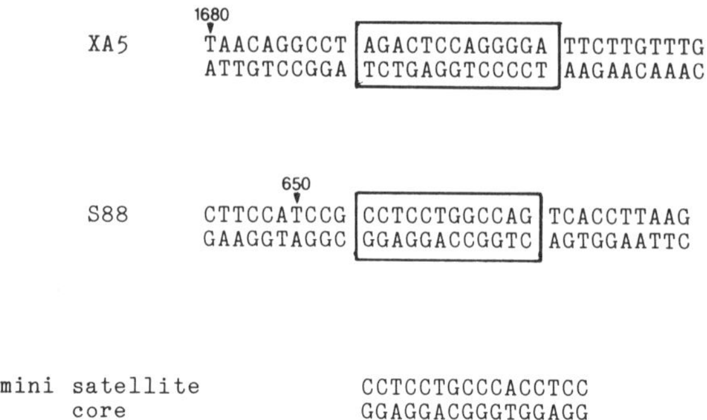

Figure 4

Nucleotide sequence of target sites of insertion mutations at the *aprt* locus. Boxed areas indicate nucleotides deleted by the insertion events. The human minisatellite core sequence (Jeffreys et al. 1985) is presented for comparison.

al. 1987). Clearly the sequences inserted into S88 and XA5 are very different. Southern blot analysis, using fragments of the mutant genes bearing the insertions as probes, revealed that the insertion in the γ-radiation-induced mutant has a sequence that is highly dispersed throughout the hamster genome, whereas the insertion in the spontaneous strain is unique (Fig. 5).

The unique nature of the insert in the spontaneous mutant allowed cloning of the donor sequence from which the insert originated and, consequently, further characterization of this transposition event. This revealed that the donor region is largely unique, although the mobilized fragment is flanked by a long alternating purine pyrimidine stretch dominated by a 46-bp dG-dT run. This donor region is not altered in the mutant strain in which the insertion occurred, and it appears to be present at the same copy number as in the wild-type strain. However, comparison of the fragment inserted into *aprt* with this donor region revealed two single base-pair alterations, which appear to be the result of polymorphisms of the donor sequence.

DISCUSSION

Genome rearrangements account for a minority (about 10%) of mutations at *aprt* locus. The pattern of the deletions observed (Fig. 1) indicates that this low frequency may be, in part, due to some sort of constraint on the *aprt* locus. It is obvious that large deletions frequently extend upstream of *aprt*, although never downstream, suggesting an essential gene or function downstream from the locus. Although this constraint may limit the frequency of deletions occurring, it is very useful for cloning deletion junctions, since at least a small fragment of *aprt* always remains to identify recombinant bacteriophage bearing the novel junctions.

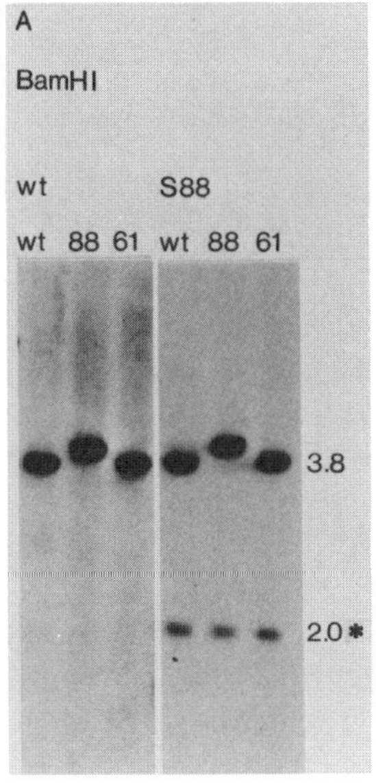

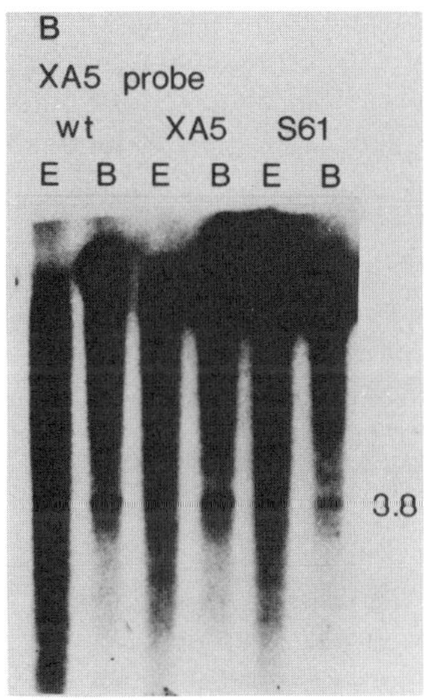

Figure 5

Southern blot analysis of digested genomic DNA probed with mutant *aprt* gene fragments bearing insertions. (*A*) Blots probed with wild-type *aprt* or the fragment bearing the insert from spontaneous mutant S88; (*B*) blots probed with the mutant gene fragment from XA5. (E) For *Eco*RI, (B) for *Bam*HI. Fragment sizes are indicated in kilobases. The starred band represents the novel fragment detected by the insertion in S88.

Clues to the mechanism by which these deletions are formed are provided by sequence analysis of the cloned junctions. We propose that the short direct repeats at the deletion termini may be involved in some sort of nonhomologous recombination event. Our data further suggest that inverted repeats, or the secondary structures that could form, might act as a recognition signal for an enzyme that cuts (not necessarily at the structures) at preferred tri- or tetranucleotide sequences. The question we are unable to resolve is what brings the other terminus into proximity to allow the exchange event. The deletion events we observe at the *aprt* locus are somewhat similar to those characterized in the *lacI* system of *Escherichia coli* (Albertini et al. 1982), but they differ from the more complex alterations found at deletion termini in mutant human globin genes (Collins and Weissman 1984; Jennings et al. 1985).

Quite unexpectedly, novel joint formation during a gene amplification event also was responsible for the generation of a deletion. In addition to the deletion, this complex rearrangement produced an inverted repeat. Together with previous reports of apparently similar structures at novel joints in arrays of amplified genes (Ford and Fried 1986), this observation indicates that these structures may play a role or are intermediates in the process of gene amplification.

We can only speculate on the mechanisms of insertion on the basis of the small number analyzed. Events at the target sites appear to be similar in that they generate small deletions. However, the fragments inserted are very different—unique in the case of the spontaneous mutant and highly repetitive for the γ-radiation-induced insertion. The highly repetitive nature of the fragment inserted in the γ-radiation-induced mutant combined with the inverted repeat adjacent to this insert (Breimer et al. 1986) make it tempting to suggest that the fragment may be some sort of remnant of a much larger fragment that was then imprecisely excised. On the other hand, the unique nature of the spontaneous insertion allows more concrete analysis of the mechanism involved. Our data suggest, in this case, that the donor sequence was duplicated before transposition into *aprt*. We cannot say what initiated this event except to speculate that the simple sequence (dG-dT) repeat (which is prone to form Z DNA) may have caused a distortion at the junction with the mobilized sequence (Singleton et al. 1984), providing some sort of focus for a nonreciprocal exchange event at the *aprt* target site.

Clearly, more examples of these events need to be characterized before detailed mechanisms can be formulated. However, the rearrangements examined thus far provide clear predictions of cellular functions and chromosomal structures that may govern their generation.

ACKNOWLEDGMENTS

We would like to thank Takeshi Seno and Dai Ayusawa for critically reading the manuscript.

REFERENCES

Albertini, A.M., M. Hofer, M.P. Calos, and J.H. Miller. 1982. On the formation of spontaneous deletions: The importance of short sequence homologies in the generation of large deletions. *Cell* **29:** 319.

Alitalo, K., M. Schwab, C.C. Lin, H.E. Varmus, and J.M. Bishop. 1983. Homogeneously staining chromosomal regions contain amplified copies of an abundantly expressed cellular oncogene (c-*myc*) in malignant neuroendocrine cells from a human colon carcinoma. *Proc. Natl. Acad. Sci.* **82:** 1707.

Breimer, L., J. Nalbantoglu, and M. Meuth. 1986. The structure and sequence of mutations induced by ionizing radiation at selectable loci in Chinese hamster ovary cells. *J. Mol. Biol.* **192:** 669.

Canaani, E., O. Dreazen, A. Klar, D. Rechavi, D. Ryan, J.B. Cohen, and D. Givol. 1983. Activation of the c-*mos* oncogene in a mouse plasmacytoma by insertion of an endogenous A-particle genome. *Proc. Natl. Acad. Sci.* **80:** 7118.

Collins, F.S. and S.M. Weissman. 1984. The molecular genetics of human hemoglobin. *Prog. Nucleic Acid Res. Mol. Biol.* **31:** 315.

Ford, M. and M. Fried. 1986. Large inverted duplications are associated with gene amplification. *Cell* **45:** 425.

Haluska, F., Y. Tsujimoto, and C. Croce. 1987. Mechanisms of chromosome translocation in B- and T-cell neoplasia. *Trends Genet.* **3:** 11.

Jeffreys, A.J., V. Wilson, and S.L. Thein. 1985. Hypervariable minisatellite regions in human DNA. *Nature* **314:** 67.

Jennings, M.W., R.W. Jones, W.G. Wood, and D.J. Weatherall. 1985. Analysis of an inversion within the human beta globin gene cluster. *Nucleic Acids Res.* **13:** 2897.

Kunkel, L.M. et al. (including 73 contributors from 25 different centers). 1986. Analysis of deletions in DNA from patients with Becker and Duchenne muscular dystrophy. *Nature* **322:** 73.

Lowy, I., A. Pellicer, J.F. Jackson, G.-K. Sim, S. Silverstein, and R. Axel. 1980. Isolation of transforming DNA: Cloning the hamster *aprt* gene. *Cell* **22:** 817.

Nalbantoglu, J. and M. Meuth. 1986. DNA amplification-deletion in a spontaneous mutation of the hamster *aprt* gene: Structure and sequence of the novel joint. *Nucleic Acids Res.* **14:** 8361.

Nalbantoglu, J., O. Goncalves, and M. Meuth. 1983. Structure of mutant alleles at the *aprt* locus of Chinese hamster ovary cells. *J. Mol. Biol.* **167:** 575.

Nalbantoglu, J., G. Phear, and M. Meuth. 1987. DNA sequence analysis of spontaneous mutations at the *aprt* locus of hamster cells. *Mol. Cell. Biol.* **7:** 1445.

Nalbantoglu, J., D. Hartley, G. Phear, G. Tear, and M. Meuth. 1986. Spontaneous deletion formation at the *aprt* locus of hamster cells: The presence of short sequence homologies and dyad symmetries at deletion termini. *EMBO J.* **5:** 1199.

Shapiro, J.A., ed. 1983. *Mobile genetic elements.* Academic Press, New York.

Singleton, C.K., M.W. Kirkpatrick, and R.D. Wells. 1984. S1 nuclease recognizes DNA conformational junctions between left-handed helical $(dT\text{-}dG)_n(dC\text{-}dA)_n$ and contiguous right handed sequences. *J. Biol. Chem.* **259:** 1963.

Zieg, J., C.E. Clayton, F. Ardeshir, E. Giulotto, E.A. Swyryd, and G.R. Stark. 1983. Properties of single-step mutants of Syrian hamster cells resistant to N(phosphonacetyl)-L-aspartate. *Mol. Cell. Biol.* **3:** 2089.

Mutation at the Dihydrofolate Reductase Locus

LAWRENCE A. CHASIN,* ADELAIDE M. CAROTHERS,† CARLOS CIUDAD,* DEZIDER GRUNBERGER,† PAMELA J. MITCHELL,* RONALD STEIGERWALT,† AND GAIL URLAUB*

*Department of Biological Sciences
†The Institute for Cancer Research
Columbia University
New York, New York 10027

OVERVIEW

Mutants deficient in the enzyme dihydrofolate reductase (DHFR) have been isolated in a Chinese hamster ovary (CHO) cell line that is hemizygous at this locus. Most mutations induced by UV light, benz[a]pyrene diolepoxide (B[a]PDE), and 2-(*N*-acetoxy-*N*-acetylamino)fluorene (AAAF) appear to be point mutations in the *dhfr* gene, in that they show DNA restriction patterns identical to those of the wild-type gene. A single-base mismatch detection method using RNA heteroduplexes was used to map most of these mutations to sites within the six exons of the gene. The mutants can be classified into three categories on the basis of their *dhfr* mRNA phenotypes. The first class has normal levels of mRNA of normal size. The second class contains shorter *dhfr* mRNA molecules because of the skipping of an exon during the maturation process. These mutations were shown to be single- or double-base substitutions occurring at exon-intron joints. The third class contains greatly reduced levels of mRNA of normal size. The *dhfr* mRNA in some of these mutants was shown to be as stable as that of wild-type cells (in the presence of actinomycin D), and the mutant *dhfr* genes are transcribed at wild-type rates. It is therefore proposed that single-base changes have altered the stability of *dhfr* pre-mRNA molecules in these mutants.

INTRODUCTION

The mammalian gene specifying the housekeeping enzyme DHFR has been the subject of somatic-cell genetic studies of gene amplification, cell-cycle and growth-phase regulation, and mutation to enzyme deficiency. Over the last few years, we have used a variety of mutagens to induce DHFR-negative mutants in a line of CHO cells that harbor only a single allele of the *dhfr* gene. In this paper, we summarize our recent results on the mapping of point mutations, both spontaneous and mutagen-induced, in the *dhfr* gene and on the RNA phenotypes of these point mutations.

DHFR plays an important metabolic role in virtually all cells by converting folate and dihydrofolate to the active reduced form of this cofactor, tetrahydrofolate. Tetrahydrofolate, in turn, plays an essential role in the de novo biosynthesis of glycine, purine nucleotides, and thymidylate. DHFR-negative mutants are viable only if they are supplied with alternative sources of these end products in the medium. Such mutants can be readily selected from hemizygous cells (in which one of the two diploid copies of the gene has been deleted) on the basis of their relative resistance to tritiated deoxyuridine suicide (Urlaub and Chasin 1980) or their inability to accumulate a fluorescent ligand of the enzyme (Urlaub et al. 1985).

The Chinese hamster *dhfr* gene is approximately 25 kb long and is organized into six exons and five introns. The introns range in size from 380 to 9400 bp; the exons are assembled into three major size species of mRNA that differ principally at their 3′ end due to differential polyadenylation (Carothers et al. 1983). The smallest mRNA contains about 1150 bases, of which 561 code for the 21,000-dalton monomeric enzyme. Most of the remaining mRNA consists of an untranslated 3′ tail. The 5′ region of the gene contains a dual promoter element resulting in a major transcriptional start 63 bases upstream of translation initiation and a minor start 44 bases further upstream. These two start sites are reflected in a repeated 29-base sequence that is conserved among hamster, mouse, and human and that contains binding sites for the transcriptional factor Sp1 (Dynan et al. 1986; Mitchell et al. 1986b). Within 200 bases of the *dhfr* cap site is another promoter that initiates transcription in the opposite direction to produce a 4-kb polyadenylated RNA whose function is unknown (Mitchell et al. 1986b). The 5′ end of the hamster *dhfr* gene contains several DNase-I-hypersensitive sites, and none of the CpG sequences tested in this region are methylated. In contrast, the remaining 22 kb of the gene contains no hypersensitive sites and is methylated (Mitchell et al. 1986b).

RESULTS AND DISCUSSION

Mutagenesis at the *dhfr* Locus

The rate of spontaneous mutation of the single *dhfr* gene present in the hemizygous CHO cell line UA21 is relatively low, 1.3×10^{-7} (Mitchell et al. 1986a). Physical and chemical mutagens can increase this rate by two to three orders of magnitude. Our results with two forms of irradiation show that UV light is a much more efficient mutagen than ionizing radiation (Table 1). We

Table 1
Mutation Rates at the *dhfr* Locus in CHO Cells

Mutagen	Mutation rate ($\times 10^7$)	Reference
None	1.3	Mitchell et al. (1986a)
γ-rays	100	Urlaub et al. (1983)
UV light	5000	P.J. Mitchell et al. (unpubl.)
B[a]PDE	200	A.M. Carothers et al. (unpubl.)
AAAF	150	Carothers et al. (1986)

All induced mutation rates represent mutants per survivor after a transient exposure to the mutagen. Survival was 20–50% in the various experiments. The spontaneous rate was estimated by fluctuation analysis.

have previously shown that γ-rays cause predominantly large deletions and inversions at the *dhfr* locus (Urlaub et al. 1986). While UV light can also lead to large deletions (Urlaub et al. 1986), it results more often in point mutations (see below). It is reasonable to think that the gross changes in structure induced by ionizing radiation may limit the yield of viable mutants. The results in Table 1 also show that the two aromatic hydrocarbon carcinogens, B[a]PDE and AAAF, are effective mutagens in this mammalian system, as they are in bacteria (Ames et al. 1973).

Distribution of Point Mutations within the *dhfr* Gene

A number of mutants induced by UV light (15), B[a]PDE (15), and AAAF (29) were analyzed by Southern blotting, using a series of probes (Carothers et al. 1986) that covers 35 kb within and surrounding the 25-kb *dhfr* gene. Unlike γ-ray-induced mutants, the majority of these mutants yielded restriction fragments indistinguishable from those of the parental cells. Although this type of analysis does not rule out the existence of small (50–100 bp) deletions, the results suggest that these three mutagens mainly cause point mutations. To map the location of these putative point mutations at the *dhfr* locus, we used an RNA heteroduplex detection assay recently developed by Winter et al. (1985). This method takes advantage of the fact that RNase A and T1 will cleave double-stranded RNA molecules preferentially at the site of a mismatch between the two strands. Uniformly labeled RNA molecules representing specific overlapping portions of wild-type *dhfr* mRNA were synthesized from vectors carrying bacteriophage promoters (Melton et al. 1984). These RNAs were each hybridized to total RNA from a given DHFR-deficient mutant. After treatment with RNase, the protected probe fragments were fractionated on denaturing polyacrylamide gels and autoradiographed. From the sizes of the new fragments generated from wild-type-mutant

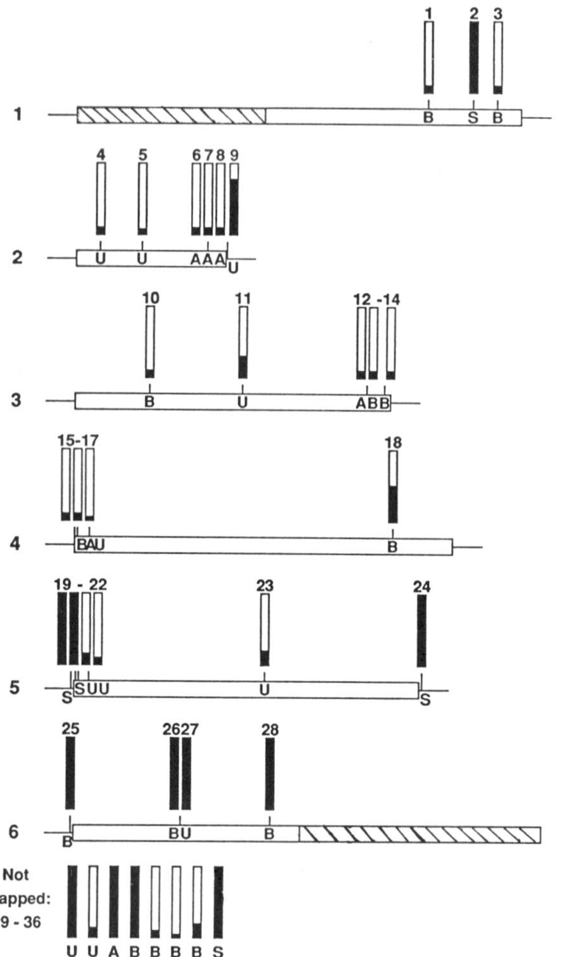

Figure 1

Location of point mutations and *dhfr* mRNA levels. The boxes indicate the six exons of the Chinese hamster *dhfr* gene and the lines represent intron sequences, most of which have been omitted. The shaded areas represent the 5′ and 3′ untranslated regions; the latter has been truncated in the figure. The position of mutations was determined on the basis of sensitivity of mismatches in RNA heteroduplex molecules to RNase cleavage (except for mutants 19, 20, and 24, which were regionally sequenced). The mutagen used to induce each mutation is indicated by a one letter-abbreviation: (A) AAAF; (B) B[a]PDE; (S) spontaneous; (U) UV light. The vertical bar above each mutation is used to show the content of *dhfr* mRNA relative to parental cells; a full bar represents 100% of that level.

heteroduplexes, the site of the mismatch could be deduced, usually to within five bases. Of 35 mutants analyzed, 28 could be mapped to 1 of the 6 *dhfr* exons. The sites of these mutations are shown in Figure 1.

There are several possible explanations for the failure of seven mutations to be mapped in this way. Because only heteroduplexes containing mRNA sequences are analyzed, this strategy necessarily overlooks mutations that occur in introns or flanking regions of the gene. It is also likely that the RNase treatment will not efficiently cleave at certain mismatched sequences, so some base-substitution mutations will be missed even though they occur in exons. Finally, mutations in other genes that affect the activity of the *dhfr* gene cannot be ruled out.

It is clear that the great majority of point mutations leading to DHFR deficiency are located within exons, and further, within the protein-coding portion of the mRNA sequence. This location is consistent with the fact that none of these mutants contains more than 2% residual DHFR enzyme (most have no detectable activity). Indeed, the selection regimen applied may be excluding leaky mutants, since each mutant is screened for its inability to grow in the absence of exogenous glycine, hypoxanthine, or thymidine; it is known that even a few percent of the wild-type level of DHFR is sufficient for prototrophy (Urlaub and Chasin 1980; Crouse et al. 1983). Nevertheless, the results suggest that it is difficult to destroy *dhfr* gene activity by single-base mutations in flanking regions or deep in introns. We did find base-substitution mutations in the universally conserved GT and AG dinucleotides that are located at the borders of introns. These mutations result in exon-skipping and have been described previously (Mitchell et al. 1986a); they are also included in Figure 1.

In several cases, the location of a mutation has been confirmed by finding a restriction site that has been destroyed or created by the nucleotide change and/or by cloning and sequencing the mutant gene fragment. Thus, mutation 10 in Figure 1 was found to destroy an *Msp*I site; subsequent sequencing revealed that the mutation is due to the deletion of a C in the recognition sequence. This mutation was induced by B[a]PDE; apparently, the reaction of the carcinogen with G can lead to a deletion opposite the adducted G, as well as base substitution. Mutation 13 destroys a *Sac*I site by virtue of a transversion of G→T induced by B[a]PDE. Mutation 16 creates a *Bst*EII site in exon 4; examination of the wild-type sequence shows that the mutation was probably the transversion of G→T induced by AAAF. Mutation 9 was induced by UV light; in this case, a double-base substitution has occurred. The GG that spans the exon 2/intron 2 joint has been changed to an AA, thus destroying the ability of this exon to be properly spliced. It is likely that this mutation arose as a result of CC dimerization with the subsequent incorporation of two As opposite the dimer. Three of the spontaneous mutations have been previously sequenced; mutations 19, 20, and 24 represent the base substitutions G→A, G→T, and G→C, respectively (Mitchell et al. 1986a).

Overall, the mutations are distributed among all six exons. However, the distribution does not appear to be random. There is a tendency for mutations to be located closer to the edges of exons. Moreover, several preferred sites are evident. Three mutations, all induced by AAAF, are clustered near the 3′ end of exon 2, four mutations (two spontaneous and two UV-induced) are clustered at the 5′ end of exon 5, and two mutations destroy the same *Sac*I site in exon 3. Perhaps the most striking specificity seen is among the spontaneous mutants. Four out of the five mutants tested affect exon 5, with each resulting in the skipping of this exon during the maturation of *dhfr* mRNA.

RNA Phenotypes of Point Mutants

Most of the point mutations fall within the protein-coding sequences of the *dhfr* gene, and almost always they exhibit no detectable DHFR activity. Thus, it is reasonable to assume that these mutations have resulted in amino acid substitutions or nonsense mutations that render the enzyme inactive. Each of the mutants was analyzed for the quality (size) and quantity of *dhfr* mRNA present, using Northern blotting and/or RNase protection techniques. This analysis revealed that at least five of the mutants produced mRNAs that were lacking a particular exon. Thus, mutants 19, 20, 21, and 36 (Fig. 1) skip exon 5, joining exon 4 directly to exon 6. Mutant 9 skips exon 2. The phenotype of mutant 25 is less firmly established, but it appears to be splicing exon 5 to a position of a few bases downstream from the normal splice acceptor site in the 5′ region of exon 6. Although incorrect exons are spliced together in most of these mutants, the level of mRNA produced is similar to that of the normal mRNA in the parental cells, suggesting that the splicing of incorrect partners is proceeding efficiently. The fact that single-base substitutions at either end of an exon can effectively preclude that exon from participation in splicing suggests that the splicing machinery considers each exon as an entity. Cryptic sites within the long introns in this gene are apparently not used when one splice site is mutated; perhaps the exons are held close to one another in preparation for splicing, so that an exon 1 to exon 3 or exon 4 to exon 6 splice is made almost as easily as the normal joining.

The RNA analysis of these mutants revealed a second class that affects *dhfr* mRNA. In this case, *dhfr* mRNA of normal size is present, but the amount of this mRNA per cell has been reduced, often dramatically. A surprisingly high proportion of the mutants analyzed fall into this category; over half contain *dhfr* mRNA at less than 20% of the parental level, as indicated in Figure 1. In many mutants, the level of *dhfr* mRNA is barely detectable, being reduced to approximately 2% of the parental level. If the proportion of *dhfr* mRNA in the parental cells is the same as the proportion of DHFR protein, then the parental cells contain approximately 20 copies per cell, and the low level mRNA mutants have less than 1 copy. These measurements were carried out on preparations of total RNA (nuclear plus cytoplasmic), ruling out the accumulation of large amounts of unprocessed transcripts in these mutants.

How do point mutations in exons cause a low level of mRNA? A priori there are at least three possibilities: decreased transcription rates, mRNA lability, and pre-mRNA lability. Transcription of the *dhfr* gene in the mutant cells was measured using a nuclear run-on assay (McKnight and Palmiter 1979). Four mutants were analyzed, and all were indistinguishable from the parental cells in transcription rate.

An estimate of mRNA stability was made in the case of two UV-induced mutants that contained about 5% residual *dhfr* mRNA. Actinomycin D was added to cultures of the mutants and to parental cells to inhibit new RNA synthesis, and polyadenylated RNA was prepared after various intervals. The quantity of *dhfr* mRNA was then assayed by Northern analysis, comparing the values obtained with those for adenine phosphoribosyl transferase mRNA as a standard. A half-life of 7.5 hours was found for both the parental cells and the mutants, suggesting that these mutations did not affect the stability of mature *dhfr* mRNA. Although this half-life is in good agreement with that found previously by other methods in mouse cells carrying amplified copies of the *dhfr* gene (Hendrickson et al. 1980; Leys et al. 1984), we cannot rule out the possibility that the actinomycin treatment interfered with the estimation of mutant mRNA stability. In support of the idea of altered mRNA stability is the finding that the first three mutations whose sequence could be deduced all involve the introduction of nonsense codons into the reading frame. The deletion of the C in mutant 10 shifts the reading frame and introduces a downstream stop codon. In mutant 13, a GAG codon has been converted to UAG. In mutant 16, the conversion of GAA→UAA can account for the creation of the *Bst*EII site. It is possible that our mutant isolation protocol is selecting for nonsense mutants by demanding the virtual complete absence of DHFR activity and that premature termination leads to mRNA lability. Several nonsense mutations in the human β-globin gene result in low mRNA (and nuclear pre-mRNA) levels (for review, see Orkin and Kazazian 1984); the mechanism of this destabilization is not known. An alternative explanation is that single-base mutations may greatly influence the stability of *dhfr* primary transcripts, an idea that has not yet been tested. It is difficult to understand how a single base change per se could have so dramatic an effect on the stability of such a large molecule. One possibility is that these mutations are removing constraints imposed by secondary structure and allowing incomplete splicing reactions to take place at cryptic sites. Abnormal structures (e.g., dead-end lariats) that result may be rapidly degraded.

ACKNOWLEDGMENTS

This work was supported by U.S. Public Health Service grants GM-22629 (L.A.C.) and CA-39547 (A.M.C. and D.G.). The assistance of Cindy Diller and Caroline Monod is appreciated.

REFERENCES

Ames, B., W. Durstor, E. Yamasaki, and D. Lee. 1973. Carcinogens are mutagens: A simple test system combining liver homogenates for activation and bacteria for detection. *Proc. Natl. Acad. Sci.* **70:** 2281.

Carothers, A.M., G. Urlaub, N. Ellis, and L.A. Chasin. 1983. Structure of the dihydrofolate reductase gene in Chinese hamster ovary cells. *Nucleic Acids Res.* **11:** 1997.

Carothers, A.M., G. Urlaub, R.W. Steigerwalt, L.A. Chasin, and D. Grunberger. 1986. Characterization of mutations induced by 2-(N-acetoxy-N-acetyl) aminofluorene in the dihydrofolate reductase gene of cultured hamster cells. *Proc. Natl. Acad. Sci.* **83:** 6519.

Crouse, G.F., R.N. McEwan, and M.L. Pearson. 1983. Expression and amplification of engineered mouse dihydrofolate reductase minigenes. *Mol. Cell. Biol.* **3:** 257.

Dynan, W.S., S. Sazer, R. Tjian, and R.T. Schimke. 1986. Transcription factor Sp1 recognizes a DNA sequence in the mouse dihydrofolate reductase promoter. *Nature* **319:** 246.

Hendrickson, S.L., J.-S.R. Wu, and L.F. Johnson. 1980. Cell cycle regulation of dihydrofolate reductase mRNA in mouse fibroblasts. *Proc. Natl. Acad. Sci.* **77:** 5140.

Leys, E.J., G.F. Crouse, and R.E. Kellems. 1984. Dihydrofolate reductase gene expression in cultured mouse cells is regulated by transcript stabilization in the nucleus. *J. Cell Biol.* **99:** 180.

McKnight, G.S. and R.D. Palmiter. 1979. Transcriptional regulation of the ovalbumin and conalbumin genes by steroid hormones in chick oviduct. *J. Biol. Chem.* **254:** 9050.

Melton, D.A., P.A. Krieg, M.R. Rebagliatti, T. Maniatis, K. Zinn, and M.R. Green. 1984. Efficient *in vitro* synthesis of biologically active mRNA and RNA hybridization probes from plasmids containing a bacteriophage SP6 promoter. *Nucleic Acids Res.* **12:** 7035.

Mitchell, P.J., G. Urlaub, and L.A. Chasin. 1986a. Spontaneous splicing mutations at the dihydrofolate reductase locus in Chinese hamster ovary cells. *Mol. Cell. Biol.* **6:** 1926.

Mitchell, P.J., A.M. Carothers, J.H. Han, J.D. Harding, E. Kas, L. Venolia, and L.A. Chasin. 1986b. Multiple transcriptional start sites, DNase I-hypersensitive sites, and an opposite-strand exon in the 5′ region of the CHO *dhfr* gene. *Mol. Cell. Biol.* **6:** 425.

Orkin, S.H. and H.H. Kazazian. 1984. The mutation and polymorphism of the human β-globin gene and its surrounding DNA. *Annu. Rev. Genet.* **18:** 131.

Urlaub, G. and L.A. Chasin. 1980. Isolation of Chinese hamster cell mutants deficient in dihydrofolate reductase activity. *Proc. Natl. Acad. Sci.* **77:** 4216.

Urlaub, G., J. McDowell, and L.A. Chasin. 1985. Use of fluorescence-activated cell sorter to isolate mutant mammalian cells deficient in an internal protein, dihydrofolate reductase. *Somatic Cell Mol. Genet.* **11:** 71.

Urlaub, G., E. Kas, A.M. Carothers, and L.A. Chasin. 1983. Deletion of the diploid dihydrofolate reductase locus from cultured mammalian cells. *Cell* **33:** 405.

Urlaub, G., P.J. Mitchell, E. Kas, V.L. Funanage, T.T. Myoda, and J.L. Hamlin. 1986. Effect of gamma rays at the dihydrofolate reductase locus: Deletions and inversions. *Somatic Cell Mol. Genet.* **12:** 555.

Winter, E., F. Yamamoto, C. Almoguera, and M. Perucho. 1985. A method to detect and characterize point mutations in transcribed genes: Amplification and overexpression of the mutant c-Ki-*ras* allele in human tumor cells. *Proc. Natl. Acad. Sci.* **82:** 7575.

Application of Two-dimensional Electrophoresis to the Detection of Human Somatic Mutations at the Protein Level

ERNEST H.Y. CHU,* MICHAEL BOEHNKE,† AND SAMIR M. HANASH‡
Departments of Human Genetics,*
Biostatistics,† and Pediatrics‡
University of Michigan
Ann Arbor, Michigan 48109

OVERVIEW

A subclone of a human diploid lymphoblastoid cell line, thymidine kinase-6 (TK-6), with consistently high cloning efficiency has been utilized to assess the feasibility of detecting somatic mutations in vitro. By two-dimensional polyacrylamide gel electrophoresis, a panel of polypeptide spots representing products of 263 loci per gel was screened. Human somatic mutation rates in vitro were estimated by measuring the proportion of protein variants among cell clones isolated at various times during exponential growth of a cell population. Whereas no spontaneous mutations have been observed in 200,000 allele tests, a total of 34 confirmed protein variants have been identified following treatment of cells with ethylnitrosourea corresponding to an induced rate of 1.6×10^{-4} per allele per cell generation. Both electromorphs exhibiting altered electric charge and/or apparent molecular weight and nullimorphs have been found. Repeat occurrence of mutation at nine polymorphic loci suggests nonrandomness of mutation in human somatic cells. These observations provide a basis for a two-dimensional electrophoresis approach for the detection of genetic damage resulting from exposure to mutagens.

INTRODUCTION

Most of the earlier mutation studies with cultured mammalian cells were based on a handful of selectable markers, mainly those conferring drug resistance (for review, see Howard-Flanders 1981; Chu et al. 1984). In an attempt to obtain a better estimate of mutation over a wide range of loci in somatic cells, Siciliano and co-workers (1983) isolated hundreds of random clones of Chinese hamster ovary cells, with or without prior mutagen treatment, and screened for different classes of mutational events at 44 well-defined loci. The approach that we have been evaluating is the use of

Banbury Report 28: Mammalian Cell Mutagenesis

two-dimensional polyacrylamide gel electrophoresis (2-D PAGE) to monitor simultaneously several hundred polypeptide gene products resolved on a single gel. Applications of this technique in the field of human genetics include the detection of (1) variants due to polymorphism, (2) alterations associated with genetic disorders, and (3) mutations that are either spontaneous or induced. We have previously utilized 2-D PAGE to detect genetic variants in plasma, erythrocyte lysates, platelets, and lymphocytes (Rosenblum et al. 1983a,b, 1984; Hanash et al. 1986a,b). At present, 2-D PAGE is one of the best prospects for improving the efficiency of estimating human germ-line mutation rates (Neel et al. 1984). The technique permits detection of mutations, leading to proteins with altered electrophoretic properties although not necessarily altered function (Klose et al. 1975, 1982; Zeindl et al. 1982; Marshall et al. 1983).

Our current studies are aimed at utilizing 2-D PAGE to establish human in vitro somatic mutation rates, both spontaneous and induced, for a variety of proteins of different nature, using the same panel of protein markers of lymphoid cells as used in studies of human germ-line mutation (Neel et al. 1984). Our results to date demonstrate the maintenance of continuous exponential growth of a line of human lymphoblastoid cells in culture, a low rate of spontaneous mutation at the loci assayed, and nonrandom occurrence of mutation. A detailed description and analysis of these results will be presented elsewhere (E.H.Y. Chu et al., in prep.).

RESULTS AND DISCUSSION

We have used a human diploid lymphoblastoid cell line, TK-6, obtained from Dr. William Thilly of the Massachusetts Institute of Technology. The cell line is heterozgyous at the *tk* locus. The methods for cell culture and single-cell cloning were described previously (Thilly et al. 1980). We showed that a subclone has a plating efficiency of about 70% in liquid medium as measured by the limiting dilution method. We have further established that the line can be maintained in exponential growth for an indefinite period. Depending on inoculum size and the schedule of medium renewal, the average generation time can be adjusted experimentally from 12 to 24 hours. Daily cell counts entered into a computer program serve to determine the rate of cell growth.

Our basic experimental scheme involves the following steps: (1) initiate and continuously propagate a recloned TK-6 cell population as a mass culture, (2) at the beginning of the experiment and at 3-week intervals, isolate 120 single cell clones, and allow each clone to reach a population size of 10×10^6 to 12×10^6 cells for 2-D PAGE analysis and cold storage in the cell bank, (3) at each sampling time, obtain additional cells from the proliferating mass culture for a determination of the mutant frequencies at three specific loci, as shown by hypoxanthine-guanine phosphoribosyl transferase deficiency or resistance to the purine analog 6-thioguanine, thymidine kinase deficiency or resistance to the thymidine analog trifluorothymidine, and Na^{+}/K^{+} adenosine triphosphatase (ATPase) defect or resistance to ouabain. However, results on spontaneous and induced rates at these specific loci will not be included in this presentation.

Using the techniques of O'Farrell (1975), TK-6 cells are solubilized in order to dissociate multimeric proteins into their constituent polypeptides, and these are then separated in two dimensions: the first on the basis of charge by isoelectric focusing and the second on the basis of molecular weight by electrophoresis in the presence of SDS. The conditions used in our current studies are based on those developed and tested for the detection of genetic variants using human lymphocytes (Neel et al. 1984). The studies presented here are based on the analysis of polypeptide spots corresponding to 263 loci on each two-dimensional gel that were selected for visual scoring by the criteria of their size, well-defined morphology, low variability among gels, and separation from other spots in the surrounding areas. Although the results presented here were derived from visual scoring of gels, much progress has been made toward automation of gel scoring. For example, in one experiment, four chemically induced structural mutations detected in TK-6 cells by visual analysis were also detectable by computer-image analysis of gels corresponding to mutant clones (Hanash et al. 1987).

To estimate mutation rate based on protein variation on two-dimensional gels, we have taken a deterministic approach by measuring the proportion of mutants (mutant fraction) at various times during the proliferative growth of a cell population (Armitage 1952, 1953; Li et al. 1985).

In one experiment (Table 1), we isolated and analyzed 384 nonmutagenized cell clones but observed no mutation in about 200,000 allele tests (384 gels × 263 loci × 2 alleles). Taking into account the number of generations and assuming equal viability of mutant and nonmutant cells, there were approximately 12 million opportunities for mutation. This suggests an upper probability bound on spontaneous somatic mutation rate in vitro of 2.5×10^{-7} per allele per cell division that is the largest rate that still results in a probability of 0.05 for no mutation. Two possible explanations for the

Table 1
Estimation of Spontaneous Rate of Electrophoretic Mutation in the TK-6 Human Lymphoblastoid Cell Line in Culture

Sampling time (i)	A	B	C	D
Number of generations (T)	21	46	67	94
Number of clones	77	100	91	116
Number of mutants	0	0	0	0

Average generation time: 23.00 ± 0.43 hours (mean ± S.E.). Number of opportunities for mutation: $2 \times \text{spots} \times \Sigma_i T_i \times \text{clones}_i = 12$ million.

absence of mutants are (1) the human spontaneous somatic mutation rate in vitro for the type of mutations detectable by 2-D PAGE is very small, and (2) the assumption of equal viability between clones that carry detectable mutation and those that do not (wild type) is incorrect. We are currently conducting experiments to test these explanations.

In contrast, a total of 34 confirmed protein variants have been identified in another experiment in which an exponentially growing population of TK-6 cells were exposed to *N*-ethyl-*N*-nitrosourea (ENU) in five successive daily exposures at 10 μg/ml for 40 minutes. Table 2 summarizes the distribution of electrophoretic mutants at four different sampling times. Of the 34 mutants presumably induced by ENU, 20 exhibited charge alterations of the polypeptide involved, 3 showed change in apparent molecular weight, 7 showed both a charge as well as an apparent-molecular-weight alternation, and 4 were nullimorphs (see the following). The electrophoretic mutants usually show a diminished intensity (to about 50%) of the common (parental) polypeptide and the appearance at a neighboring location of a new variant polypeptide of equal intensity. Figure 1 illustrates some examples of electrophoretic mutants of TK-6 cells induced by ENU.

In a separate study (Hanash et al. 1986a) 2-D PAGE analysis of lymphoid cell proteins revealed that 14 of 106 polypeptides scored in 40 families were polymorphic. Of the 14 polymorphic polypeptides, 9 were among the 263 loci scored for mutation in the TK-6 cell line. At the beginning of the analysis, we knew that the individual from whom the TK-6 cells were derived was heterozygous at two such polymorphic loci (spots 81/82 and 284/285). After the completion of this experiment, a third locus (spots 28/29) was found to also exist in a heterozygous state. A total of four nullimorphs were identified thus far, i.e., in each mutant, one allelic product of the two polypeptides of the heterozgyous gene was not detected. A nullimorphic mutant at spot 81/82 is shown in Figure 1. The disappearance or nondetection of one of the two allelic products at heterozgyous loci could be due to a number of reasons, such as nonsense mutation, deletion or chromosome rearrangement involving the gene, and chromosome loss. However, compared to electrophoretic

Table 2
Estimation of Rate of Electrophoretic Mutations Induced by Ethylnitrosourea in the TK-6 Human Lymphoblastoid Cell Line in Culture

Sampling time	A	B	C	D	Total
Day	10	38	57	77	
Number of clones	112	101	106	95	414
Number of mutants	8	4	9	13	34

Average generation time: 18.47 ± 0.02 hours. Number of allele tests: 414 × 263 × 2 = 218,000. Induced mutation rate: 1.6×10^{-4}/allele/cell division.

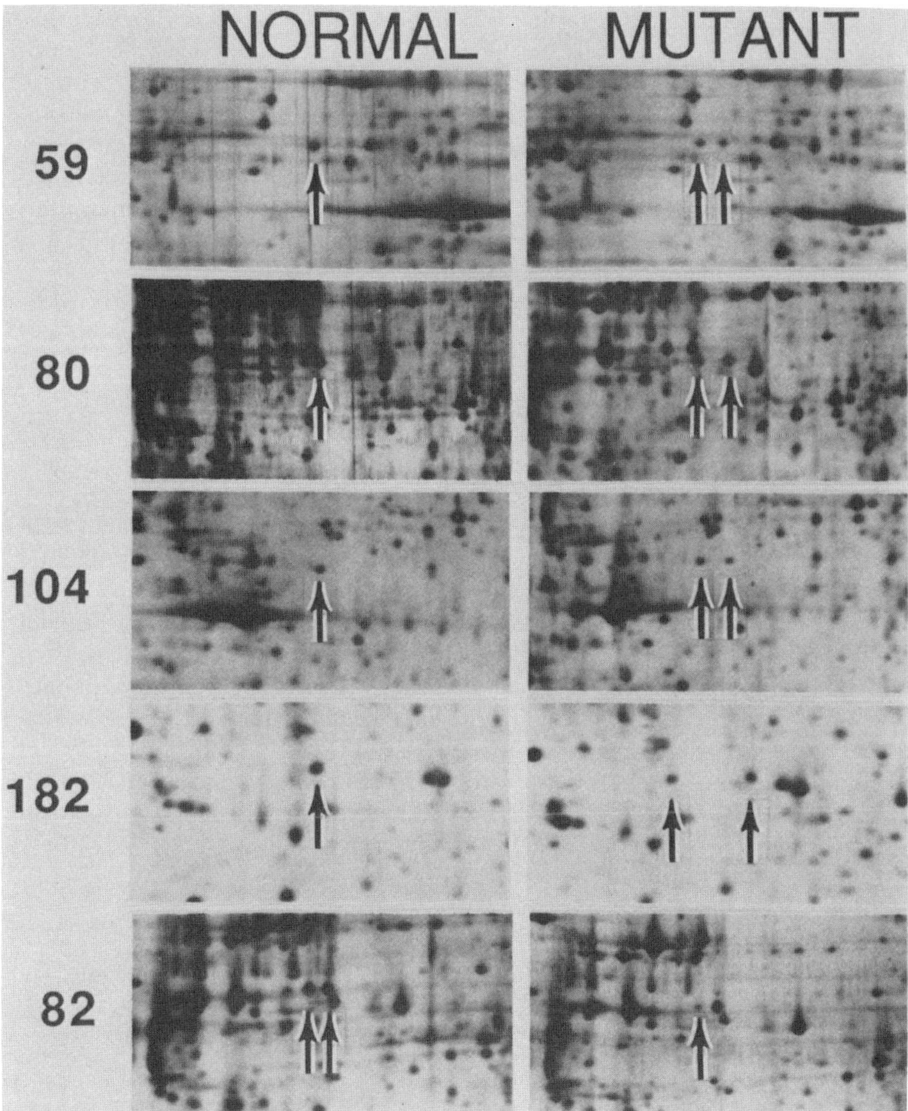

Figure 1

Enlargements of sections of two-dimensional gels from five independent clones of ethylnitrosourea treated TK-6 cells each exhibiting a mutation. Four polypeptide spots (59, 80, 104, and 182) show new charge variation and the fifth spot (82) is a nullimorph.

mobility mutations that are likely due to nucleotide base changes, nullimorphs seem to occur in these cells at an appreciably higher frequency than expected.

The existence of polymorphic loci implies that these genes may be hypermutable in the germ cells. In this particular chemical mutagenesis experiment with TK-6 cells, 4 electrophoretic variants (285, 82, 129-1, 129-2) were found among the 9 previously identified polymorphic loci. Assuming that these are independent mutations and that mutation is at random, the probability of 4 of the 34 mutations occurring at the 9 previously identified polymorphic loci is less than 0.029 (i.e., $p < 0.029$). Furthermore, mutations occurred more than 3.4 times as frequently at the 9 known polymorphic loci as at the remaining 254 loci.

Finally, multiple mutations were observed at the loci encoding polypeptides 28/29, 70, 95/121, and 129; specifically three mutations at 70 and two each at 28/29, 95/121, and 129. Assuming that these are independent mutations and that mutation is at random, the probability of observing five or more repeat mutations among 34 mutations at 263 loci is less than 0.042 (i.e., $p < .042$). The methods for calculating these probabilities will be presented elsewhere (S.M. Hanash et al., in prep.).

The above results are clearly preliminary and further experiments are either under way or planned to improve the estimate of spontaneous and induced rates of human somatic mutation in vitro. The expected results will allow us an opportunity to make a comparison with the rates of germinal mutation in human populations, especially since the same battery of protein markers of lymphoid cells are used. The experimental design and techniques we have been developing, including the biological, biochemical, and biostatistical procedures, and the automated analysis of two-dimensional gels, should be applicable to future studies on the rates of human somatic mutations in vitro and in vivo, as well as to the assessment of genetic hazards of human exposure to environmental agents.

ACKNOWLEDGMENTS

The research was supported by a grant (CA-26803) from the National Cancer Institute. The authors gratefully acknowledge the invaluable assistance of B.J. Lamb, W. Niezgoda, S. Pivirotto, and G. Sundling.

REFERENCES

Armitage, P. 1952. The statistical theory of bacterial populations subject to mutation. *J. R. Stat. Soc. B.* **14:** 1.

———. 1953. Statistical concepts in the theory of bacterial mutation. *J. Hyg.* **51:** 162.

Chu, E.H.Y., I.-C. Li, and J. Fu. 1984. Mutagenesis studies with cultured mammalian cells: Problems and prospects. In *Mutation, cancer and malformation* (ed. E.H.Y. Chu and W.M. Generoso), p. 315. Plenum Press, New York.

Hanash, S.M., L. Baier, M. Gatleau, and D. Welch. 1986a. Genetic variants detected among one hundred and six lymphocyte polypeptides observed in two-dimensional gels. *Am. J. Hum. Genet.* **39:** 317.

Hanash, S.M., B.B. Rosenblum, J.V. Neel, L.J. Baier, and D. Markel. 1986b. Genetic analysis of thirty three platelet polypeptides detected in two-dimensional polyacylamide gels. *Am. J. Hum. Genet.* **38:** 352.

Hanash, S.M., E.H.Y. Chu, R. Kuick, M. Skolnick, J. Neel, J. Strahler, S. Pivirotto, and W. Niezgoda. 1987. Detection of human somatic cell structural gene mutation by two-dimensional electrophoresis. *Proteins* **2:** 13.

Howard-Flanders, P. 1981. Mutagenesis in mammalian cells. *Mutat. Res.* **86:** 307.

Klose, J. 1975. Protein mapping by combined isoelectric focusing and electrophoresis of mouse tissues. *Humangenetik* **26:** 231.

———. 1982. Genetic variability of soluble proteins studied by two-dimensional electrophoresis on different inbred mouse strains and on different mouse organs. *J. Mol. Evol.* **18:** 315.

Li, I.-C., S.-C.H. Wu, J. Fu, and E.H.Y. Chu. 1985. A deterministic approach for the estimation of mutation rates in cultured mammalian cells. *Mutat. Res.* **149:** 127.

Marshall, R.R., A.S. Raj, F.J. Grant, and J.A. Heddle. 1983. The use of two dimensional electrophoresis to detect mutations induced in mouse spermatogonia by ethylnitrosourea. *Can. J. Genet. Cytol.* **25:** 457.

Neel, J.V., B.B. Rosenblum, C.F. Sing, M.M. Skolnick, S.M. Hanash, and S. Sternberg. 1984. Adapting two-dimensional gel electrophoresis to the study of human germline mutation rates. In *Methods and applications of two dimensional gel electrophoresis of proteins* (ed. J.E. Celis and R. Bravo), p. 259. Academic Press, New York.

O'Farrell, P.H. 1975. High resolution two dimensional electrophoresis of proteins. *J. Biol. Chem.* **250:** 4007.

Rosenblum, B.B., S.M. Hanash, and J.V. Neel. 1983a. High resolution separation of plasma proteins. Application to genetic analysis. In *Proceedings of the 5th International Prospective Biology Colloquim* (ed. G. Siest et al.), p. 91. Masson, Paris.

Rosenblum, B.B., J.V. Neel, and S.M. Hanash. 1983b. Two-dimensional electrophoresis of plasma proteins reveals "high" heterozygosity indices. *Proc. Natl. Acad. Sci.* **80:** 5002.

Rosenblum, B.B, J.V. Neel, S.M. Hanash, N. Yew, and J. Jospeh. 1984. Identification of genetic variants in erythroid lysate by two-dimensional gel electrophoresis. *Am. J. Hum. Genet.* **36:** 601.

Siciliano, M.J., G.M. Adair, E.N. Atkinson, and R.M. Humphrey. 1983. Induced somatic cell mutations detected in cultured cells by electrophoresis. *Isozymes Curr. Top. Biol. Med. Res.* **10:** 41.

Thilly, W.G., J.G. DeLuca, E.C. Furth, H. Hoppe IV, D. Kaden, J.J. Krolewshi, H.L. Liber, T.R. Skoplk, S.A. Slapikoff, R.J. Tizard, and B.W. Penman. 1980. Gene-locus mutation assays in diploid human lymphoblast lines. *Chem. Mutagens* **6:** 331.

Zeindl, E., K. Sperling, and J. Klose. 1982. Mutagenicity testing on non-selectively cloned Chinese hamster ovary cells with the protein-mapping method. *Mutat. Res.* **97:** 67.

Molecular Analysis of the *aprt* Locus

On the Complexity of Mutagenic Events at the Mouse Lymphoma *tk* Locus

MARCIA L. APPLEGATE* AND JOHN C. HOZIER*†
*Department of Biological Sciences and
†Institute of Molecular Biophysics
Florida State University
Tallahassee, Florida 32306

OVERVIEW

The mouse lymphoma mutagen assay system detects forward mutations at the heterozygous thymidine kinase (*tk*) locus in L5178Y TK$^{+/-}$3.7.2C cells. We have sublocalized the mouse *tk* gene to chromosome 11, bands B–E, by in situ hybridization of a mouse *tk* cDNA probe to L5178Y chromosomes. This chromosomal segment is involved in a heterogeneous set of rearrangements associated with loss of TK activity in slow-growing (small-colony) mutants. We are proceeding with in situ analysis of TK-deficient mutants to reveal positional changes or deletions of the active *tk* allele. Southern blot analysis of DNA isolated from 25 TK$^{-/-}$ mutants reveals major deletions involving the functional *tk* allele of all small-colony and most large-colony mutants studied. Among small-colony mutants, this loss of the functional *tk* allele is often found associated with other cytogenetically visible events such as translocations. In contrast, in vitro observations of a high frequency of cytogenetically visible or large deletion mutations at hemizygous loci are comparatively rare. Gene inactivation at heterozygous regions of the genome may therefore include lesions, such as deletions involving adjacent genes, that might be lethal at a hemizygous region. Some mutations at heterozygous loci may also arise by mechanisms requiring the presence of a homologous chromosome, such as reciprocal recombination and gene conversion, that are precluded at hemizygous loci. These extensive mutational events detected at the heterozygous mouse *tk* locus appear to be similar to chromosome rearrangements observed in human malignancies and germ-line mutations.

This manuscript has been reviewed by the Health Effects Research Laboratory, U.S. Environmental Protection Agency, and approved for publication. Approval does not signify that the contents necessarily reflect the views and policies of the agency, nor does mention of trade names or commercial products constitute endorsement or recommendation for use.

Banbury Report 28: Mammalian Cell Mutagenesis

INTRODUCTION

The history and basic principles of the mouse lymphoma L5178Y TK$^{+/-}$3.7.2C mutagen assay system have been described previously (Clive et al. 1979; Clive; Moore et al.; both this volume), and we include only a very brief outline here. TK is an enzyme in the pyrimidine salvage pathway that converts thymidine to thymidine monophosphate for use in DNA synthesis. The gene coding for TK has been localized to chromosome 11 in mouse cells by somatic-cell genetics (Kozak and Ruddle 1977). The mouse lymphoma mutagen assay quantitates mutation at the heterozygous *tk* locus in mouse L5178Y TK$^{+/-}$3.7.2C cells. Cells with a $tk^{-/-}$ genotype are induced by incubation of the heterozygote with mutagenic test substances and are selected by growth in the presence of trifluorothymidine (TFT), a pyrimidine analog that allows cells carrying the forward mutation $tk^{+/-} \rightarrow tk^{-/-}$ to survive, but kills cells that retain the $tk^{+/-}$ genotype. TFT, when phosphorylated by TK, irreversibly inhibits thymidylate synthetase, preventing DNA replication. The synthesis of thymidine monophosphate in TK$^{-/-}$ cells proceeds via the de novo biosynthetic pathways.

When cloned in soft agar, mutants recovered in the mouse lymphoma L5178Y TK$^{+/-}$3.7.2C assay system show a bimodal size distribution that is characteristic for different mutagens. Large-colony mutants have growth properties that are similar to those of the parental heterozygote line, whereas small-colony mutants show slower growth kinetics. In addition, whereas large-colony mutants do not show visible karyotypic abnormalities at chromosome 11, approximately 90% of small-colony mutants have deletions, translocations, or other rearrangements involving this chromosome early in their clonal history. Although some of these are unstable, particularly those involving dicentrics, about 60% of small-colony mutants have abnormal copies of chromosome 11 20 to 30 generations after cloning (Hozier and Sawyer 1981; Hozier et al. 1985; Moore et al. 1985).

We have been interested in defining the cytogenetic and molecular genetic features of the *tk* gene that would allow us to analyze mutations involving the functional *tk* allele in a heterozygous genetic environment. Our overall strategy includes detailed study of the chromosome structure, gene organization, level of expression, and transcript abnormalities of a representative population of spontaneous and induced TK$^{-/-}$ mutants.

To define more clearly the extent of genetic lesions in small- and large-colony mutants, we have designed methods for the high-resolution analysis of early metaphase L5178Y 3.7.2C chromosomes. In addition, we have studied *tk* gene structure by Southern blot analysis of DNA from the parental $tk^{+/-}$ heterozygote and from a selected set of chemically or radiation-induced mutants. Combining the molecular and cytogenetic approaches, we have begun in situ molecular hybridization studies of the *tk* gene in L5178Y $TK^{+/-}$3.7.2C and $TK^{-/-}$ cells.

RESULTS

Cytogenetic Abnormalities among Small-colony $TK^{-/-}$ Mutants

In L5178Y $TK^{+/-}$3.7.2C cells, the centromeric regions of the two chromosome 11 homologs are easily distinguishable allowing assignment of chromosomal abnormalities to a specific homolog (Hozier et al. 1982). Because all of the visible deletion and translocation events seen at chromosome 11 in karyotypes of small-colony $TK^{-/-}$ mutants involve one homolog, designated 11b, we have hypothesized that this chromosome contains the tk^{+} allele. To characterize in more detail the lesions observed in small-colony karyotypes, we have developed techniques that allow resolution of up to about 520 bands in mouse chromosomes (Sawyer and Hozier 1986). Figure 1 shows that the range of break points observed at medium-to-high resolution for all mutants studied is not confined to a single band when multiple cytogenetically abnormal mutants are compared (although breakpoints observed for a given mutant are consistent from cell to cell). This breakpoint heterogeneity among various small-colony $TK^{-/-}$ mutants indicates that the rearrangement events associated with inactivation at the *tk* locus extend over a fairly large segment of the distal end of chromosome 11. Because the breakpoints observed at chromosome 11 in small-colony mutants occur over a segment containing several high-resolution bands (Sawyer et al. 1987; see below), it seems unlikely that genetic damage in small-colony mutants is the result solely of intragenic events at the *tk* locus.

Localization of the Mouse *tk* Gene to the Distal Portion of Chromosome 11 by In Situ Hybridization

To test whether the *tk* gene in mouse L5178Y 3.7.2C cells maps within the segment of chromosome 11 delineated by mutant breakpoint analysis, we performed in situ hybridization to mitotic chromosomes using a tritium-labeled mouse *tk* cDNA probe (pMTK4) that was cloned and provided to us by P.F. Lin and co-workers (Lin et al. 1985). Figure 2 shows a summary of ^{3}H grains counted over chromosome 11 homologs of 47 cells following autoradiographic exposure and banding of hybridized chromosomes. The grains at the distal end of chromosome 11 represent 19.6% of the total number scored. They are similarly distributed over both copies of chromosome 11 of the $TK^{+/-}$3.7.2C heterozygote, indicating the presence of TK sequences on both homologs.

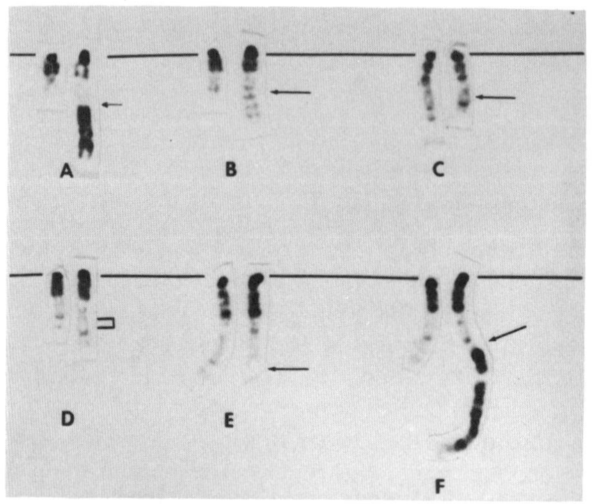

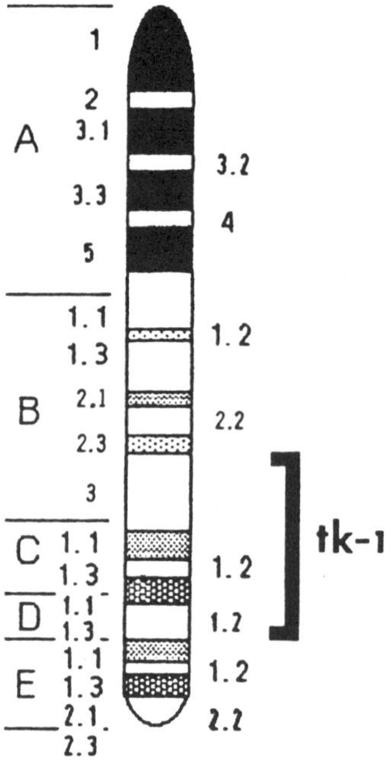

Figure 1

Examples of six different small colony $TK^{-/-}$ mutants (*A–F*). Arrows indicate breakpoints in translocations (*A,B,F*) (*F* is mutant C.342 as described in Fig. 3), an interstitial deletion (*C*) and a terminal deletion (*E*). The bracket in *D* shows the extent and location of an interstitial insertion. Note that all rearrangements involve chromosome 11b, which has the larger centromere of the homologous pair and is presumed to contain the functional allele in the $TK^{+/-}$3.7.2C heterozygote (Hozier et al. 1982). The bracketed segment on the high-resolution chromosome 11 ideogram shown in the right-hand panel represents the range of breakpoints observed among various small-colony mutants.

A single small-colony $TK^{-/-}$ mutant (C.342), containing a translocation between a chromosome 2 and chromosome 11b, has been analyzed by in situ hybridization (Fig. 3). No grains appear over the chromosome 11b portion of

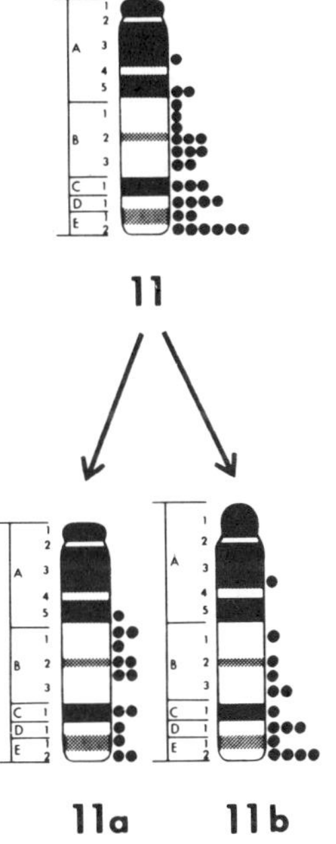

Figure 2

Distribution of grains over chromosome 11 homologs of L5178Y $TK^{+/-}$ 3.7.2C cells following autoradiographic exposure of slides hybridized in situ with the mouse TK cDNA probe pMTK4 labeled with [^{3}H]dCTP. G-banded chromosomes from 47 cells were analyzed. Of 149 total grains, 29% are on chromosome 11, and of these, 65% appear at or distal to band B3. The grains are shown mapped to chromosome 11 (*top*) and divided as observed between the two homologs (*bottom*). Chromosomes 11a and 11b are distinguishable due to a centromeric heteromorphism (Hozier et al. 1982).

the translocation product. In contrast, a highly significant proportion of grains appears at the distal end of chromosome 11a. Although in situ hybridization analysis of the mutational event in this cell line is complicated by an insertion or terminal addition of two extra bands of unknown origin at the distal region of chromosome 11a, we believe that the data are consistent with loss of *tk* gene sequences from 11b in the chromosome rearrangement. Therefore, analysis of both breakpoint heterogeneity and in situ hybridization patterns leads to the possibility of extensive and apparently complex rearrangements in the vicinity of the *tk* gene in some small-colony mutants.

Southern Blot Analysis of *tk* Gene Structure in L5178Y $TK^{+/-}$ and $TK^{-/-}$ Cell Lines

To study the effects of mutation at the *tk* locus itself in small- and large-colony mutants, we have performed Southern blot analysis of DNA isolated from L5178Y $TK^{+/-}$3.7.2C and derivative mutant cell lines. DNA from each was digested with a set of restriction enzymes, size separated by gel electrophoresis, blotted, and hybridized with the mouse *tk* cDNA probe, pMTK4. To identify TK-coding sequences, hybridizing bands were compared with similar

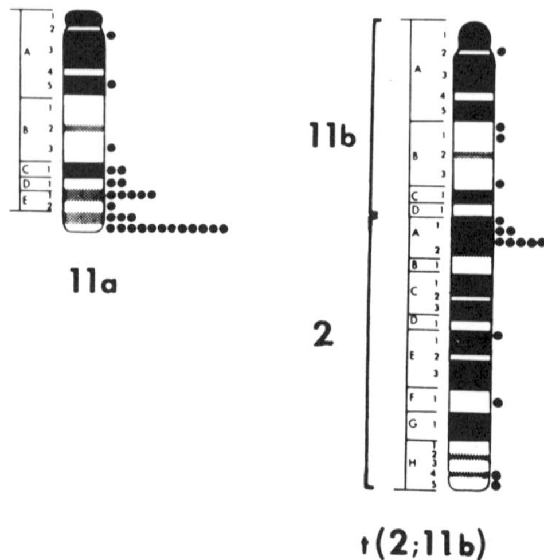

Figure 3
In situ hybridization of chromosomes from the t(2;11) small-colony mutant C.342 with pMTK4; 97 cells were analyzed. Of 282 grains, 28 are on chromosome 11a, with 25 of these distal to band C1.

digests of DNA from the cosmid pMTK116, which contains a full-length mouse genomic *tk* gene capable of transforming mouse LTK$^-$ cells to the TK$^+$ phenotype (Lin et al. 1985). In addition to functional gene-coding sequences, Southern blots of mouse cell DNA reveal sequences hybridizing to *tk* cDNA, presumably pseudogene sequences, that do not appear within the *tk* gene cloned in pMTK116. However, they do appear in the TK$^{-/-}$ cell line LTK$^-$, which contains no functional cytosolic *tk* gene sèquences (Lin et al. 1985; and Figs. 4, 5, and 6 below).

By comparing Southern blots of normal and mutant cell DNAs cleaved with a large panel of restriction enzymes, we were able to identify a probable restriction site heteromorphism distinguishing the two *tk* alleles. The restriction enzyme *Nco*I cuts DNA from L5178Y TK$^{+/-}$3.7.2C cells into five fragments (Fig. 4). Two of these, at 6.4 and 4.4 kb, appear in pMTK116 DNA cut with *Nco*I, indicating that they represent functional *tk* gene sequences. Two others at 5.5 and 2.1 kb that appear in *Nco*I-digested LTK$^-$ cell DNA are TK-like sequences, possibly derived from pseudogenes. In DNA from 24 of the 25 large- and small-colony mutants that we have analyzed, we find a loss of the 6.41-kb *Nco*I fragment (M.L. Applegate et al., unpubl.) that we have mapped to the 3′ end of the *tk* gene cloned in pMTK116 (Figs. 4 and 5). The absence of this band reflects deletion of *tk* sequences from the TK$^+$ allele, and indicates that the band is heteromorphic in the two alleles of

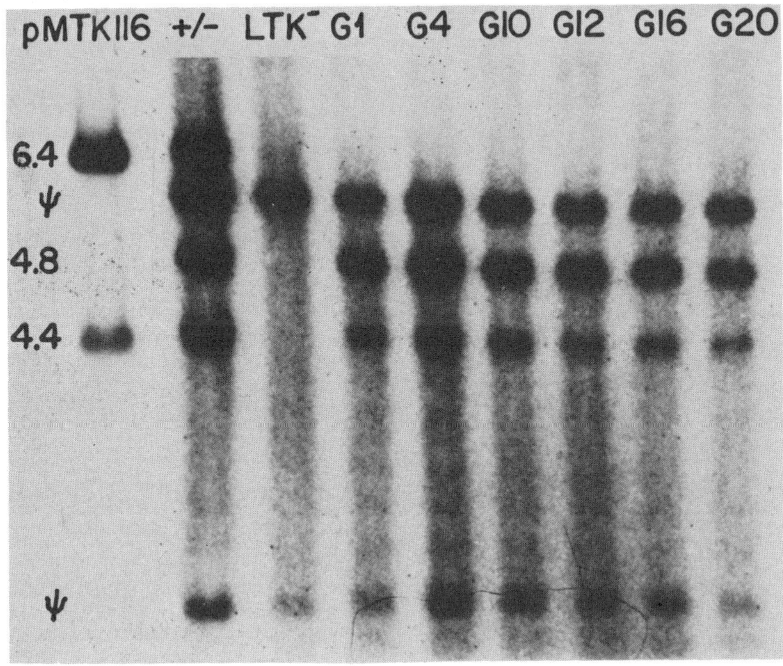

Figure 4

Southern blot of L5178Y TK$^{+/-}$3.7.2C, LTK^{-}, and six γ-radiation-induced mutant genomic DNAs, as well as DNA from cosmid pMTK116, digested with *Nco*I and probed with the TK cDNA probe pMTK4. The 6.4- and 4.4-kb bands appearing in pMTK116 represent TK-coding sequences, those in *ltk*$^{-}$ DNA (labeled ψ) are TK-like sequences, and the additional 4.8-kb band seen in the genomic digests probably represents the variant *tk*$^{-}$ allele of the 6.4-kb band at the 3′ end of the *tk*$^{+}$ allele that is deleted in 24 of 25 mutants tested.

L5178Y TK$^{+/-}$3.7.2C. The mutants whose DNAs showed deletions included five small-colony mutants recovered from bleomycin-treated cultures, six small-colony mutants induced with γ-radiation, three small-colony methyl methanesulfonate (MMS)-induced mutants, and one each of small-colony pyrimethamine and ethyl methanesulfonate (EMS)-induced mutants, as well as five large-colony methotrexate mutants, one large-colony EMS mutant, and one large-colony mutant induced by pyrimethamine. The mutant cell line whose DNA showed an apparently normal pattern of bands, consistent with a small intragenic mutation, was a large-colony, EMS-treated clone. We believe that all of the deletion events observed represent total loss of the *tk*$^{+}$ allele, since partial deletions would be expected to show alterations in the sizes of restriction fragments generated by enzymes other than *Nco*I. We have not observed any such changes in other digests (Fig. 6).

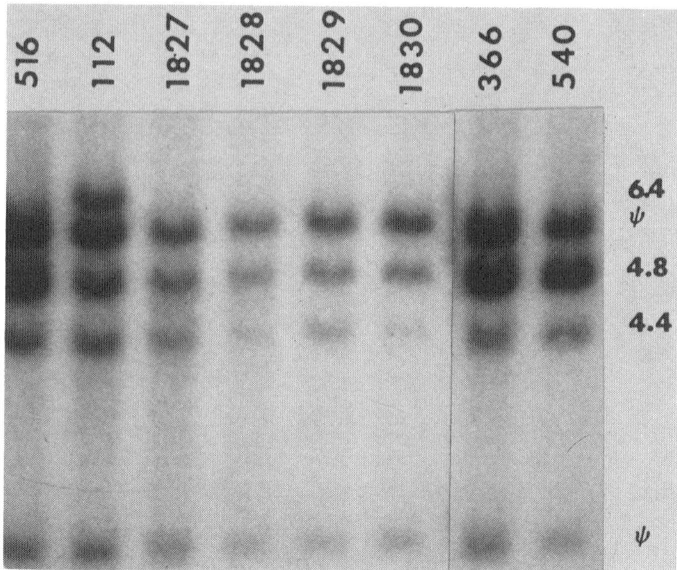

Figure 5

*Nco*I digests of several mutant DNAs showing the apparently normal digest pattern of the large-colony EMS mutant 112. Cell lines 516, 1827–1830, and 112 are large-colony mutants. The other two mutant lines, 366 and 540F are small-colony mutants.

Figures 4 and 5 indicate a fifth band in *Nco*I digests of L5178Y $TK^{+/-}$3.7.2C DNA, at 4.8 kb, which probably represents the heteromorphic counterpart in the tk^- allele of the 6.4-kb band that is deleted from the tk^+ allele in the mutants. When blots of *Nco*I digests are probed with a 5.0-kb *Hin*dIII fragment from the 3′ end of the gene, both the 6.41-kb band and the 4.9-kb *Nco*I band hybridize, indicating that they share homology within this region (data not shown). The 4.9-kb band probably arises from the presence of an *Nco*I site unique to the tk^- allele, which cuts the 6.4-kb *Nco*I fragment into a 4.8- and a 1.5-kb band in this allele. The 1.5-kb band is not visible on Southern blots because it does not contain *tk* exon fragments. Although bands of this size could also be generated by a small-deletion event, we do not see the size changes in bands using other restriction enzymes that such a deletion would be expected to produce.

Thus, for a large proportion of a selected group of mutants, loss of the functional *tk* allele is associated with gene inactivation. Among some small-colony mutants (e.g., C.342; see Fig. 3), this general allelic loss forms a background upon which other chromosome events, such as translocations, can occur.

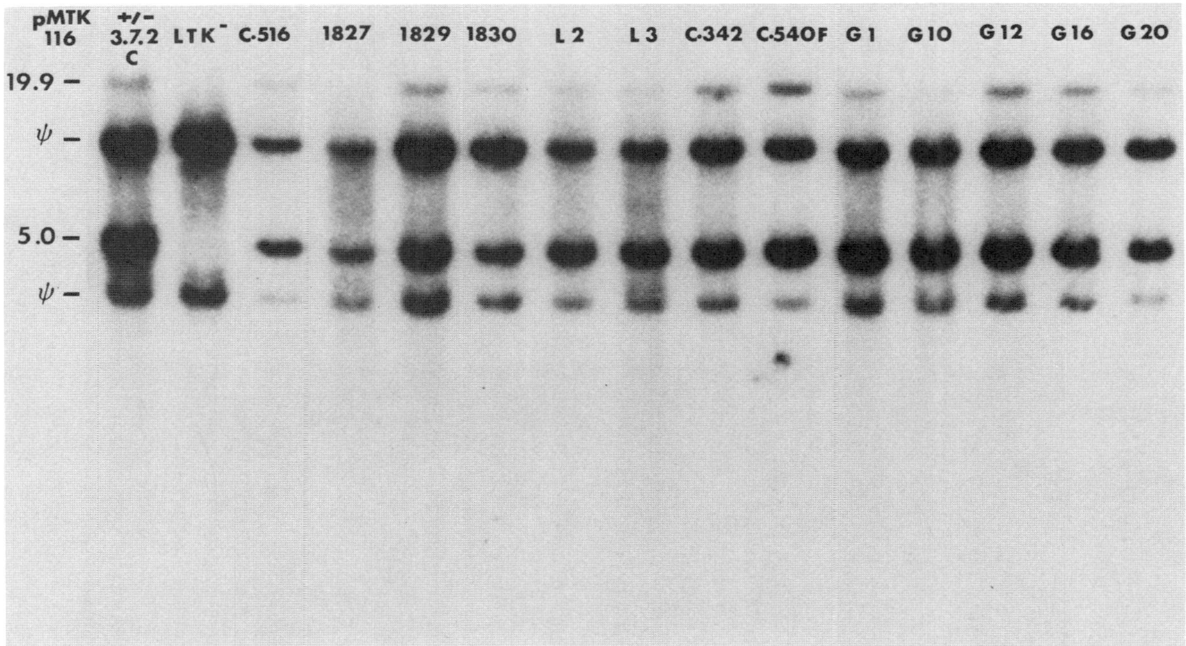

Figure 6

*Hin*dIII digests of DNA from several mutants. The 19.9- and 5.0-kb fragments are present in *Hin*dIII-digested pMTK116 and are indicated by horizontal bars. These represent coding sequences. The two bands marked ψ are present in *ltk*$^-$ DNA and may represent pseudogenes. None of these mutants show band-size changes expected to accompany intragenic deletions. Mutants 516, 1827, 1829, and 1830 are large-colony mutants; all the others are small-colony mutants

DISCUSSION

We have presented evidence that all of the small-colony and most of the large-colony $TK^{-/-}$ mutants that we have studied harbor deletions spanning the active *tk* allele of the parental heterozygote, L5178Y $TK^{+/-}$ 3.7.2C. John Little and collaborators (Little et al., this volume, and references therein) have reported similar results from their studies of the human *tk* gene in a lymphoblastoid cell line heterozygous at the *tk* locus. Many of these mutants were induced with chemical or physical agents that produce a very poor mutagenic response at hemizygous loci (for review, see Moore et al., this volume). The data therefore support the hypothesis that heterozygous loci may allow recovery of deletion mutations that have lethal effects at hemizygous sites when the deletions extend into genes that flank the selective locus (Clive et al. 1980; Evans et al. 1986). Alternatively, heterozygous test loci may allow recovery of mutations arising as primary lesions or as consequences of cellular DNA repair processes mediated by mechanisms requiring the presence of a pair of homologous chromosomes, i.e., reciprocal or nonreciprocal homologous recombination. To distinguish between simple deletion events and those resulting in, for example, deletion of the tk^{+} allele accompanied by duplication of the nonfunctional allele, we are currently quantitating the amounts of TK-specific sequences in the heterozygote and mutant cell line.

One of the reasonable expectations that we hold for cytogenetic and molecular level analyses of mutants recovered in mammalian in vitro mutagenesis assay systems is that these studies will provide insight into intragenic and chromosomal alterations observed in cancer and other human diseases. We are using the mouse lymphoma TK mutagenesis assay system to study such genetic alterations, since it provides an autosomal heterozygous test locus with cytogenetic and molecular properties that allow detailed investigation of mutation at the *tk* locus, despite the inevitably complex nature of any analysis involving a heterozygous gene.

Deletions and loss of heterozygosity are documented concomitants of acquired and inherited congenital human diseases. Visible chromosome deletions, such as chromosome-5p monosomy associated with the "cri-du-chat" syndrome, have been known for some time. Less extensive deletions not visible at the cytogenetic level (e.g., those associated with loss of globin genes in the thalassemias) have required molecular analysis. An interesting recent example of clinical importance is Duchenne muscular dystrophy, in which the very large gene (~ 1 million bp) responsible for the disease has been mapped to the human X chromosome. Molecular analysis of this muscular dystrophy gene locus has revealed heterogeneity of translocation and deletion breakpoints (Monaco and Kunkel 1987), indicating variety and complexity of mutational events associated with the disease phenotype. Loss of specific segments of the genome as predisposing factors is implicated in the etiology of retinoblastoma (Cavenee et al. 1983) and Wilm's tumor (Koufos et al. 1984).

With regard to acquired (somatic) genetic changes, Turner et al. (1985) have shown that a surprisingly large proportion ($\sim 60\%$) of human lymphocytes selected as in vivo hypoxanthine-guanine phosphoribosyl transferase (HGPRT) mutants from normal subjects have substantial alterations in gene structure, rather than simple point mutations. These alterations are complex and include deletions, exon amplification, and other rearrangements. Dracopoli et al. (1985) have shown that major (multilocus or whole chromosomal) somatic mutations are common in malignant melanomas. When specific oncogenes are studied in a wide spectrum of tumor types using molecular probes, apparent allelic alterations including deletions and/or amplifications have been observed (Yokota et al. 1986).

Although combined cytogenetic and molecular techniques are seldom used in these clinically relevant analyses of genetic change (often for technical reasons), a pattern of complexity and multigenic extent of the alterations is apparent. Technical advances such as high-resolution cytogenetics, as well as pulse-field gel electrophoresis for analysis of very large segments of the genome, will make such studies of clinical material more feasible and should lead to an understanding of the possible overall importance of large mutational events to human genetic disease. At the same time, we are interested in acquiring a detailed understanding of such changes in chromosome structure, as well as their underlying mechanisms, and a mechanistic understanding of all classes of mutagenic events at the heterozygous mouse *tk* locus to determine to what extent it meets our initial expectations of providing an in vitro mutagenesis system for analysis of complex genetic changes relevant to human disease.

ACKNOWLEDGMENTS

We thank Pin-fang Lin for the cosmid pMTK116 and the cDNA clone pMTK4. We also thank Natalie de Vassal and Greg Juhn for expert technical assistance and Amy Wadhams for providing data from her Master's thesis work. Finally, we thank Debbie LaFrance for her patience and expert typing of the manuscript.

REFERENCES

Cavenee, W.K., T.P. Dryja, R.A. Phillips, W.F. Benedict, R. Godbout, B.L. Gallie, A.L. Murphree, L.C. Strong, and R.L. White. 1983. Expression of recessive alleles by chromosomal mechanisms in retinoblastoma. *Nature* **305:** 779.

Clive, D., A.G. Batson, and N.T. Turner. 1980. The ability of the L5178Y/TK$^{+/-}$ mouse lymphoma cells to detect single gene and viable chromosome mutations: Evaluations and relevance to mutagen and carcinogen screening. In *The predictive value of short term screening tests in carcinogenicity evaluation* (ed. G.M. Williams et al.), p. 103. Elsevier, Amsterdam.

Clive, D., K.O. Johnson, J.F.S. Spector, A.G. Batson, and M.M.M. Brown. 1979. Validation and characterization of the L5178Y/TK$^{+/-}$ mouse lymphoma mutagen assay system. *Mutat. Res.* **59:** 61.

Dracopoli, N.C., A.N. Houghton, and L.J. Old. 1985. Loss of polymorphic restriction fragments in malignant melanoma: Implications for tumor heterogeneity. *Proc. Natl. Acad. Sci.* **82:** 1470.

Evans, H.H., J. Mence, M.-F. Horng, M. Ricanati, C. Sanchez, and J. Hozier. 1986. Locus specificity in the mutability of L5178Y mouse lymphoma cells: The role of multilocus lesions. *Proc. Natl. Acad. Sci.* **83:** 4379.

Hozier, J. and J. Sawyer. 1981. Cytogenetic analysis of the L5178Y/TK$^{+/-} \rightarrow$ TK$^{-/-}$ mouse lymphoma mutagenesis assay system. *Mutat. Res.* **84:** 169.

Hozier, J., J. Sawyer, D. Clive, and M. Moore. 1982. Cytogenetic distinction between the TK^{+} and TK^{-} chromosomes in the L5178Y TK$^{+/-}$3.7.2C mouse-lymphoma cell line. *Mutat. Res.* **105:** 451.

———. 1985. Chromosome 11 aberrations in small colony L5178Y TK$^{-/-}$ mutants early in their clonal history. *Mutat. Res.* **147:** 237.

Koufos, A., M.F. Hansen, B.C. Lampkin, M.L. Workman, N.G. Copeland, N.A. Jenkins, and W.K. Cavenee. 1984. Loss of alleles at loci on human chromosome 11 during genesis of Wilm's tumour. *Nature* **309:** 170.

Kozak, C.A. and F.H. Ruddle. 1977. Assignment of the genes for thymidine kinase and galactokinase to *Mus musculus* chromosome 11 and the preferential segregation of this chromosome with chinese hamster/mouse somatic cell hybrids. *Somatic Cell Genet.* **3:** 121.

Lin, P.F., H.B. Lieberman, D.-B. Yeh, T. Xu, T.-Y. Zhao, and F.H. Ruddle. 1985. Molecular cloning and structural analysis of murine thymidine kinase genomic and cDNA sequences. *Mol. Cell. Biol.* **5(11):** 3149.

Monaco, A.P. and L.M. Kunkel. 1987. A giant locus for the Duchenne and Becker muscular dystrophy gene. *Trends Genet.* **3(2):** 33.

Moore, M.M., D. Clive, J.C. Hozier, B.E. Howard, A.G. Batson, N.T. Turner, and J. Sawyer. 1985. Analysis of trifluorothymidine-resistant (TFT^{r}) mutants of L5178Y/TK$^{+/-}$ mouse lymphoma cells. *Mutat. Res.* **151:** 161.

Sawyer, J.R. and J.C. Hozier. 1986. High resolution of mouse chromosomes: Banding conservation between man and mouse. *Science* **232:** 1632.

Sawyer, J., M. Moore, and J. Hozier. 1987. High resolution cytogenetic analysis of TK$^{-/-}$ mutants of L5178Y TK$^{+/-}$3.7.2C cells: Variation in break points among sigma TK$^{-/-}$ mutants. *Environ. Mutagen.* (suppl. 8) **9:** 95.

Turner, D.R., A.A. Morley, M. Haliandros, R. Kutlaca, and B.J. Sanderson. 1985. In vivo somatic mutation in human lymphocytes frequently result from major gene alterations. *Nature* **315:** 343.

Yokota, J., Y. Tsunetsugu-Yokota, H. Battifora, C. Le Fevre, and M.J. Cline. 1986. Alterations of *myc*, *myb*, and *ras*^{Ha} proto-oncogenes in cancers are frequent and show clinical correlations. *Science* **231:** 261.

Molecular Analysis of Mutations at the *tk* and *hgprt* Loci in Human Cells

JOHN B. LITTLE, DAVID W. YANDELL, AND HOWARD L. LIBER
Laboratory of Radiobiology
Harvard School of Public Health
Boston, Massachusetts 02115

OVERVIEW

The DNA structure of spontaneous and induced mutations in a human lymphoblastoid cell line was analyzed by restriction mapping at the X-linked hypoxanthine guanine phosphoribosyl transferase (*hgprt*) locus and at the heterozygous thymidine kinase (*tk*) locus. Restriction-fragment-length polymorphisms (RFLP) were used to identify the functional *tk* allele, as well as to determine the involvement of other loci on chromosome 17. A large fraction of spontaneous mutations recovered at the autosomal *tk* locus arose by loss of the entire *tk* gene; relatively few arose from point mutations. The fraction of induced mutations that involved allele loss varied with the inducing agent. The induced mutant fraction and distribution of molecular structural changes were similar for HGPRT mutants as for normal-growth-rate TK mutants. An additional class of slow-growth-rate mutants was recovered at the *tk* locus; these mutations were almost all associated with gene loss. Loss of the functional *tk* allele was often accompanied by loss of linked loci on chromosome 17. This was not due to nondisjunction and rarely in spontaneous mutants to detectable karyotypic changes. These results emphasize the importance of large-scale structural changes involving the entire gene or multiple genes in mutations in human cells.

INTRODUCTION

Mutations in mammalian cells may result from three general types of structural changes at the DNA level: (1) localized alterations in base sequence such as results from base-pair substitutions or frame shifts, (2) larger alterations within the target gene, including intragenic rearrangements and small deletions, and (3) large-scale changes, leading to loss of the entire gene. In addition to large deletions, loss of a functional gene may result from nondisjunction, mitotic recombination, or gene conversion.

There has been considerable recent interest in the structural analysis of both spontaneous and induced mutations in mammalian cells (Thacker 1985).

Banbury Report 28: Mammalian Cell Mutagenesis

In studies of the loss of functional enzyme activity at the adenine phosphoribosyl transferase (*aprt*) locus in Chinese hamster ovary (CHO) cells, for example, spontaneous mutants have been associated almost entirely with point mutations (Meuth and Arrand 1982; Nalbantoglu et al. 1983; Grosovsky et al. 1986). Sequencing of the *aprt* gene in these mutants has indicated a high frequency of base-pair substitutions. Similarly, analysis of mutants in a hemizygous dihydrofolate reductase (*dhfr*) gene has also shown a high frequency of single-base changes (Chasin et al., this volume). Structural analyses of mutations at the X-linked *hgprt* locus have indicated that 30–50% of spontaneous or X-ray-induced mutants are associated with detectable intragenic rearrangements, the remainder arising from small-scale changes involving less than 150–200 bp (Albertini et al. 1985; Turner et al. 1985; Thacker 1986; Liber et al. 1987).

These studies, however, all involve the analysis of mutations at hemizygous or X-linked loci, where only one copy of the gene is present. Such an experimental model facilitates the analysis of structural changes by restriction mapping on Southern blots, which would be complicated by the presence of the gene on the homologous chromosome. However, mutagenic mechanisms such as homologous mitotic recombination or gene conversion are not available at such loci. Furthermore, nondisjunction or large-scale deletions would likely be lethal events at hemizygous loci; many potential mutations arising from these processes may well go undetected. Previous attempts to examine recessive mutations at heterozygous or homozygous autosomal loci have provided findings that are difficult to interpret (Nalbantoglu et al. 1983; Simon et al. 1983).

The present investigation was therefore undertaken to analyze the molecular structural changes associated with spontaneous and induced mutations at an autosomal locus in human cells. These results are compared with those for mutations at the X-linked *hgprt* locus examined in parallel experiments with the same cell line. This cell line is heterozygous for the autosomal *tk* locus. Structural changes at the functional *tk* allele were examined by analysis of an RFLP located within the *tk* locus (Yandell et al. 1986). RFLPs located elsewhere in chromosome 17 were studied to estimate the extent of the structural changes associated with the mutant phenotype.

RESULTS

These experiments were all carried out with a human B-cell lymphoblastoid cell line designated TK-6 (Skopek et al. 1978; Liber and Thilly 1982). These cells grow rapidly in suspension and can be cloned with high efficiency in 96-well microtiter plates. This cell line is karyotypically stable, near diploid (Yandell and Little 1986), and contains one functional and one nonfunctional but unrearranged *tk* gene both in their native locations on chromosome 17. Treatment with mutagens resulted in a six- to tenfold increase in mutation frequency over spontaneous levels. The experiments were designed such that most mutant clones analyzed arose independently.

The methodology employed has been described in detail elsewhere (Yandell et al. 1986 and in prep.; Liber et al. 1987). Genomic DNA from mutant clones was cleaved with restriction enzymes, electrophoresed on agarose gels, transferred to nitrocellulose filters by the Southern technique, and hybridized directly with radioactively labeled probes. The *hgprt* probe, *tk* probe (ptk11) and *erbA*I oncogene probe (*cHerbA*I) are cDNA probes, whereas the *D17S2* is an anonymous probe. Informative RFLPs were identified at the latter three loci. These polymorphisms included a *Sac*I site for the *tk* locus, a *Xho*I site for the *erbA*I locus, and a *Pst*I site for the *D17S2* probe.

For analysis of HGPRT mutants, genomic DNA was digested with three different restriction enzymes: *Bam*HI, *Hin*dIII, and *Eco*RI. Representative Southern blot hybridization patterns of DNA from wild-type TK-6 cells as well as from four X-ray-induced mutant clones cut with *Hin*dIII are shown in Figure 1. No change in the *Hin*dIII restriction pattern nor in those for other enzymes occurred in mutant clones 2 and 4. This finding indicates that the mutation resulted from a small-scale change (< 200 bp). Mutant clones 1 and 3, on the other hand, showed significant changes in the restriction pattern from the wild-type cells manifested by loss of bands for one or more enzymes. This finding indicates that a large-scale change involving structural rearrangements of the *hgprt* gene has occurred. The DNA structural changes observed in 14 spontaneous and 28 X-ray-induced HGPRT mutant clones are summarized in Table 1. About one third of the spontaneous and over 50% of the X-ray-induced mutant clones were associated with large-scale changes; most of these involved loss of the entire gene.

Representative Southern blot hybridization patterns of genomic DNA from control ($TK^{+/-}$) and mutant ($TK^{-/-}$) clones digested with *Sac*I and hybridized to the human *tk* cDNA probe are shown in Figure 2. The DNA was digested with the *Sac*I restriction enzyme. Lanes 9 through 14 represent unselected control clones ($TK^{+/-}$), whereas the banding patterns shown in lanes 1 through 8 were derived from spontaneously arising TK mutants. The 14.8 and 8.4 kb-bands are members of a polymorphic pair and correspond to the functional and nonfunctional copies of the *tk* gene respectively. Changes in this banding pattern in mutant clones reflect three classes of structural alterations. First, loss of the *Sac*I site in the active allele, as manifested by loss of the 14.8 kb polymorphic band, with no change in the restriction pattern for a number of other enzymes that map the complete *tk* gene (lanes 3, 5–8), indicates that the mutant phenotype is associated with loss of the entire functional allele. Second, the appearance of new bands that may or may not

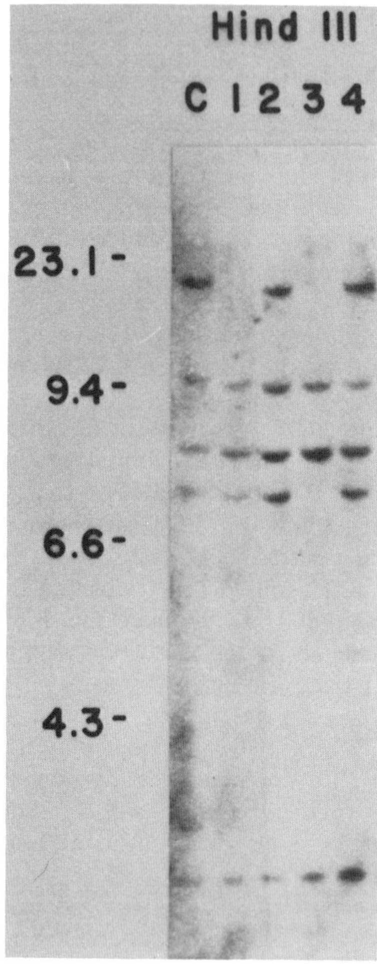

Figure 1

Southern blot hybridization patterns of genomic DNA from wild-type TK-6 cells (*C*) and four X-ray-induced mutant clones (*1–4*). The DNA was digested with *Hin*dIII and hybridized to a human *hgprt* cDNA probe. Molecular-weight markers on the left are in kilobases. The consistent 10.0-,8.0-, and 3.4-kb bands are from autosomal pseudogenes. Mutant clones in lanes *1* and *3* show loss of one or more of the bands associated with the X-linked *hgprt* gene (see text).

be accompanied by the loss of the 14.8-kb *Sac*I band (lanes 1 and 4) indicates that structural rearrangements have occurred within the functional *tk* gene. Third, mutants that show no detectable alterations in the banding pattern following digestion with *Sac*I or the other enzymes (lane 2) have resulted from point mutations; i.e., changes limited to a very small region of DNA (< 200 bp). Such small-scale changes may include base-pair substitutions and small deletions or additions.

The results of experiments in which 218 spontaneous and induced normal-growth-rate mutant clones as well as 25 unselected control clones were analyzed by this technique are presented in Table 2. As can be seen, all 25

Table 1
DNA Structural Changes Observed in Spontaneous and X-ray-induced HGPRT Mutant Clones

	Number	HGPRT enzyme activity	
		none	partial
Spontaneous mutants with restriction fragment alterations	5	5	0
Spontaneous mutants without restriction fragment alterations	9	5	4
Total	14	10	4
X-ray-induced mutants with restriction fragment alterations	15	15	0
X-ray-induced mutants without restriction fragment alterations	13	8	5
Total	28	23	5

Data from Liber et al. (1987).

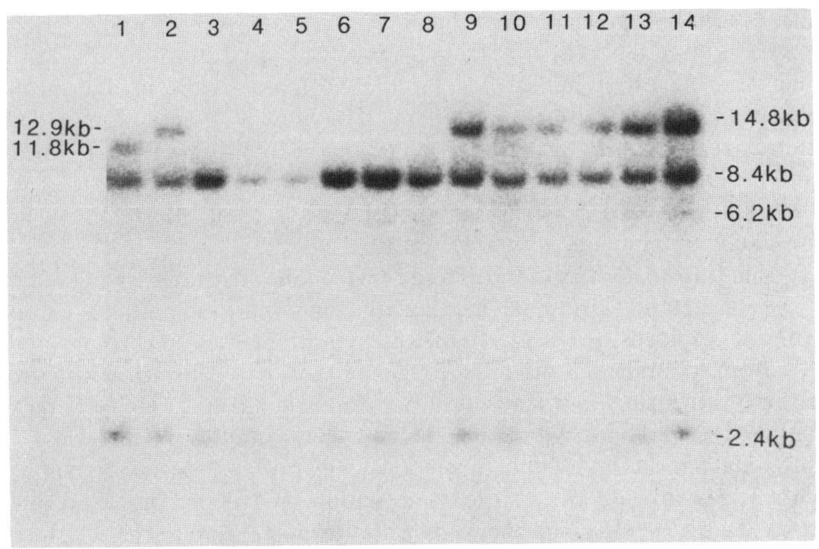

Figure 2
Hybridization patterns of genomic DNA from eight spontaneously arising ($TK^{-/-}$) mutant clones (*1–8*) and from six wild-type ($TK^{+/-}$) clones (*9–14*). The DNA was hybridized to a full-length human *tk* cDNA probe and digested with *Sac*I. The 14.8- and 8.4-kb bands are members of a polymorphic pair. The 14.8-kb band associated with the functional allele is lost in all mutant clones except that in lane *2*. In lanes *1* and *4*, new bands have appeared (11.8 and 12.9 kb, respectively), indicating an intragenic rearrangement has taken place (see text).

Table 2
Molecular Analysis of 421 TK$^{-/-}$ Mutant Clones and 25 Control Clones

	Number analyzed	Structural changes at *tk* locus: no change	rearranged	loss
Control clones (TK$^{+/-}$)	25	25	0	0(−)
Normal-growth mutants				
Spontaneously arising	51	10	5	36(71%)
X-ray-induced	56	18	3	35(63%)
UV-light-induced	20	10	1	9(43%)
Mitomycin-C-induced	67	45	4	18(27%)
EMS-induced	24	20	2	2(8%)
Slow-growth mutants				
Spontaneously arising	120	4	1	115(96%)
X-ray-induced	22	1	1	20(91%)
Mitomycin-C-induced	37	1	0	36(97%)
EMS-induced	24	2	0	22(92%)

control clones showed no structural changes at the *tk* locus. The results in Table 2 indicate that a large fraction of spontaneously arising mutants, as well as those induced by X-rays or UV light, involve loss of the entire functional *tk* allele. On the other hand, most of the normal-growth-rate mitomycin-C mutants and probably all of those induced by ethyl methanesulfonate (EMS) resulted from changes too small to be detected (point mutations). Since the induced-mutant fraction following EMS treatment was only six-fold above background, the two clones showing allele loss in the EMS treatment group were likely spontaneous mutants.

To gain information as to the extent to which the structural changes leading to loss of heterozygosity in spontaneous and induced mutants extended beyond the *tk* locus, RFLPs were identified at other sites on chromosome 17. DNA from spontaneous and induced TK mutants was hybridized with probes for these other sites and restriction patterns studied in order to determine whether loss of TK activity was associated with multilocus allele loss. The approximate locations of the chromosome-17 probes utilized are shown in Figure 3. Recent data (J.B. Little et al., unpubl.) have localized the sequence recognized by the *D17S2* probe to the short arm of chromosome 17, although this result requires confirmation. Thus, the *erbA*I and the *tk* loci are linked on the long arm of chromosome 17; involvement of both loci would suggest a large deletion or region of mitotic recombination had occurred. Involvement of the short-arm marker would suggest a nondisjunctional event.

The results of experiments in which 255 TK mutant clones were analyzed at the *erbA*I locus are shown in Table 3. The mutant clones are divided into two

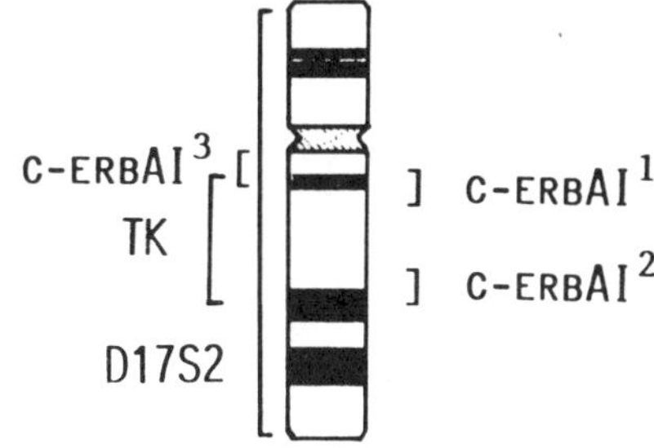

Figure 3
Approximate locations of the *tk*, *erbA*I, and *D17S2* loci on human chromosome 17. Recent unpublished data suggest that the *D17S2* locus is probably located on the short arm.

groups. The first includes those in which the Southern-banding patterns following digestion with *Xho*I were unchanged from those of wild-type cells. The second group showed loss of the *Xho*I polymorphism, indicating loss of the *erbA*I allele. Presumably, loss of the linked *tk* and *erbA*I genes in a TK mutant clone indicates that a large-scale multilocus change has taken place. As shown in Table 3, 25–50% of the mutant clones associated with loss of the entire *tk* allele also showed loss of the *erbA*I gene, indicating a multilocus change had occurred.

Of the more than 300 clones analyzed for loss of heterozygosity at the locus that hybridized to the *D17S2* probe, 6 showed allele loss. Two of these (which arose in a single experiment and may represent a single event) showed no change at the *tk* and *erbA*I alleles, whereas both of these loci were lost in the remaining mutants. Karyotypic analysis of one of these revealed two apparently normal chromosome-17 homologs, suggesting that nondisjunction and reduplication of the chromosome carrying the nonfunctional gene may have occurred. However, the significance of these findings awaits karyotypic analysis of the remainder of these clones.

Preliminary karyotyping has been carried out on 18 spontaneous TK mutant clones associated with loss of the *tk* locus. Both copies of chromosome 17 were present in all clones studied. In 16 of the 18 clones, chromosome 17 was karyotypically normal. In the remaining two, presumptive evidence for small deletions involving the long arm of chromosome 17 was found.

During the course of these experiments, it became evident that two different classes of mutant clones occur at the *tk* locus, in terms of their phenotypic characteristics. Normal-growth-rate mutants were scored in microtiter dishes 12 days after seeding. When serially passaged, the doubling time of these mutant cells was similar to that of parental TK-6 cells. A separate class of slow-growth-rate mutants was scored 20 days after seeding; these mutant cells retained a stably prolonged doubling time over many generations of replication in vitro. The mean generation time of normal-growth-rate mutants during serial passaging in culture was 15.3 ± 1.0 hour, whereas the doubling time for slow-growth-rate mutants was 33.4 ± 4.3 hours. There was no over-

Table 3
Structural Changes Observed in Thymidine Kinase Mutants at *tk* and *erbA*I Loci

Phenotype	Inducing agent	Number of clones analyzed	*tk/erbA*I[a]			
			+/+	+/−	−/+	−/−
Mutant $TK^{-/-}$						
Normal growth rate	Spontaneously arising	40	10	2	17	11
	Mitomycin-C	43	29	2	9	3
	X-ray	27	8	1	11	7
	UV-Light	15	6	3	4	2
Slow growth rate	Spontaneously arising	98	3	0	45	50
	Mitomycin-C	32	1	0	17	14

Data from D.W. Yandell et al. (in prep.).
[a] + or − before / indicates presence (+) or loss (−) of active *tk* allele (+ category includes no change or intragenic structural rearrangements, see Table 1). + or − after / indicates presence or loss of active *erbA*I allele.

lap in the distribution of doubling times among mutant clones in these two classes; thus, the two phenotypes are easily distinguishable. Approximately 70% of spontaneous and 80% of mutant clones induced by 150 rads of X-rays were of the slow-growth-rate type, whereas a smaller fraction of EMS-induced mutants were of this class. No slow-growth-rate mutants were observed at the *hgprt* locus; the induced-mutant fractions at this locus were very similar to those found for normal-growth-rate mutations at the *tk* locus.

Results on the structural analysis of slow growth-rate TK mutants are also shown in Tables 2 and 3. As can be seen in Table 2, 90% or more of spontaneous slow growth-rate mutants as well as of those induced by all of the agents studied involved loss of the entire *tk* allele. One half of these were multilocus events, as evidenced by concomitant loss of the linked *erbA*I gene (Table 3). Most striking are the results with EMS. Most and probably all normal-growth rate EMS-induced mutants resulted from small-scale intragenic changes (probably point mutations), whereas 90% of EMS-induced slow-growth-rate mutants showed loss of the entire *tk* gene. The structural analysis of X-ray- and EMS-induced mutations at the *tk* and *hgprt* loci are compared in Table 4. Most if not all HGPRT and normal-growth-rate TK mutants induced by EMS involved point mutations, in contradistinction to the high frequency of large-scale changes seen in slow-growth-rate TK mutants. These findings suggest that different mutational mechanisms are responsible for the two classes of mutations at the heterozygous autosomal *tk* locus.

DISCUSSION

Nearly 90% of all spontaneous mutant clones recovered at the autosomal *tk* locus were associated with loss of the entire *tk* allele. Nearly 50% of these involved multilocus changes, as evidenced by loss of another locus on the long arm of chromosome 17. This finding is in contradistinction to the observation that only about 35% of spontaneous mutations at the X-linked *hgprt* locus involved detectable structural alterations in the gene (Table 1) and the

Table 4
X-ray- and EMS-Induced Mutants Showing Structural Rearrangements or Allele Loss on Southern Blots

Inducing agent	*hgprt*	*tk* normal growth	*tk* slow growth
X-ray[a]	15/28(54%)	20/26(77%)	21/22(95%)
EMS[b]	2/17(12%)	4/24(17%)	22/24(92%)

[a]Dose 150 rads; mutant fraction ten-fold over spontaneous frequency.
[b]Dose 150 μM for 4 hr; mutant fraction six-fold increase over spontaneous frequency.

findings of a much lower frequency of such large-scale changes at other hemizygous loci (Meuth and Arrand 1982; Nalbantoglu et al. 1983; Grosovsky et al. 1986; Chasin et al., this volume). These results suggest that a class or classes of mutations may be recovered at heterozygous autosomal loci that are not seen at hemizygous or X-linked loci. One such class might be a large multilocus deletion, involving genes required for cell growth and survival. Presumably, such genes on the homologous chromosome would remain intact at heterozygous loci, furnishing the critical gene product, although in reduced dosage. In addition, however, other mutational mechanisms are available at autosomal loci and not at hemizgyous loci. These include homologous mitotic recombination and gene conversion. Thus, many spontaneous and induced mutants may be lethal at hemizygous or X-linked loci, either because of large-scale deletions involving many genes or because "repair" mechanisms involving homologous chromosomes cannot operate.

The observations that the short-arm marker was rarely involved in mutant TK clones, and that recognizable karyotypic changes were uncommon in mutant clones associated with loss of the *tk* gene suggest that mutlilocus changes as identified by molecular structural analyses are infrequently associated with nondisjunction or cytogenetically detectable deletions or rearrangements. These findings lead us to hypothesize that these multilocus changes, particularly those associated with slow-growth-rate mutants, may result from mitotic recombination. This hypothesis is supported by the dichotomy in the spectrum of changes observed in the two classes of EMS-induced TK mutants.

The cellular and cytogenetic analyses of TK mutants in the L5178Y mouse lymphoma system have provided results consistent with our findings that essentially all slow-growth-rate mutants at the *tk* locus arose by allele loss (Moore et al. 1985; Applegate and Hozier, this volume). However, cytogenetic analysis of these mutants showed a strong correlation between visible cytogenetic damage and colony-size classification (Hozier et al. 1985; Moore et al. 1985), whereas our cytogenetic data showed no apparent relationship between the slow-growth phenotype and the visible chromosomal alterations. Consistent with our results, Evans et al. (1986) found that more mutations were induced at the heterozygous *tk* locus than at the X-linked *hgprt* locus. Studies at the *aprt* locus in rodent cells also indicate that complex genetic events are involved in the expression of recessive mutations at heterozygous loci (Bradley and Letovanec 1982; Adair et al. 1983; Simon et al. 1983).

In summary, the results described in this paper emphasize the importance of allele loss presumably by mitotic chromosomal mechanisms in mutagenesis at autosomal loci. They further suggest that in vitro models for recessive somatic mutation that are based at homozygous or hemizygous loci, although particularly useful in the analysis of changes in DNA sequence or small deletions, may ignore a large category of genetically important events. A significant fraction of spontaneous mutant clones as well as those induced by all mutagens tested involved large-scale genetic alterations. The mechanism for these large-scale changes, in particular whether they involve simple deletions or mitotic recombinational events, remains to be elucidated.

ACKNOWLEDGMENT

This research was supported by grants CA-11751, CA-34037, and ES-00002 from the U.S. National Institutes of Health.

REFERENCES

Adair, G.M., R.L. Stallings, R.S. Nairn, and M.J. Siciliano. 1983. High-frequency structural gene deletion as the basis for functional hemizygosity of the adenine phosphoribosyltransferase locus in Chinese hamster cells. *Proc. Natl. Acad. Sci.* **80:** 5961.

Albertini, R.A., J.P. O'Neill, J.A. Nicklas, N.H. Heintz, and P.C. Kelleher. 1985. Alterations of the *hprt* gene in human in vivo-derived 6-thioguanine resistant T-lymphocytes. *Nature* **316:** 369.

Bradley, W.E.C. and D. Letovanec. 1982. High-frequency nonrandom mutational event at the adenine phosphoribosyltransferase (*aprt*) locus of sib-selected CHO variants heterozygous for *aprt*. *Somatic Cell Genet.* **8:** 51.

Evans, H.H., J. Mencl, M.-F. Horng, M. Ricanti, C. Sanchez, and J. Hozier. 1986. Locus specificity in the mutability of L5178Y mouse lymphoma cells: The role of multilocus lesions. *Proc. Natl. Acad. Sci.* **83:** 4379.

Grosovsky, A.J., E.A. Drobetsky, P.J. deJong, and B.W. Glickman. 1986. Southern analysis of genomic alterations in gamma-ray-induced *aprt*$^-$ hamster cell mutants. *Genetics* **113:** 405.

Hozier, J., J. Sawyer, D. Clive, and M. Moore. 1985. Chromosome 11 aberrations in small colony L5178Y/tk$^{+/-}$ mutants early in their clonal history. *Mutat. Res.* **147:** 237.

Liber, H.L. and W.G. Thilly. 1982. Mutation assay at the thymidine kinase locus in diploid human lymphoblasts. *Mutat. Res.* **94:** 467.

Liber, H.L., K.M. Call, and J.B. Little. 1987. Molecular and biochemical analyses of spontaneous and X-ray-induced mutants in human lymphoblastoid cells. *Mutat. Res.* **178:** 143.

Meuth, M. and J.E. Arrand. 1982. Alterations of gene structure in ethyl methane sulfonate-induced mutants of mammalian cells. *Mol. Cell. Biol.* **2:** 1459.

Moore, M.M., D. Clive, B.E. Howard, G.A. Batson, and N.T. Turner. 1985. In situ analysis of trifluorothymidine-resistant (TFT^r) mutants of mouse lymphoma cells. *Mutat. Res.* **151:** 147.

Nalbantoglu, J., O. Goncalves, and M. Meuth. 1983. Structure of mutant alleles at the *aprt* locus of Chinese hamster ovary cells. *J. Mol. Biol.* **167:** 575.

Skopek, T.R., H.L. Liber, B.W. Penman, and W.G. Thilly. 1978. Isolation of a human lymphoblastoid line heterozygous at the thymidine kinase locus: Possibility for a rapid human cell mutation assay. *Biochem. Biophys. Res. Commun.* **84:** 411.

Simon, A.E., M.W. Taylor, and W.E.C. Bradley. 1983. Mechanism of mutation at the aprt locus in chinese hamster ovary cells: Analysis of heterozygotes and hemizygotes. *Mol. Cell. Biol.* **3:** 1703.

Thacker, J. 1985. The molecular nature of mutations in cultured mammalian cells: A review. *Mutat. Res.* **150:** 431.

———. 1986. The nature of mutants induced by ionizing radiation in cultured hamster cells. III. Molecular characterization of HPRT deficient mutants induced by gamma-rays or alpha-particles showing that the majority have deletions of all or part of the *hprt* gene. *Mutat. Res.* **160:** 267.

Turner, D.R., A.A. Morley, M. Haliandros, R. Kutlaca, and B.J. Sanderson. 1985. In vivo somatic mutants in human lymphocytes frequently result from major gene alterations. *Nature* **315:** 343.

Yandell, D.W. and J.B. Little. 1986. Chromosome 14 marker appearance in a human B lymphoblastoid cell line of nonmalignant origin. *Cancer Genet. Cytogenet.* **20:** 231.

Yandell, D.W., T.P. Dryja, and J.B. Little. 1986. Somatic mutations at a heterozygous autosomal locus in human cells occur more frequently by allele loss than by intragenic structural alterations. *Somatic Cell Mol. Genet.* **12:** 255.

Molecular Analysis of Human *hprt* Mutations

RICHARD A. GIBBS AND C. THOMAS CASKEY
Baylor College of Medicine and
Howard Hughes Medical Institute
Houston, Texas 77030

OVERVIEW

Procedures for the rapid detection of point mutations can greatly improve both the diagnosis of human genetic diseases and the characterization of somatic mutations in mammalian cells. The ribonuclease-A (RNase-A)-cleavage technique has been employed for the molecular diagnosis of mutations at the human hypoxanthine phosphoribosyl transferase (*hprt*) locus that lead to the Lesch-Nyhan (LN) syndrome. RNase-A cleavage and denaturing gradient gel electrophoresis, in combination with the polymerase chain reaction (PCR) procedure for the amplification of specific nucleotide sequences, offer the potential for detection of all possible point mutations.

INTRODUCTION

DNA probes have now been obtained for all the commonly employed selectable markers in mammalian cell mutagenesis test systems, and there are a large number of reports describing Southern blot analysis (Southern 1975) of spontaneous and induced mutations at these loci. These studies have confirmed many predictions of the molecular basis of mutations, which stem from earlier observations made at the protein and cellular levels. For example, ionizing radiation frequently induces gross changes in gene structure and yields substantial alterations in Southern blot patterns (Thacker 1986; Gibbs et al. 1987). Conversely, mutagens that are known to produce mainly single-base changes rarely generate Southern blot changes (Fuscoe et al. 1983; Stankowski et al. 1986). A further dissection of the spectrum of mutations (mutational spectra) that are induced by different mutagens is difficult to achieve by the Southern blot technique as restriction-enzyme-fragment-length alterations of less than 50–200 bp are difficult to resolve. Consequently, new procedures have been developed to facilitate the rapid detection of single nucleotide changes. These techniques are RNase-A cleavage and denaturing gradient gel electrophoresis, which can detect single-base changes without prior knowledge of the precise location of the mutations. We have employed RNase-A cleavage to diagnose the molecular defects in LN patients (a human, total HPRT-deficiency syndrome), and we are currently developing strategies to utilize denaturing gradient gel electrophoresis and a procedure for the amplification of DNA sequences in order to improve the sensitivity of assays for point mutations. The potential for pulsed-field gel electrophoresis to improve mutational assays by resolving very large DNA changes, in the range of 20–1000 kb, has been illustrated elsewhere and will not be considered here (Burmeister and Lehrach 1986; Kenwrick et al. 1987).

DISCUSSION

RNase-A-cleavage Analysis of LN Patients

Myers et al. (1985a) and Winter et al. (1985) first described the RNase-A-cleavage procedure that enables identification of single-base mismatches in RNA:DNA and RNA:RNA heteroduplexes, respectively. In each case, a radiolabeled RNA probe is hybridized to a complementary RNA or DNA strand and treated with RNase A to digest single-stranded regions of RNA and some internal mismatches. Following digestion, the RNase-A-cleavage products are analyzed on denaturing polyacrylamide-7 M urea gels, and from the length of the protected RNA fragments, the position of internal cleavage sites can be determined. Overlapping RNA probes allow the unambiguous localization of the RNase-A-sensitive sites, relative to the ends of the duplex. The efficiency of the cleavage of single-base mismatch sites varies from 0% to 100%, depending on the individual bases involved and, to a lesser extent, the sequence context of the mismatch. Overall, between one third and one half of all possible single-base changes will generate RNase-A-sensitive sites. More substantial mismatches, resulting from deletions or rearrangements, offer greater potential for cleavage because of the more extensive single-stranded regions in the heteroduplexes. The RNase-A-cleavage procedure thus provides an assay for exact complementarity of an RNA probe with a particular nucleotide sequence.

The LN syndrome is a severe human neurological defect that is correlated with a total deficiency of HPRT (Lesch and Nyhan 1964; Kelley et al. 1967). Many LN cases arise as a result of de novo mutations within the *hprt* locus (Haldane 1935). Approximately 15% of LN patients reveal altered *hprt* gene structures by Southern blot analysis, including partial and complete gene deletions, intragenic duplications, and rearrangements. The remaining 85% of LN cases appear normal by Southern blot analysis and, presumably, are the result of point mutations, small deletions, or rearrangements that inactivate the *hprt* gene (Yang et al. 1984). The predominance of "small" lesions and the heterogeneity of the mutations make the molecular analysis of the LN syndrome a challenge for which the RNase-A-cleavage assay is eminently suited.

The human *hprt* gene is comprised of 9 exons dispersed over 44 kb of genomic DNA, whereas the HPRT-peptide-coding region is represented by only 651 bp (Brennand et al. 1983; Patel et al. 1986). Consequently, the *hprt* mRNA provides a convenient target for RNA probing. However, since *hprt* mRNA is present in a low abundance, 0.01% of the poly(A)$^+$ (Melton et al. 1981), it is difficult to detect in total cellular RNA. We have adapted commercially available polyuridylic acid (poly[U])-bound affinity paper (mAP: Amersham-Searle) (Wreschner and Herzberg 1984) to improve the sensitivity of the *hprt* probing, as shown in Figure 1. Total cellular RNA is prepared by a standard guanidinium isothiocyanate procedure (Chirgwin et al. 1979). The RNA is hybridized to a radiolabeled, antisense *hprt* RNA probe generated from *hprt* cDNA fragments that are cloned into in vitro transcription vectors. The hybrids are then partially purified by passage through the poly(U) affinity paper, which binds the polyadenylated fraction of the cellular RNA but does not retain ribosomal RNA, transfer RNA, or any radiolabeled RNA probe that is not hybridized to an HPRT message. The filters are washed in 0.5 M NaCl to remove any nonspecifically bound contaminants, after which the probe:HPRT hybrids are eluted under conditions that will melt the poly(U):poly(A) hybrid regions but which leave the more extensively paired probe:HPRT duplexes intact. Subsequently, the hybrids are digested with RNase-A and analyzed by gel electrophoresis, essentially as described by Winter et al. (1985).

The use of the poly(U) paper step improves the clarity of the signal from the *hprt* mRNA and eliminates the background RNA "ladders" that are often seen in RNA protection procedures. There is no need to purify the RNA probes by gel electrophoresis prior to the assay. Any DNA template that is not removed following the in vitro transcription reactions to prepare probes

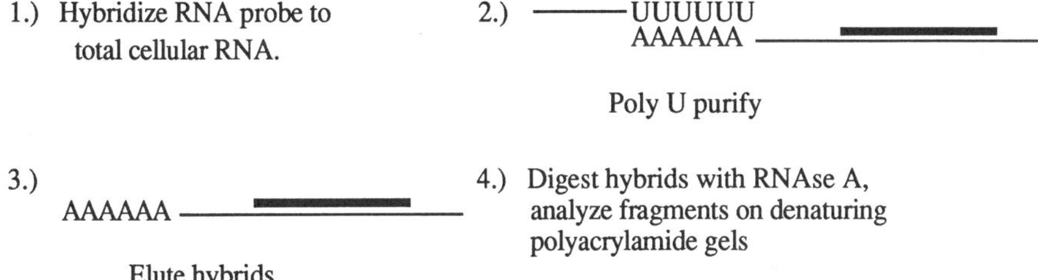

Figure 1

Flow diagram for the RNase-A-cleavage assay. (*1*) Antisense, radioactive RNA probe is incubated with total cellular RNA in 80% formamide and 0.4 M NaCl to enable the formation of specific RNA:RNA hybrids. (*2*) The mix is passed through poly(U)-bound affinity paper, which retains mRNA, including mRNA bound specifically to the radioactive probe. Nonspecifically bound contaminants are washed free from the paper, then (*3*) the hybrids are eluted. (*4*) The hybrids are digested with RNase A to remove single-stranded regions and digest some internal mismatches, and the radiolabeled probe fragments are analyzed on denaturing polyacrylamide gels.

will not give spurious RNA protection, since the RNA:DNA hybrids will not be retained by the poly(U) paper. The only disadvantages of poly(U) paper chromatography appear to be that extra time is required (~ 1 hr per assay), and some signal is lost during the elution process. Approximately 50 μg of total cellular RNA is the smallest amount of starting material used in this procedure to detect *hprt* mRNA.

To date, RNase-A-cleavage analysis has identified the known single-base mismatch positions in *hprt* mRNA from a LN patient, $\text{HPRT}_{\text{Kinston}}$ (Gibbs and Caskey 1987), and a partial HPRT-deficiency case, $\text{HPRT}_{\text{London}}$ (Wilson et al. 1983) (Fig. 2). We have also identified cleavage of a G:A mismatch in hybrids formed between an RNA probe and murine ornithine transcarbamylase (*otc*) mRNA (Veres et al. 1987). In each of these cases, the efficiency of cleavage was less than 100%. In contrast, Winter et al. (1985) observed 100% cleavage of U:C, A:G, and A:A mismatches in RNA:RNA duplexes. The different cleavage efficiencies may be due to the sequence context of the mutations, and the extensive study of Myers et al. (1985a) has identified individual mismatches in RNA:DNA hybrids that vary in cleavage efficiency from 0% to 100%, depending on the neighboring nucleotides. The sequence-dependent context effects may extend beyond the immediate neighbors to the mismatch sites, and so, many sequence combinations will need to be analyzed to derive a precise set of consensus-sequence cleavage rules.

RNase-A-cleavage sites have been identified in 5 unrelated LN cases among 14 individuals who had normal HPRT Southern blot patterns. In each of the five cases, the RNase-A-cleavage sites were different, corroborating the earlier observation from Southern blot analysis that LN mutations are heterogeneous (Yang et al. 1984). The RNA fragment lengths show that three of these cases are either point mutations or very small deletions (<5 bp), whereas the remaining two cases have deletions of approximately 40 and 85 bases in the RNA (Fig. 3).

The LN cases in whom RNase-A-sensitive sites were not identified are most likely to have point mutations in the HPRT-coding region that do not give mismatches that are sensitive to RNase-A digestion. It is also possible that these events are the result of insertion mutations, which do not always permit cleavage of the radiolabeled probe (P. Patel, in prep.), or mutations outside the HPRT-coding region, thereby affecting the ultimate expression of *hprt* mRNA.

The identification of RNase-A-cleavage sites can lead to improved molecular diagnosis for members of LN patients' families. We are currently sequencing the RNase-A-sensitive sites from the *hprt* mRNA in the five patients described here, using the PCR procedure (see below) to enrich for the sequences of interest. The nucleotide-sequence information will then be used to design a strategy to identify directly the nucleotide alterations within the

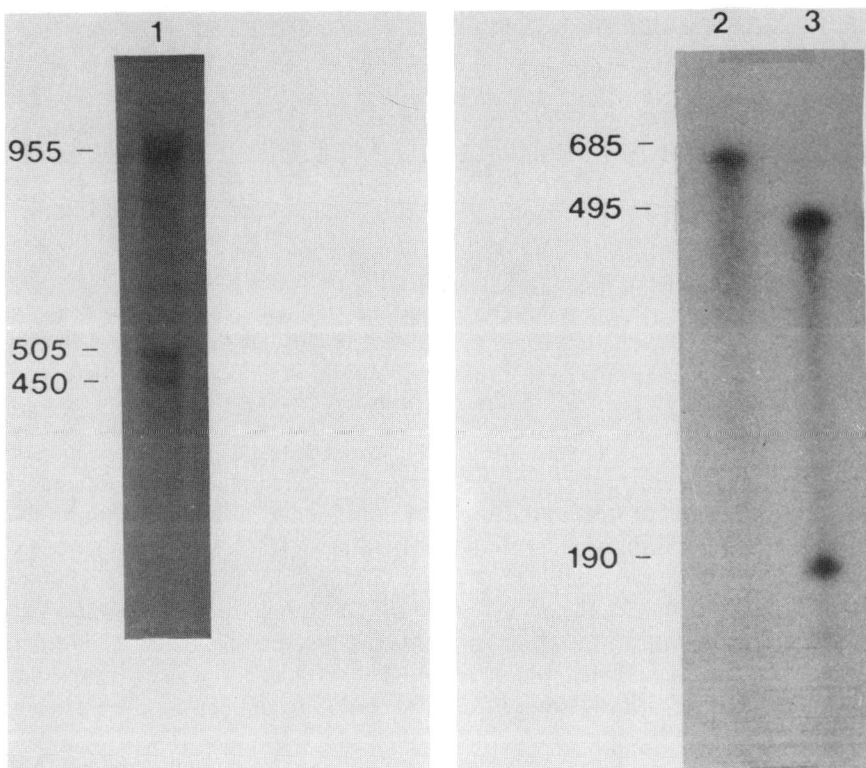

Figure 2

RNase-A-cleavage analysis of human *hprt* mRNAs. The mRNA samples are from $HPRT_{London}$ (*1*), a partial HPRT-deficiency case (Wilson et al. 1983), and from RJK 896 (*2*) and RJK 894 (*3*), two LN patients (Gibbs and Caskey 1987). The samples were analyzed according to the methods outlined in Figure 1. The 955 base fragment (lane *1*) corresponds to full-length protection of an RNA probe that is complementary to the 651-base-HPRT-peptide-coding region and approximately 300 bases of flanking sequence. From cleavage of a G:U mismatch within the RNA:RNA hybrids, 505 and 450 base fragments are produced. The 685-base fragment (lane *2*) indicates protection of an RNA probe complementary to approximately 500 bases of the HPRT-peptide-coding region and 185 bases of the 3′-flanking region. Two bands (lane *3*) were generated from internal cleavage of this probe when hybridized to RNA from RJK 894.

DNA of the LN patients' family members. For example, if an RNase-A-sensitive site is caused by a nucleotide change that alters a restriction endonuclease recognition site, then a standard Southern blot analysis can be used to identify carriers among the relatives. Alternatively, an allele-specific oligonucleotide (ASO) can be synthesized and then used to discriminate the normal DNA sequence from the mutant by differential hybridization under stringent washing conditions (Kidd et al. 1983).

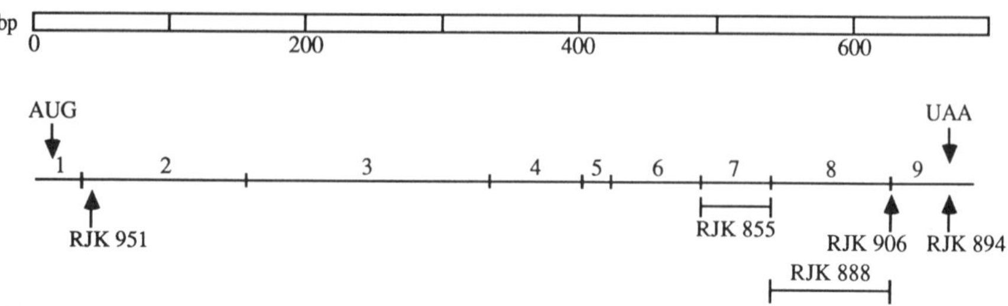

Figure 3

Mutations identified in human *hprt* mRNA by the RNase-A-cleavage procedure. The RNase-A-sensitive sites in the *hprt* mRNA of five LN patients were mapped using the procedure outlined in Figure 1.

The variation of the RNase-A-cleavage procedure described here can be used to study somatic mutations and to improve mammalian cell mutagenesis assays. Collections of *hprt* mutations can be screened to identify mutational "hot spots," to determine the relative frequency of RNA-splicing mutations, and to discriminate those molecular lesions induced by mutagens that cannot be distinguished by Southern blot analysis. A detailed characterization of "spontaneous" mutational spectra in cultured cells is of particular interest, since weak mutagens may be identified by their characteristic molecular lesions, even if quantitative measures of mutation frequency cannot be easily made.

Denaturing Gradient Gel Electrophoresis and the PCR

Mullis and Faloona (1987) have described a PCR procedure for the amplification of specific nucleotide sequences from a complex mixture. First, double-stranded DNA is thermally denatured, and a pair of synthetic oligonucleotide primers that straddle the DNA sequences of interest are hybridized and extended with the large fragment of *Escherichia coli* DNA polymerase I (Klenow). The two primers are chosen so that the extension product of each is able to serve as a template for the other primer in subsequent rounds of denaturation, annealing, and polymerization. Thus, after many reaction cycles, a number of copies of the DNA sequence between the two primers are produced. Important features of this procedure are (1) that the total amount of the target sequence increases exponentially with the number of PCR rounds performed, (2) that the reaction has a high degree of specificity due to the need to coordinate two independent priming and extension events, and (3) that after several rounds of PCR, a reaction product with defined 5′ and 3′ ends will predominate. These fragments have discrete lengths because they result from primed synthesis using a product of a previous PCR cycle as a template. The power of the PCR procedure is quite extraordinary, and unique mammalian DNA sequences can be amplified 2×10^5-fold by just 20 reaction cycles in less than 2 hours, using total genomic DNA as starting material. The PCR procedure has been employed elsewhere for the enrichment of DNA sequences from β-globin genes to facilitate mutation detection by ASO probing and to enable parts of the β-globin gene to be rapidly sequenced by circumventing the more conventional λ-cloning and screening procedures (Saiki et al. 1985, 1986). We have used PCR to amplify specific nucleotide sequences from the mRNA of the mouse *otc* gene, consequently enabling us to identify the precise nucleotide change associated with an RNase-A-cleavage site (Veres et al. 1987). Currently, we are employing a series of oligonucleotide primers to simplify the cloning of regions of *hprt* cDNA by PCR (Fig. 4).

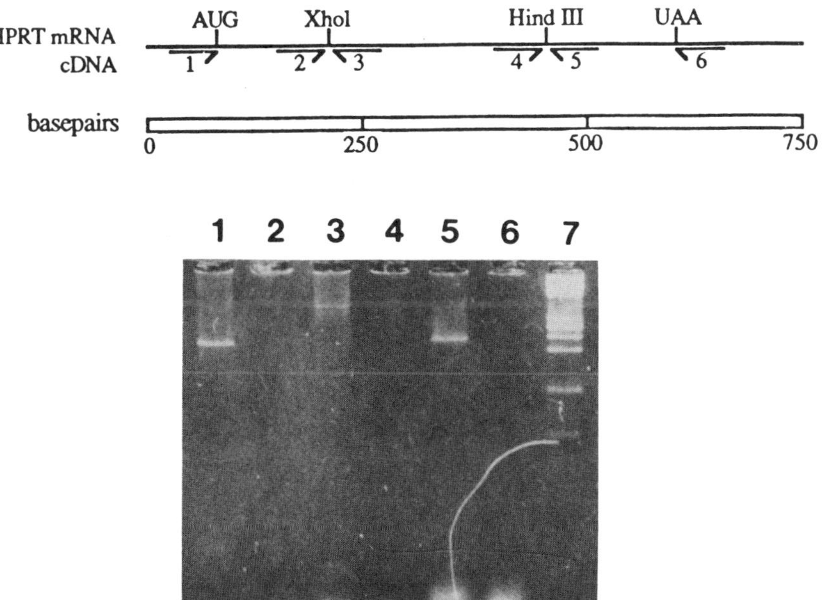

Figure 4

Strategy for rapid cloning of *hprt* mRNA via the PCR procedure. Six synthetic oligonucleotide primers, complementary to the *hprt* cDNA, have been prepared. Primer 1 has an *Eco*RI-restriction-endonuclease recognition site included as a tail on the 5′ end. Primers 2–5 flank natural 6-base restriction-endonuclease recognition sites, and primer 6 contains a single-base mismatch to the *hprt* cDNA sequence, which gives rise to an *Eco*RI-restriction-endonuclease recognition site. Primers 1:3, 2:5, and 4:6 each allow a portion of the *hprt* cDNA to be amplified by PCR (see text) and then cloned into vectors for DNA sequencing. Amplification products from ten-PCR cycles, using the three primer sets of *hprt* cDNA, have been analyzed by electrophoresis on a 4% NuSieve agarose gel (SeaKem). The starting material in each reaction was 200 ng of cloned *hprt* cDNA. Lanes *1*, *3*, and *5* show discrete bands representing the PCR products of primers 4:6, 2:5, and 1:3, respectively. Lanes *2*, *4*, and *6* are samples before PCR. Each lane represents 10% of the total reaction mixture. Lane *7* is *Hae*III-digested ϕX174 DNA.

The PCR procedure is well suited to act in concert with other techniques for mutation analysis. Denaturing gradient gel electrophoresis is a procedure that enables the resolution of single-DNA-base changes by exploiting the difference in melting temperatures (T_ms) of closely related DNA duplexes. When electrophoresed through a gradient of increasing concentrations of formamide and urea, the mobility of a particular DNA fragment decreases abruptly as the duplex begins to denature to single strands. The T_m is highly dependent on DNA sequence composition; hence, the precise position in the gel where the mobility decreases is dependent on the DNA base sequence. Potentially, all DNA-base-sequence changes can be detected by this procedure; however, the precise location of the base substitution within the duplex can influence ultimate mobility changes (Myers et al. 1985d).

Myers et al. (1985b,c) have demonstrated that when a DNA duplex is linked to a highly GC-rich DNA sequence prior to electrophoresis, the likelihood of detecting point mutations is dramatically increased. This is because the observed decrease in electrophoretic mobility is dependent on the formation of partially denatured molecules, which in turn depends on the presence of domains of different T_m levels within a single duplex. The GC-rich "clamps" provide a high-melting domain, so that all the rest of any duplex will fall in relatively low-melting regions, and single-DNA-base changes that alter the T_m level of the low-melting domains will change the electrophoretic properties of the molecule. The GC clamps can be introduced by a DNA cloning step, utilizing a cloning vector with a naturally occurring GC-rich region, but this has the disadvantage of being no less time consuming than conventional approaches to obtain DNA-sequence information, and so it is not suitable for the rapid screening of mutations directly from mammalian cells. As an alternative, we are now pursuing a strategy of incorporating a poly(C) "tail" on the 5′ terminus of one of the oligonucleotides used in the PCR amplification reaction. After several reaction cycles, the sequences between the PCR primers will be both amplified and covalently linked to a 100% GC-rich region, providing the highest possible melting domain as a clamp for the denaturing gradient gel procedure. A careful selection of oligonucleotide primers can thus allow unique mammalian DNA sequences to be amplified for denaturing gradient gel electrophoresis and offers a maximum likelihood of detection of point mutations in these regions.

SUMMARY

The RNase-A cleavage assay has been employed to detect human *hprt* mutations that could not be identified by Southern blot analysis. Some of these mutations had previously been identified as single-base substitutions, which demonstrates that the assay has the power to detect *hprt* point mutations. The RNase-A cleavage assay has increased the fraction of LN patients that can be diagnosed at the molecular level from 15% to 50%. The demonstration of the utility of the RNase-A-cleavage procedure is an important development in mutation screening, and it is likely that RNase-A-cleavage analysis of somatic mutations will lead to a greater refinement of the classification of the mutational spectra of many environmental mutagens.

Further strategies for detecting point mutations are being developed. Currently, denaturing gradient gel electrophoresis offers the greatest potential for the rapid detection of all point mutations within a defined region of a mammalian cell genome. The PCR procedure for amplification of DNA sequences offers the possibility for simultaneous enrichment of total cellular DNA preparations for specific target sequences and the introduction of GC-rich clamps to increase the likelihood of producing an electrophoretically detectable variation.

ACKNOWLEDGMENTS

This work was supported in part by grant number DK31428 from the National Institutes of Health. R.A.G. is supported by a postdoctoral fellowship from the Muscular Dystrophy Association, and C.T.C. is an investigator with the Howard Hughes Medical Institute. We thank K. Nguyen for technical assistance. Special thanks to Grant MacGregor for useful comments on the manuscript and to David Adcock for artwork.

REFERENCES

Brennand, J., D.S. Konecki, and C.T. Caskey. 1983. Expression of human and Chinese hamster hypoxanthine-guanine phosphoribosyltransferase cDNA recombinants in cultured Lesch-Nyhan and Chinese hamster fibroblasts. *Proc. Natl. Acad. Sci.* **79:** 1950.

Burmeister, M. and H. Lehrach. 1986. Long-range restriction mapping around the Duchenne muscular dystrophy gene. *Nature* **324:** 582.

Chirgwin, J.M., A.E. Przybyla, R.J. MacDonald, and W.J. Rutter. 1979. Isolation of biologically active ribonucleic acid from sources enriched in ribonuclease. *Biochemistry* **18:** 5294.

Fuscoe, J.C., R.G. Fenwick, D.H. Ledbetter, and C.T. Caskey. 1983. Deletion and amplification of the HGPRT locus in Chinese hamster cells. *Mol. Cell. Biol.* **3:** 1086.

Gibbs, R.A. and C.T. Caskey. 1987. Identification and localization of mutations at the Lesch-Nyhan locus by ribonuclease A cleavage. *Science* **236:** 303.

Gibbs, R.A., J. Camakaris, G.S. Hodgson, and R.F. Martin. 1987. Molecular characterization of ^{125}I decay and X-ray-induced HPRT mutants in CHO cells. *Int. J. Radiat. Biol.* **51:** 193.

Haldane, J.B.S. 1935. The rate of spontaneous mutation of a human gene. *J. Genet.* **31:** 317.

Kelley, W.N., F.M. Rosenbloom, J.F. Henderson, and J.E. Seegmiller. 1967. A specific enzyme defect in gout associated with overproduction of uric acid. *Proc. Natl. Acad. Sci.* **57:** 1735.

Kenwrick, S., M. Patterson, A. Speer, K. Fischbeck, and K. Davies. 1987. Molecular analysis of the Duchenne muscular dystrophy region using pulsed field gel electrophoresis. *Cell* **48:** 351.

Kidd, V.J., R.B. Wallace, K. Itakura, and S.L.C. Woo. 1983. Alpha 1-antitrypsin deficiency detection by direct analysis of the mutation in the gene. *Nature* **304:** 230.

Lesch, M. and W.L. Nyhan. 1964. A familiar disorder of uric acid metabolism and central nervous system function. *Am. J. Med.* **36:** 561.

Melton, D.W., D.S. Konecki, D.H. Ledbetter, J.F. Hejtmancik, and C.T. Caskey. 1981. *In vitro* translation of hypoxanthine guanine phosphoribosyltransferase mRNA: Characterization of a mouse neuroblastoma line that has elevated levels of hypoxanthine guanine phosphoribosyltransferase protein. *Proc. Natl. Acad. Sci.* **78:** 6977.

Myers, R.M., Z. Larin, and T. Maniatis. 1985a. Detection of single base substitutions by ribonuclease cleavage at mismatches in RNA:DNA duplexes. *Science* **230:** 1242.

Myers, R.M., S.G. Fischer, L.S. Lerman, and T. Maniatis. 1985b. Nearly all single base substitutions in DNA fragments joined to a GC-clamp can be detected by denaturing gel electrophoresis. *Nucleic Acids Res.* **13:** 3131.

Myers, R.M., S.G. Fischer, T. Maniatis, and L.S. Lerman. 1985c. Modification of the melting properties of duplex DNA by attachment of a GC-rich DNA sequence as determined by denaturing gradient gel electrophoresis. *Nucleic Acids Res.* **13:** 3111.

Myers, R.M., N. Lumelsky, L.S. Lerman, and T. Maniatis. 1985d. Detection of single base substitutions in total genomic DNA. *Nature* **313:** 495.

Mullis, K.B. and F.A. Faloona. 1987. Specific synthesis of DNA *in vitro* via a polymerase catalyzed chain reaction. *Methods Enzymol.* (in press).

Patel, P.I., P.E. Framson, C.T. Caskey, and A.C. Chinault. 1986. Fine structure of the human hypoxanthine phosphoribosyltransferase gene. *Mol. Cell. Biol.* **6:** 393.

Saiki, R.K., T.L. Bugawan, G.T. Horn, K.B. Mullis, and H.A. Erlich. 1986. Analysis of enzymatically amplified β-globin and HLA-DQα DNA with allele-specific oligonucleotide probes. *Nature* **324:** 163.

Saiki, R.K., S. Scharf, F. Faloona, K.B. Mullis, G.T. Horn, H.A. Erlich, and N. Arnheim. 1985. Enzymatic amplification of beta-globin sequences and restriction analysis for diagnosis of sickle cell anemia. *Science* **230:** 1351.

Southern, E.M. 1975. Detection of specific sequences among DNA fragments separated by gel electrophoresis. *J. Mol. Biol.* **98:** 503.

Stankowski, L.F., K.R. Tindall, and A.W. Hsie. 1986. Quantitative and molecular analysis of ethyl methanesulfonate- and ICR 191-induced mutation of AS52 cells. *Mutat. Res.* **160:** 133.

Thacker, J. 1986. The use of recombinant DNA techniques to study radiation-induced damage, repair and genetic change in mammalian cells. *Int. J. Radiat. Biol.* **50:** 1.

Veres, G., R.A. Gibbs, S.E. Scherer, and C.T. Caskey. 1987. The molecular basis of the sparse fur mouse mutation. *Science* **237:** 415.

Wilson, J.M., A.B. Young, and W.N. Kelley. 1983. Hypoxanthine-guanine phosphoribosyltransferase deficiency. *N. Engl. J. Med.* **309:** 900.

Winter, E., F. Yamamoto, C. Almoguera, and M. Perucho. 1985. A method to detect and characterize point mutations in transcribed genes: Amplification and overexpression of the mutant c-Ki-*ras* allele in human tumor cells. *Proc. Natl. Acad. Sci.* **82:** 7575.

Wreschner, D.H. and M. Herzberg. 1984. A new blotting medium for the simple isolation and identification of highly resolved messenger RNA. *Nucleic Acids Res.* **12:** 1349.

Yang, T.P., P.I. Patel, A.C. Chinault, J.T. Stout, L.G. Jackson, and C.T. Caskey. 1984. Molecular evidence for new mutation in the HPRT locus in Lesch-Nyhan patients. *Nature* **310:** 412.

Molecular Analysis of Mutations at the *hgprt* Locus in Human T Lymphocytes

J. PATRICK O'NEILL, JANICE A. NICKLAS, TIMOTHY C. HUNTER, LINDA M. SULLIVAN, AND RICHARD J. ALBERTINI

Genetics Laboratory
University of Vermont
Burlington, Vermont 05401

OVERVIEW

Human T lymphocytes from peripheral blood samples have been employed to measure the in vivo "spontaneous" frequency and the in vitro γ-irradiation-induced frequency of mutation at the hypoxanthine-guanine phosphoribosyl transferase (*hgprt*) locus. Mutant colonies were studied by Southern blot analysis to determine the frequency and extent of gross structural alterations in the *hgprt* gene. In-vivo-derived mutants were isolated from 8 individuals, and 15 of 164 (9.1%) mutants showed *hgprt* changes. These alterations included partial and total gene deletion as well as DNA insertions, duplications, and restriction-enzyme-site changes. Procedures have been developed to quantify the in vitro induction of mutations by γ-irradiation in human T lymphocytes. Southern blot analysis of the *hgprt* gene has revealed alterations in some of the mutants, including partial and total gene deletion. The T cell receptor (TCR) β-gene rearrangement patterns have been used to define the clonality of individual mutants and to differentiate sibling colonies of a single induced mutant from multiple identical mutation events ("hot spots"). The combined in vivo/in vitro characterization of mutations at the *hgprt* locus should help to define the molecular nature of both spontaneous and γ-irradiation induced mutagenic events in human somatic cells.

INTRODUCTION

The development of conditions for the in vitro propagation and single-cell cloning of human T lymphocytes provides an exciting new avenue for the study of mutation frequency and mutagenic mechanisms in the human population. The in vivo mutant frequency at the *hgprt* locus can be measured by selection for cells resistant to 6-thioguanine (TG^r) by a T-cell cloning assay (Albertini et al. 1982; Morley et al. 1983, 1985; Albertini 1985; Henderson et al. 1986; O'Neill et al. 1987). Southern blot analyses (Southern 1975) of

Banbury Report 28: Mammalian Cell Mutagenesis

genomic DNA from these mutants with an *hgprt* cDNA probe have revealed a variety of DNA alterations in the *hgprt* gene, and such studies should provide insight into the nature of spontaneous mutation events occurring in vivo (Albertini et al. 1985; Turner et al. 1985; Messing et al. 1986; Bradley et al. 1987; Nicklas et al. 1987). In addition, the use of (TCR) gene probes has allowed determination of the in vivo mutation frequency (Nicklas et al. 1986, 1987). The distinction between mutant and mutation frequency is discussed elsewhere in this volume, particularly in terms of its importance in human mutagenicity monitoring (Albertini et al., this volume). A second aspect of the ability to propagate T lymphocytes in cell culture is their utilization for studies of mutation induction in vitro. Preliminary reports suggest that this can be accomplished, although the optimal conditions for mutant expression and selection have not been rigorously defined (Sanderson et al. 1984; Vijayalaxmi and Evans 1984). Study of in vitro mutation induction by mutagenic agents should provide information on the heterogeneity of the human population response to a specific agent and may define the existence of individuals with unusual susceptibility, both increased and decreased as compared to "normal." In addition, molecular characterization of in-vitro-induced mutations may define the type of alteration produced by a given agent and thus provide information on the mechanism of mutagenicity of that agent.

This paper summarizes our analysis of structural alterations in in-vivo-derived HGPRT mutants of human T lymphocytes. It also presents progress in the development of a quantitative assay for the in vitro induction of mutations by γ-irradiation and initial molecular characterization of these induced mutants. A unique advantage of the T-lymphocyte assay is the ability to differentiate sibling mutants from mutation hot spots by the use of the TCR gene probes. This in vivo/in vitro approach should allow description of the molecular nature of mutation in these human somatic cells, both sponanteous and induced.

RESULTS

In-vivo-derived *hgprt* Mutations

The optimal conditions for T lymphocyte cloning and TG^r mutant selection have been described in another report (O'Neill et al. 1987) and will not be discussed here. In-vivo-derived nonselected and TG-selected colonies have been propagated in vitro and analyzed by Southern blot analysis as described previously (Albertini et al. 1985; Nicklas et al. 1986). We have recently completed the analysis of 50 nonselected and 164 TG-selected colonies isolated from 8 male individuals. Details of this study will be described by

Nicklas et al. (1987). Of the 164 mutant colonies studied, 15 (9.1%) showed *hgprt* changes and these are listed in Table 1. Intepretations of the gene alterations are based on the restriction fragment patterns reported by Patel et al. (1984) and Yang et al. (1984). In most cases, the mutants were analyzed with two restriction endonucleases, *Bam*HI and *Hin*dIII. For some blots, an *Msp*I subfragment of the *hgprt* cDNA probe was employed, and, therefore, no information on exon 1 was obtained. No mutants from individuals 1 and 2 showed *hgprt* structural alterations. Mutants from the other 6 individuals showed a variety of structural alterations. Our conclusion on the specific alteration is listed in the last column of Table 1 for these 15 in-vivo-derived mutations. Four mutants (1–4) showed essentially total deletions of the *hgprt* gene, one showed a deletion at the 5′ end of the gene (exons 2–3, mutant 5), and three showed apparent insertions at the 3′ end of the gene (exons 6–9, mutants 6–8). Three mutants showed ill-defined duplications of gene sequences (mutants 9–11) and four showed apparent restriction enzyme-site changes or rearrangement/deletion of gene sequences (mutants 12–15). The data in Table 1 demonstrate the feasibility of defining the type of mutations that arise in vivo in humans as a result of spontaneous events. These results show that less than 10% of the spontaneous mutations appear to contain large DNA alterations at this analytical level. However, the gross alterations that are found appear to occur throughout the gene, including complete deletions. We have now collected approximately 100 TG^r mutants from each of 3 individuals, and analysis of the *hgprt* gene by Southern blot and other methods is in progress. This approach should allow better definition of the DNA alterations involved in spontaneous mutation events in human somatic cells.

In-vitro-induced *hgprt* Mutations

Related to the above studies is the development of a methodology to quantify and characterize mutation events induced in vitro in T lymphocytes. For initial studies, we chose γ-irradiation of G_0-phase cells because of the wealth of cytogenetic data available for this treatment. The existence of radiation-exposed human populations offers the opportunity of comparing the in-vitro-induced alterations in the *hgprt* gene with those alterations induced by in vivo exposure (Messing et al. 1986).

In vitro culture conditions have been developed that allow quantification of γ-irradiation-induced mutagenicity at the *hgprt* locus. The culture conditions are similar to those described recently (O'Neill et al. 1987) in terms of media and amount of T-cell growth factor (TCGF) for mutant selection, except that a TG selection density of 1×10^4 cells/well was found to yield best results with in vitro proliferating T lymphocytes.

Table 1
hgprt Structural Alterations in *in-vivo*-derived T-lymphocyte Mutants

Individual	MF ($\times 10^{-6}$)[a]	Fraction altered[b]	Enzymes[c] *Bam*HI	*Hin*dIII	*Eco*RI	*Msp*I	*Pst*I	Conclusion
(1) LS 28	24.0	0/13						
LS 39	13.9	0/3						
(2) LS 40	1.2	0/11						
(3) P 1	3.8	3/13						
		C5	−25	—	—	−9.4	—	exon 6–9 (6)
			+35			+16.5		(addition)
		E2	−8	n.c.[d]	n.c.	—	—	exon 1 (12)
			+19					(*Bam*HI site)
		E6	−3.7	—	—	−1.35	—	1–9 del. (1)
			−8			−1.8		
			−22			−3.8		
			−25			−9.4		
(4) LS 50	1.7	1/17						
LS 51	5.6	1/8						
		B12E9	−3.7	−7	—	—	—	2–9 del. (2)
			−22	−7.2				
			−25	−17				
		A5B6	n.c.	—	−8	—	—	exon 6–9 (13)
					+0.8			(*Eco*RI site)
					+0.65			
(5) J 38	2.2	2/7						
J 38-2	3.1	2/16						exon 2–3 (14)
		A9G2	—	−7	—	—	—	(*Hin*dIII site
				+5.2				or rearrangement)

		B11F2	—	−7 +4	—	—	—	exon 2–3 (15) (*Hind*III site or rearrangement)
		A19A1	−25 +30	−17 +23	—	+5.4	+6.4	exon 7–9 (7) (addition)
		B15G4	−22 −25	−7 −17				2–9 del. (3)
(6) J 40	3.1	2/21						
J 40-2	2.1	1/9						
		A13D1	−22	−7	—	−4.1 +4.0	−1.2 −2.2	2–3 del. (5)
		A7F7	n.c.	+7.5	—	n.c.	n.c.	dupl. (9)
		B6B2	+2.2	—	—	—	—	dupl. (10)
(7) J 49	9.3	2/22						
J 49-2	7.4	0/9						
		A17B4	−25 +28	−17 +20	—	—	—	exon 6–9 (8) (addition)
		A6H12	+40	—	—	—	—	dupl. (11)
(8) J 51	8.6	1/25						
		A13A8	—	−17	−8 −8.3	—	−1.2 −2.2 −3.8 −4.9 −5.0	2–9 del. (4) (addition)

[a]TG[r] mutant frequency.
[b]Fraction of mutant colonies showing altered restriction fragment patterns on Southern blot analysis.
[c]Restriction enzymes employed.
[d]n.c. indicates no change in fragment pattern observed.

Cells from one individual were irradiated with a ^{60}Co source, stimulated with phytohemagglutinin (PHA), and plated for cytotoxicity determination 36–40 hours later. The cells were also subcultured and plated for mutant selection at 2–3-day intervals. The mutant frequency was calculated as described previously (Albertini et al. 1982; O'Neill et al. 1987). Figure 1 shows the cytotoxicity of γ-irradiation toward T lymphocytes. Cell cloning was determined prior to the first cell division induced by PHA. The cells were also subcultured in order to define the phenotypic expression time. Figure 2 shows that 4–7 days is sufficient time for the mutant expression, as both expression times yield similar dose-response relationships (Fig. 3). Seven days expression was chosen as optimal for this cell system. Figure 4 shows that the induction of TG^r mutants by γ-irradiation of G_0-phase T lymphocytes is linear with the log surviving fraction as reported for a variety of other mammalian cells. Some effect of selection cell density was observed (Fig. 5) and was thought to reflect metabolic cooperation (Subak-Sharpe et al. 1969). However, recent studies show this to be the result of suboptimal amounts of TCGF that resulted in mutant recovery of less than 100% at a selection cell density of 1×10^4 cells/well. Improvements in TCGF activity determinations have alleviated this problem and increased the quantitative nature of the assay.

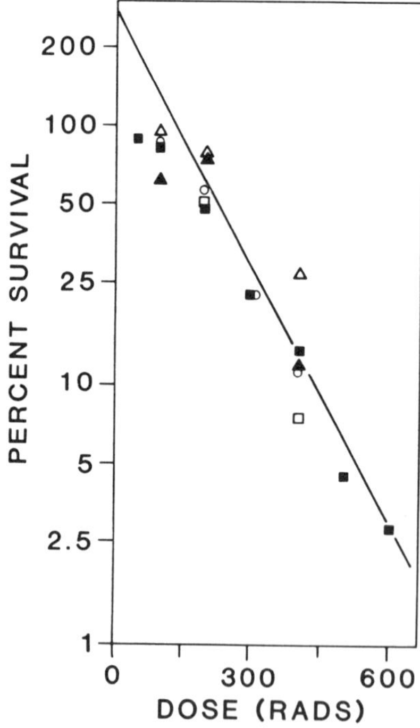

Figure 1

Cytotoxicity of γ-irradiation of T lymphocytes. Cells were exposed to irradiation as described, incubated with PHA for 40 hr, and plated in 96-well microtiter dishes at 2 cells/well for cloning determinations. Cloning efficiencies are expressed relative to the unirradiated control culture. The absolute cloning efficiencies were 0.61 (△), 0.24 (▲), 0.28 (□), 0.36 (■), and 0.44 (○). Five independent determinations with the same individual are presented.

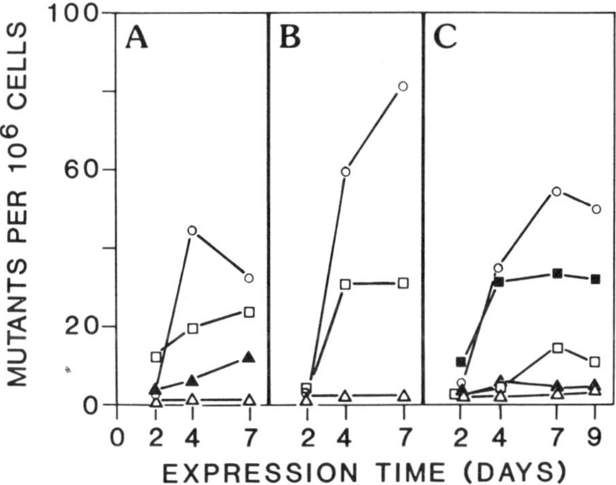

Figure 2

Expression time of the TG^r mutant phenotype in γ-irradiated T lymphocytes. Cultures were irradiated, incubated with PHA for 40 hr and plated in 96-well microtiter dishes at 2 cells/well for cloning determination and at 1×10^4 cells/well in 10 μM TG for mutant selection (time of irradiation designated day 0 and 40 hr time designated day 2). Thereafter, cells were subcultured as described and plated on the indicated days. Cultures received 0 (△), 100 (▲), 200 (□), 300 (■) or 400 (○) rads irradiation. *A*, *B*, and *C* represent three independent determinations with the same individual.

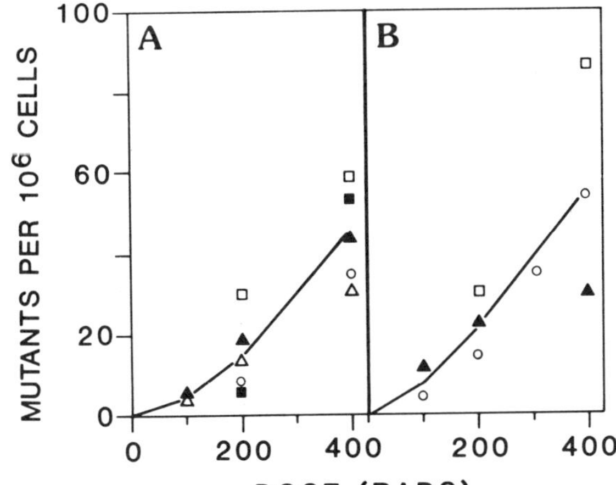

Figure 3

Dose response for mutant frequency with irradiated cells. The induced mutant frequency is presented as a function of irradiation dose in cultures at day 4 (*A*) and day 7 (*B*) of expression.

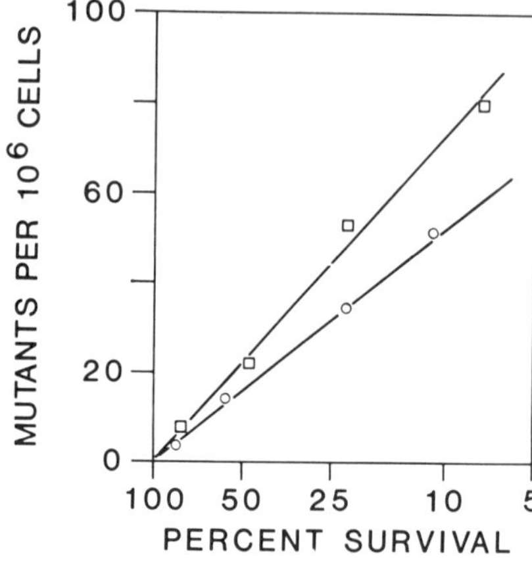

Figure 4

Relationship between mutant frequency and cytotoxicity with irradiated cells. The relationship between induced mutant frequency and cytotoxicity is presented for two independent determinations with the same individual. The mutant frequency was measured on day 7 of phenotypic expression.

Mutants induced by 300 rads of irradiation were isolated and studied by Southern blot analysis. Figure 6 shows the results obtained with TG^r mutants from one individual selected after 8 days of expression as described in Figures 1–5. The four unselected colonies showed the expected pattern for a *Hin*dIII digest (lane A). Eight of the mutants show the same unaltered pattern and are also labeled A. One mutant shows a complete deletion (lane B) and five mutants show a loss of the 17-kb fragment and a gain of a 23-kb fragment (lane C). These latter five appear to contain identical alterations in exons

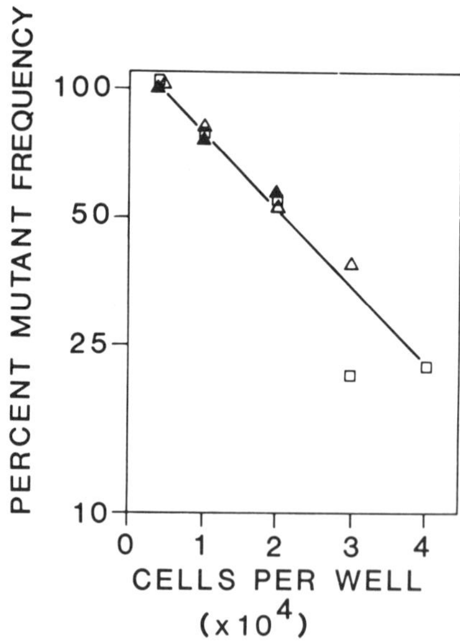

Figure 5

The effect of selection cell density on measured mutant frequency. Cultures were exposed to 300 rads of irradiation and plated for mutant selection on day 7 of expressions at 5×10^3 to 4×10^4 cells/well. The mutant frequency is given in percent relative to the frequency obtained with 5×10^3 cells/well. The symbols represent three independent determinations with the same individual.

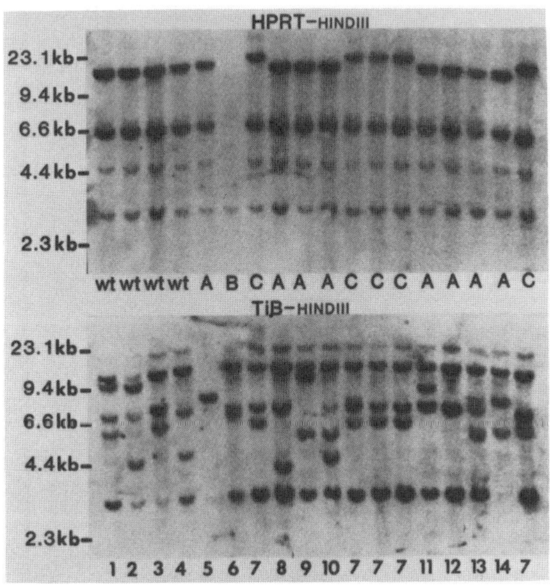

Figure 6

Southern blot analyses of the *hgprt* and TCR β-genes. Four unselected and 14 TG-selected colonies were isolated from cultures plated on day 8 of expression after exposure to 300 rads of γ-irradiation. The TG^r mutant frequency was 62.6×10^6 (nonirradiated control, 1.1×10^{-6}) *Hin*dIII digests were probed with the *hgprt* gene (top) and TCR β-gene (bottom) probes. The *Hin*dIII *hgprt* fragments are 17 (exons 5–9), 7 (exons 2–3), and 4.9 kb (exon 4). The 4 unselected colonies are designated *wt*, and the 14 TG^r-selected colonies are designated *A*, *B*, or *C* for the HGPRT blot. The TCR blot lanes are arbitrarily designated 1–14.

5–9. Interpretation of these results (eight mutants with no alteration and five with the same alteration) would normally be impossible because phenotypic expression of in-vitro-induced mutants requires cell division. The eight mutants with no DNA alteration could be eight independent events or eight siblings of a single event. Similarly, the five mutants could be siblings or five independent events indicative of a mutagenic hot spot. The advantage of T lymphocytes is the ability to differentiate between these two possibilities by employing the TCR gene rearrangement patterns as an independent measure of clonality. The bottom panel of Figure 6 shows TCR β-gene probe results. The four unselected clones have four unique TCR rearrangements (1–4). The eight mutants with no DNA alteration show eight unique rearrangements (5, 8–14) indicative of eight independent mutagenic events. The HGPRT deletion mutant (lane *B*) also shows a unique TCR rearrangement (6). However, the five mutants with the 17 kb loss/23 kb gain (HGPRT lane *C*) have identical TCR rearrangements (7) consistent with their being siblings of a single mutagenic event. These 14 mutants then appear to be the result of 10 independent mutation events, only 2 of which caused gross *hgprt* gene alterations. The TCR rearrangement analysis is an important tool for the study of in vitro mutation induction in that it provides an independent measure of clonality. Its use will allow for more complete description of the types of *hgprt* gene changes induced by γ-irradiation, especially the possibility of mutation hot spots, since the independent nature of the events can be determined.

DISCUSSION

The use of human T lymphocytes for measurement of in-vivo- and in-vitro induced mutations at the *hgprt* locus offers several unique opportunities for the study of the mechanisms responsible for gene alteration. It is the only assay that allows determination of the frequency of mutations existing in vivo in somatic cells of humans.

Either the autoradiography (Albertini 1985) or the cloning assay (Henderson et al. 1986; O'Neill et al. 1987) will provide a measure of the *mutant* frequency, and the use of the TCR gene rearrangement analysis will yield the *mutation* frequency (Nicklas et al. 1986, 1987). The use of Southern blot analysis of the *hgprt* gene gives information on the molecular nature of these in-vivo-derived mutations in humans (Albertini et al. 1985; Turner et al. 1985; Bradley et al. 1987; Nicklas et al. 1987). This approach will allow definition of the nature of spontaneous mutation in these human somatic cells.

The use of human T lymphocytes in in vitro assays of mutation induction provides new possibilities for the study of mutagen susceptibility and mechanism in humans. Comparison of the in vitro induction of TG^r mutants with blood samples from many individuals may allow definition of the heterogeneity of response to that agent in the human population. Individuals with increased or decreased susceptibility could then be discovered, and molecular analysis of the nature of the changes in the *hgprt* gene might elucidate the mechanisms involved in the induction of mutations by that agent. Of course, molecular analysis of the *hgprt* mutations induced in any individual does shed light on the mechanism. The T-lymphocyte cloning assay is invaluable for any attempt to extend in vitro mutagenicity results into an in vivo genetic risk prediction. This is the only assay capable of allowing a comparison of the specific gene alteration induced by an agent both in vitro and in vivo in humans. For example, comparison of the DNA alterations induced in vitro by γ-irradiation (as described in this paper) to those putatively induced in vivo (Messing et al. 1986) could provide definitive evidence for the in vivo induction of mutations in humans.

Finally, the use of the T-lymphocyte should provide the needed perspective for use of the wealth of data on in vitro induction of mutations in other mammalian cell systems in human risk assessment.

ACKNOWLEDGMENTS

Research supported by National Cancer Institute grant 2-RO1-CA-30688-04A1 and Department of the Environment contract DE-AC02-83ER60147-A-003. We thank Inge Gobel for preparing this manuscript.

REFERENCES

Albertini, R.J. 1985. Somatic gene mutations *in vivo* as indicated by the 6-thioguanine resistant T-lymphocytes in human blood. *Mutat. Res.* **150:** 411.

Albertini, R.J., K.S. Castle, and W.R. Borcherding. 1982. T-cell cloning to detect the mutant 6-thioguanine resistant lymphocytes present in human peripheral blood. *Proc. Natl. Acad. Sci.* **79:** 6617.

Albertini, R.J., J.P. O'Neill, J.A. Nicklas, N.H. Heintz, and P.C. Kelleher. 1985. Alterations of the *hprt* gene in human 6-thioguanine resistant T-lymphocytes arising *in vivo*. *Nature* **316:** 369.

Bradley, W.E.C., J.L. Gareau, A.M. Seifert, and K. Messing. 1987. Molecular characterization of 15 rearrangements among 90 human in vivo somatic mutants shows that deletions predominate. *Mol. Cell. Biol.* (in press).

Henderson, L., H. Cole, J. Cole, S.E. James, and M. Green. 1986. Detection of somatic mutations in man: Evaluation of the microtitre cloning assay for T-lymphocytes. *Mutagenesis* **1:** 195.

Messing, K., A.M. Seifert, and W.E.C. Bradley. 1986. *In vivo* mutant frequency of technicians professionally exposed to ionizing radiation. In *Monitoring of occupational genotoxicants* (ed. M. Sorsa and H. Norppa), p. 87. A.R. Liss, New York.

Morley, A.A., K.J. Trainor, J.L. Dempsey, and R.S. Seshadri. 1985. Methods for study of mutations and mutagenesis in human lymphocytes. *Mutat. Res.* **147:** 363.

Morley, A.A., K.J. Trainor, R. Seshadri, and R.B. Ryall. 1983. Measurement of *in vivo* mutations in human lymphocytes. *Nature* **302:** 155.

Nicklas, J.A, J.P. O'Neill, and R.J. Albertini. 1986. Use of T-cell receptor gene probes to quantify the *in vivo hprt* mutations in human T-lymphocytes. *Mutat. Res.* **173:** 67.

Nicklas, J.A., T.C. Hunter, L.M. Sullivan, J.K. Berman, J.P. O'Neill, and R.J. Albertini. 1987. Molecular analyses of *in vivo hprt* mutations in human T-lymphocytes. I. Studies of low frequency "spontaneous" mutants by Southern blots. *Mutagenesis* **2:** (in press).

O'Neill, J.P., M.J. McGinniss, J.K. Berman, L.M. Sullivan, J.A. Nicklas, and R.J. Albertini. 1987. Refinement of a T-lymphocyte cloning assay to quantify the *in vivo* thioguanine resistant mutant frequency in humans. *Mutagenesis* **2:** 87.

Patel, P.I., R.I. Nussbaum, P.E. Framson, D.H. Ledbetter, C.T. Caskey, and A.C. Chinault. 1984. Organization of the *hprt* gene and related sequences in human genome. *Somatic Cell Mol. Genet.* **10:** 483.

Sanderson, B.J.S., J.L. Dempsey, and A.A. Morley. 1984. Mutations in human lymphocytes: Effect of X- and UV-irradiation. *Mutat. Res.* **140:** 223.

Southern, E.M. 1975. Detection of specific sequences among DNA fragments separated by gel electrophoresis. *J. Mol. Biol.* **93:** 503.

Subak-Sharpe, J.H., R.R. Burk, and J.P. Pitts. 1969. Metabolic cooperation between biochemically marked mammalian cells in tissue culture. *J. Cell. Sci.* **4:** 353.

Turner, D.R., A.A. Morley, M. Haliandros, R. Kutlaca, and B.J. Sanderson. 1985. *In vivo* somatic mutations in human lymphocytes frequently result from major gene alterations. *Nature* **315:** 343.

Vijayalaxmi and H.J. Evans. 1984. Measurement of spontaneous and X-irradiation induced 6-thioguanine resistant human blood lymphocytes using a T-cell cloning technique. *Mutat. Res.* **129:** 283.

Yang, T.P., A.C. Chinault, J.T. Stout, and L.G. Jackson. 1984. Molecular evidence for new mutation at the *hprt* locus in Lesch-Nyhan patients. *Nature* **310:** 412.

Comments

Hsie: Earlier, Barry Glickman asked me about apples and oranges. If I recall correctly, David DeMarini mentioned yesterday that cisplatin apparently induced largely, if not exclusively, small colonies in the mouse lymphoma cells, and thus, in the apple system. Then, in your orange system, you mentioned that you induced almost exclusively base-pair substitution and frameshift mutations. It seems to me that doesn't exactly fit. How would you reconcile the potential apples' and oranges' similarities and differences?

Glickman: I think I can reconcile that. Let me begin by saying that having an obvious explanation doesn't make it the right one, but I think that there is an explanation.

If you look at our gamma-ray data, the criticism that we have received, and we have received, indeed, much criticism for using the hemizygous strain, is that you cannot get deletions because they can only be found in one direction. There are two answers to that. The flippant one is, "We have a few deletions. It's very hard to sequence a gene that's not there." In our cloning system, that is certainly true because we have based it upon homologous recombination.

But, in the gamma-ray work, we know we are missing classes of events. However, we do have some rearrangements, and we do have some deletions. The deletions are interesting. The three deletions that we have appear to have a common endpoint within *aprt*. The other end extends well outside the gene. Regarding rearrangements, of the six rearrangements we have now cloned, we have 8 of the 12 pieces and are going back and looking to find out from where they broke. That way, we will understand a little bit better in that data set exactly what we can see and what we can't see.

Elliot Drobetsky repeated those experiments with D423, which is hemizygous. We know we have three times as much mutation, and we know from blotting that there are many more deletions and rearrangements. My point is that mutation is not random. At the deletion endpoints and at break points, they reflect some quality of DNA that we don't know much about.

So, what I would propose is, in the case of the work with the large and small colonies, because of the window through which you are looking, you see a certain image. All of us who are doing mutagenesis work have chosen a given window. Whether it be one system or another, we are all trying to look at the same thing, but we don't even know yet whether we are looking in the same house with our windows. So, in the case of the proflavin, we are looking again through this window through which we can see, because of the system, this subset of data. We will have to look at the hemizygous strain, and hopefully, when more people have their shuttle vector data, the *aprt* and *dhfr* data, and so forth, the windows will have become large enough to understand how the pieces fit. Generally, I think that is all that is really the matter. Again, mutation is not random; we see a subset.

Hsie: How about the possibility of the existence of multifunctions from a single agent? Perhaps proflavin is no exception. It could be multifunctional, in such a way, when the mutant hunters pick up the nice looking big colonies, and in this case, that may turn out to be a base substitution that would have been preferentially picked. Mark Meuth is shaking his head. Maybe he didn't do what I thought.

Meuth: The spontaneous collections were obtained from fluctuation tests, where most of the positives have single colonies, so we just take what's there. They're not necessarily big or small.

Sarasin: Barry, you insisted on the fact that in some of your mutants you have several mutations at several base pairs apart from each other. How do you explain that? Do you think they represent independent events or they are due to a common mechanism?

Glickman: In the case of UV, the calculation is that there are 0.025 dimers per target site. We know that the dimers aren't randomly distributed. We don't know what dimers are involved, I should add, but I don't think that it's a case at any time that the damage is compounded.

Repair is an important component. Yesterday, we were talking about spontaneous mutation and the role of repair. For example, we can be very deceived. We know that in the lack of excision repair the spontaneous mutation frequency is a little higher. In the absence of, let's say, *recA* or *umuc*, there's not much of a difference of frequency. Yet, when you actually look at the spectrum of spontaneous mutants, you discover that, while the mutation frequency has not changed, the distribution of mutation has. And so, in the case of a Uvr^- strain, some point mutants have gone up and some of the frameshifts have gone down. So, what you are constantly seeing is a change of one lesion for another, because of the biochemistry, and one mutational endpoint for another. I am a firm believer of the idea that damage goes through a series of pathways. You can say you started out with this damage, but is that different when the next step is an apurinic site? It's still a lesion. Repair processes make mistakes. I believe that these multiple changes reflect very complex lesions; as an example, cisplatin can do intrastrand and interstrand crosslinks.

Chasin: Mark, you went very quickly over your spontaneous mutants. I don't want to carry this analogy too much further, but since in this case both you and Barry are looking at apples for spontaneous mutants. Do you see the same spectrum? Or are there differences that might indicate that the spontaneous mutants aren't so spontaneous but may be arising due to subtle differences in culturing conditions?

Meuth: We characterized several spontaneous mutations resulting from small events, i.e., either point mutations or small deletions, which were localized and mapped by loss of restriction endonuclease sites. We didn't sequence whole mutant genes. Consequently, we may not be looking at the same sites as Barry, but the mix of base-pair changes we saw was different. We had no particular bias in transitions, as many GC→AT as AT→GC. We saw a couple of transversions as well, and small deletions were about one-third of the spontaneous small events that we sequenced. So, the spectra appear to be different.

As Elliot Drobetsky in Barry's lab communicated the manuscript about the UV hot spot to me, we were able to do the experiment of hybridizing oligonucleotides specific for the hot spot to blots of digests from our mutant strains. We end-labeled an oligonucleotide in which the potential hot spot was in the middle of the sequence. If you have a mismatch in that sort of situation, the oligo melts off at a much lower temperature. Out of 60 or so mutants that we screened using this technique, none of them appear to have lost that site. So it appears that this site is not hot in our collection of spontaneous mutants.

Glickman: We have done the oligomeric probing of the same mutants. Because the system was new, many of our mutants were cloned two and three times independently and checked out. The point is, in the position we are calling 241, we have 7 out of 30. I don't know if you had any, but it isn't one you would have normally picked up, except that it also has a phenotype. So I think we would have known it it had been in yours. It's leaky, for one thing.

In any case, we become concerned about that. By oligomeric probing, we know our mutants are what we think they are. They also came up in the UV collection at the same site. However, it is interesting. The same person did both experiments, in the sense that Elliot Drobetsky did your collection and my collection, and there is a difference.

We have a new collection of about 200 independent mutants at which we are going to look. We also have collected mutants, not with the azaadenine but with a diaminopurine, and it is not in that collection.

There are, however, in terms of that question of spontaneous mutations, some rather amusing differences from *E. coli* that we have learned the very hard way. Whether you have 0.3 ml or 2.0 ml in a tube, you get a different spontaneous spectrum. In the case of CHO cells, I can tell you of an unfortunate accident—if you connect an oxygen tank instead of an air tank to your incubator, you end up increasing the "spontaneous" by a factor of ten. Clearly, very small differences in how you do things are going to have an impact on what you see. I confess, in *E. coli,* we weren't really prepared for that. We found out when a new student came in and filled up the tubes to the top. We discovered something very important. We said, "One day we're going to do anaerobic and see what that does." Well, it's very interesting, there were almost no mutants. So, little differences can make a big difference, and big differences, like the filling of the tube, make enormous differences.

We don't know the cloning, the media, or the subtleties yet of the CHO system, but we can be very sure that this difference we are finding reflects a subtlety that we have to learn to understand.

Chasin: I think we should be aware of the possibility of culturing differences and not start throwing up our hands and asking, "What's going on?" when we see disparate results for spontaneous mutations. In your response, actually, you mentioned some things that were really not subtle—one was substituting oxygen for air. Another was that you see a different spectrum with a different selective agent.

Glickman: It's very important.

Chasin: Another point was that your hot spot produces a leaky mutant. So, depending on how stringent the selection is, leaky mutants (and the hot spot) could be missed.

Glickman: No. That has to do with HAT resistance as opposed to azaadenine. It has to do with growth in another medium as opposed to the actual selection. We have to be very careful.

First of all, you know that some mutants are going to escape just simply because your selection isn't designed to catch them. So you have to start with some threshold where you say, "This is what we are going to look at." We have agreed essentially on the same threshold, all of us. We have a variety of mutant sites now. We have probably 70 very different substitution sites, so we have that big a window. When we have expanded our collections—and we can all think that way—when you have expanded your collection, you learn a little bit about the system, then perhaps you have to think of alternative selection mechanisms and so forth. Let's face it, we are at the beginning of the study, and we are asking very simple questions. When you talk about subtleties, I think that for spontaneous, it is clear that we've got very careful things to consider. However, when you do UV, your dominant effect of the treatment is probably okay. What I do and what you do in another system and so forth, I think those answers are going to be the same. However, the spontaneous is a very subtle question that we don't understand.

Adair: I would like to follow up on this and point out some encouraging similarities and some disturbing differences in the *aprt* mutant data from our three laboratories. In comparing these data, one should keep in mind that Mark and Barry are both working with mutants isolated from D422, whereas most of my work has been with mutants isolated from AT3-2. I have also looked at smaller data sets from several other hemizygous lines. If you take Mark's data, among 120 mutants, he sees 10 single site losses, 1 site gain, 8 deletions, and 1 insertion. In AT3-2, among 161 mutants, I see 15 site losses, 4 site gains, 8 deletions, and 1 insertion. In another set of 56 mutants isolated from another hemizygote, I see essentially the same thing as in AT3-2. So, with two different CHO cell lines, two different labs, and presumably slightly different selection conditions, we are seeing about the same spectrum. If you look at the individual site-loss and site-gain mutants, our collections show very similar distributions.

In addition to these data for the parental line, I have also looked at repair-deficient lines. The numbers there are comparable. I have looked at 142 mutants from one repair complementation class and 169 from another, so the sample sizes are fairly large. Again, we see a lot of similarities and a few differences, but basically we see most of the same site losses and site gains, and a few unique sites that may have to do with the repair deficiencies.

The most disturbing thing to me so far in comparing our data sets concerns the hot spot, because Barry gets lots and lots of independent mutants at this 241 site. As it turns out, one should be able to pick up such mutations as *mbo*2 site gains. The original sequence at this site is TCCTTC. The first C is the mutational hot spot. Barry gets lots of C→T transitions at this site either spontaneous or with UV. A C→T transition will create an *mbo*2 recognition site, so one should be able to pick up such mutations at the Southern blot level.

In 161 spontaneous mutants of AT3-2, another 142 and 169 mutants in the two repair-deficient lines, and another 56 spontaneous mutants that we have looked at in other heterozygotes, we haven't picked up any spontaneous *mbo*2 site gains at this hot-spot site. After UV mutagenesis, we have picked up one in one repair-deficient class and two in the other repair-deficient class, so we know we pick up the phenotype under our selection conditions. We know they are UV-induced in a repair-deficient background, but we have not seen them spontaneously in either our repair-proficient or repair-deficient lines. Thus, there seems to be a real difference in terms of lab-to-lab variation in mutability of the hot spot that I don't quite understand.

Hutchinson: Just to follow up on that point, if you look at those numbers, the implication is that Barry is selecting mutants under very restrictive conditions. Then he might pick up, for example, single-base substitutions that only change the amino acid, whereas the other two experiments might not. That is, in these other two experiments, a protein with a single change in an amino acid might still have enough activity so that the mutation is not detected. This type of selection has been seen in experiments with prokaryotes.

Glickman: That's possible. There are some possibilities here, because you don't know that that change you found is really at that site, and we would have to probe it with the oligomer. However, the interesting thing here is, let's face it, normally when two people are working on opposite sides of the country they have no communications. We have excellent communications. We have had the same person do these experiments.

Hutchinson: To the best of your knowledge and belief, the selection criteria are the same?

Glickman: There are differences. Now that we have found this difference, we have started to look at it. We don't use the same serum. We don't use the same company on a number of sources. Mark and I have laughed a little bit at the difference between Toronto and Montreal water. Now, that is always the joke, of course, that one comes up with when you have something to explain. However, there are differences.

We were forced for a long time to ignore the difference simply because we were trying to get the data together, but with oligomeric probing, it is quite simple to try a few conditions, to try a few different sera, and to just take out a few mutants. If it's 7 out of 30 or 0, that's not a big number in order to get an answer. With oligomeric probing, we can do it now.

Hutchinson: However, can't you interchange mutants strains and determine what they look like under different conditions?

Glickman: It would be interesting to send you a strain and to see if it grows properly. It could be a simple explanation.

Chu: It may be a big order. First, I want to know if you have examined cell clones isolated under nonselective conditions. In other words, random clones that have nothing wrong with *aprt*, and yet you probe with *aprt* probes and see if the mutations still occurred? That means a kind of control. Second, are the sequences known? Have you seen nucleotide changes in introns? Third, would it now be possible for you to do a counter-selection for reversions showing changes as predicted by what we know from prokaryotes? For the same kind of reversions, would they have occurred either spontaneously or in response to specific agents?

Glickman: The first question, we haven't looked at random clones. We have cloned and sequenced, I think, seven independent clones for the wild-type sequence without finding changes. I don't know how we would probe random clones very easily, but we have looked at several.

In terms of the question of introns, we have never seen a change in an intron other than at the splice junction site. However, changes do occur in introns, and our evidence for that is that we have now sequenced a number of other CHO alleles and that is very interesting. The other allele for which we have the full sequence is 37 base substitutions, of which only 3 produce amino acid changes and about half of which are an intron sequence. We also have a number of other bits of sequences from cDNA, which do not give us the intron information but which do tell us that the variety of alleles available in CHO is quite large: at 33%, 5% heterogeneity without much effect on the protein seems to be the rule.

On reversion, we have looked at a number of sites where we thought reversion would be easy to do, picked some, and then tried to establish the limits for reversion, either in cases where we have tried to do things like homologous recombination or tried to take an amber mutant to see if we get a suppressor, and so forth. So far, our score of getting revertants is very low. We don't have any decent data. We just know that it's a good idea.

Davidson: We have some data on reversion in the *gpt* system. We find basically two different classes of revertants. Most of them occur at frequencies below 10^{-7}. Among the first group that we sequenced, three of them returned to the correct base at the original site. One of them was a different mutation, a reversion by a different alteration within the codon, putting in an entirely different amino acid. Then we have a class of mutants with reverted frequencies of approximately 1–10%, and these seemed to revert to gene amplification rather than by an alteration within the coding sequence.

My guess is that we will find, and you probably will also, all the types of different revertants that one would expect. Then I think that one of the interesting questions is whether or not any of these turn out to be suppressors.

Glickman: I didn't answer that question properly. I realize that what we were looking at was trying to develop reversion assays for very specific kinds of events that occurred at low frequency. What we do know, for example, is the duplication of 15 bp reverts at a high frequency. If we measure the stability of the mutation, if we look at reversion frequency, we have lots that have a high reversion frequency. We expect at some point to sequence those revertants to see whether they fit any kind of model.

When you asked the reversion system, my point of view was based upon the question: Do we have reversion assays? The qualification of reversion assays would be low spontaneous rates, because what you are trying to do is have something that doesn't revert very easily that you could use as an assay for a specific change after treatment with whatever you are interested in. So, there is the conflict. You take something that is very stable, you try to get revertants, and you find that it is hard. So, these are two different answers to the question.

Adair: I would like to respond to Ernie's (Chu) questions, also.

With regard to looking for unselected changes at the *aprt* locus, Mike Siciliano and I recently tried one experiment where we looked for electrophoretic mobility shifts at *aprt* after ENU mutagenesis, without selection for $APRT^-$ phenotypes. Out of 200 clones, we didn't see any, so we don't have anything positive there.

With regard to reversion, one of the things we have done is to take most of the restriction site-loss mutants that we have obtained and also a few of the site-gain mutants and have attempted to select revertants. In most cases we see spontaneous reversion at very low frequencies, 10^{-7} to something less than 5×10^{-9}. In several cases where we haven't obtained spontaneous revertants, we have been able to obtain them after EMS mutagenesis. In most cases, we see the reappearance of the original restriction pattern, suggesting reversion at the original site of mutation. An exception is the *pst*I site, which is at a 3′ splice junction. It appears fairly hot for mutation; we get lots of independent isolates of mutants at that site. With these mutants, we sometimes see reversion at the original site with regeneration of the original wild-type restriction pattern, and in other cases might see a change in the band, but it appears to be slightly smaller than the original; these may involve a different sort of event. In several cases, where we have been unsuccessful in obtaining revertants, we have been able to show that a very small deletion was responsible for what originally appeared to be a single site loss at the Southern blot level.

Calos: Barry, when you say you sequenced the whole gene, how much are you sequencing?

Glickman: In almost every case, everything.

Calos: 4 kb?

Glickman: 3.2, 3.3.

Calos: Have you ever found two mutations that are separated by more than 100 bases?

Glickman: No. I should qualify that in the following way: That statement "we sequenced everything" is true for the spontaneous, it's true for the UV, but it is not yet true for benzo[a]pyrene or cisplatin. As you go through the gene, when you find it, those are the ones you have, and then the next batch will be run through the next set of primers and the next set, and so there may very well be a surprise down the road in the latter two collections.

Chasin: Doesn't the answer to Michele's question really answer Ernie's question about unselected mutants? You were not getting other changes.

Glickman: But if we have sequenced, let's say, 85% of things, we have never seen a single event that didn't give an amino acid change. We have seen some things that didn't get an amino acid change when they were neighboring bases, and the two of them together gave one change. So, I believe that the number of such events must be very low.

Hutchinson: Mark, regarding your finding that your inserts are duplicates of sequences which remain in the genome. A specific question: Do you know anything about the flanking sequences in the original sites? More generally, could you expand a bit on why you chose your particular model.

Meuth: We have the sequence. Do you mean of the donor sequence surrounding it?

Hutchinson: Yes.

Meuth: We have the sequence of a 4-kb fragment bearing the mobilized sequence. There is nothing really notable about it. There are no direct or inverted repeats.

Hutchinson: No interesting sequence homologies?

Meuth: The inserted sequence has no significant homology with the target site or with any other sequence in the various data bases. There are some inverted repeats occurring in the donor sequence, but nothing really at the termini. It is possible that a much larger bit was inserted into the target site but was then imprecisely excised. Again these inverted repeats are nothing like LTRs.

Hutchinson: Why do you favor the duplicated model?

Meuth: The inserted sequence has obviously been duplicated because there are now three copies in the mutant strain as opposed to two in the wild type. The only question is when it was duplicated. An alternative explanation is that the transposition occurred during G2 and originated from the chromatid which was then segregated at mitosis. Therefore there may not really have been a duplication directly connected with the event.

Dixon: I have a question that concerns how you actually go about doing the experiments to pick the spontaneous mutants. This relates to whether you can state how many events are occurring and when they are occurring. Do you have a whole set of 100 cultures that have been started from one cell and then grown to look for mutants? What is it that you are actually mechanically doing?

Meuth: We set up a large number of replica cultures with a small number of cells, maybe 100, in each. We grow them to about 2 million cells before plating out the entire culture under selective conditions. Since the mutations are rare, random events, you have a very large percentage of cultures with no mutants, whereas most of the positives have single colonies. From those multiple colonies, only one mutant is taken for characterization.

Dixon: When you actually look at the numbers of mutants in the dishes at the end, do the numbers make sense in terms of the mutants arising from random spontaneous events?

Meuth: Yes.

Glickman: There is no other explanation for the data than that they are independent. That's the first thing with which we had to deal. It must be that. Something's wrong. That's not possible.

Dixon: Have you then, rather than taking just one from each dish, actually confirmed that they are all the same on each dish, as they should be if they arose from one event?

Meuth: No, I haven't done that.

Glickman: We took three from one dish. They were all different. I should say that at the time we did it, it was an induced one.

Dixon: So, it's not clear if it is following what you would predict?

Adair: In cases where we had a low frequency of mutants in a particular replicate culture, and have picked several, we saw the same mutation. In cases where there were higher numbers of mutants on plates in a set, or where I saw very morphologically distinct colonies and have picked more than one isolate from a culture, they were sometimes different, probably representing different mutational events occurring in the population. The way we do the experiments, we are very confident that we are looking at independent events.

Dixon: I grant you that. It's just a question of whether you can tell anything about the actual mechanism by which the mutations are occurring, based on the kinetics at which they appear.

Meuth: Right. If it's an early event, they would all be the same.

Davidson: Barry, could you say a little bit more about the apparent strand specificity that you were seeing?

Glickman: It's very simple. When we looked at sequencing of the minus mutants, knowing that we were looking at GC to AT predominantly, and so forth, we came up with an interesting observation. If you ask what strand is responsible, where the mutations are, it is the nontranscribed strand. So, what we were thinking is that, because deamination occurs much more frequently in single strand than double strand, that if the transcribed strand isn't somehow hydrogen bonded because of the RNA, that leaves this strand unhappily single. That's bad for you. So, there is a paper by Dr. Fix and myself in *MGG* describing this bias, this asymmetry in mutation. Because it's *lacI* and we know how many sites are available and where they are, we can make a conclusion that says it is not due to the availability of sites.

We asked the same question in *aprt.* The question we can then ask is: Are the mutations occurring in the transcribed or nontranscribed strand? They are occurring on the nontranscribed strand, but we don't know a thing about available sites. So, it is a very soft conclusion that is just consistent with the idea that deamination occurs of the single strand.

Davidson: How can you say in which random mutation it is occurring?

Glickman: Because it is GC→AT. If it's deamination, we know which strand it is.

Meuth: But your mutants are not in methylatable CG sites.

Glickman: And the're not in methylated sites. They are not in the CTG sequence.

I wanted to add one thing that had to do with the hot spot. The hot spot that we have at position 241, we don't have it in the gamma spectrum, and we don't have it in the benzo[a]pyrene nor the cisplatin spectrum. We have it in the spontaneous, and we have it in the UV. That is the site that I mentioned as being also the hot spot sequence in the shuttle. So, we do have something interesting here, and it has to be resolved. I think that the site itself seems to be one that does mutate. The question is, I think, why does it appear in one system and not in another, in the spontaneous as opposed to the induced. However, that sequence is not something that we are picking up in every spectrum. It's not likely to be some kind of accident because of how the work is done or something that occurs during the steps.

De Marini: On site specificity, but sort of dose-related, I'm wondering, Barry, do you have any data where you collected mutants at different doses or different levels of survival and then looked at the spectra? The reason I ask is, based on Ernie Chu's work 17 years ago, he saw this difference in spectra, Fred (de Serres) and Herman (Brockman) saw this in *Neurospora*, and Don (Clive) and Martha (Moore) saw it certainly in mouse lymphoma. It might be expected to be found at the molecular level, if you get that collection of mutants throughout the dose range.

Glickman: This is a mammalian meeting so it's not fair to go back to *E. coli*, but the statement on *E. coli* would be that we have done it for UV, we have done it for ionizing radiation, and we have now done it for alkylating agents, a whole series of them. There are some things to be learned. In the spectra change with ionizing radiation in *E. coli*, you go from transitions to tranversions. In the UV when you do it, you find there are sites that behave in a linear fashion and in a dose-squared fashion. So, it does matter.

The only data we have so far in the mammalian cell is with benzo[a]pyrene. It's not much data. If you look at a low dose and a high dose, at the high dose there are very few things that revert and very few leaky ones. At low dose, which is where we were trying to get high survival like 50%, there are lots of things that revert. So, there seems to be a qualitative difference, but we haven't done any sequencing of the batch—or any looking really or even blotting—of the batch from the higher dose.

Gibbs: Is that explainable by spontaneous mutations at the low doses?

Glickman: It's a 100-fold at 50% survival, a 100-fold over background.

Gibbs: To refer back to David's comment, what about in the ionizing radiation case? I don't think you can say that there is any evidence for a qualitative difference in mutational spectra at different doses of ionizing radiation in mammalian cells.

De Marini: Well, Ernie showed it.

Glickman: That's where it's really the keenest. The data base is really good.

Hsie: I'm glad to see the beautiful thing about ENU as a simple mutagen, obviously demonstrated here again with Ernest's (Chu) experiments. Now, Larry (Thompson), since the mouse experiment using the specific locus test is also nonselective in a way, can Ernie or you tell me how the mutation rates you both generated independently compare?

Chu: We made an estimate of the frequency of ENU-induced structural gene mutants in the order of 10^{-4} per allele.

Russell: I don't know how the in vitro and in vivo exposures can be meaningfully compared. In one case, cells are bathed more or less directly in the chemical; in the other, chemical that is injected in known quantity has to reach the germ-cell target. With the maximum exposure that can be administered intraperitoneally without causing a great deal of germ-cell killing, the mutation rate is in the order of 10^{-3} per locus.

Hsie: We can detect a factor of ten.

Glickman: I guess I'm a little uncertain. I'm new to the mammalian side of things, and I only recently was exposed to the kinds of calculations that you are making. When you take a treatment like ENU, which is a one-shot deal basically, with a half-life that is quite short, it seems to me that the dosimetry you want is alkylations per nucleotide so that then you can make comparisons.

In studies that Fred de Serres was involved in with a group of us in Holland, we looked at huge numbers of organisms—mice, *Drosophila,* human cells, *Salmonella, E. coli*, and phage—and tried to estimate mutants through this whole range.

So, when you talk about exposure in terms of generation, I don't understand what you mean. Could you contextualize why you talked about per generation for something that is given in an instant?

Chu: Yes. We haven't done the dosimetry and we express the mutation frequency as a function of a dose of ENU. The calculation we have done is based on the assumption that all electromorphs in the ENU experiments were induced by the mutagen at one time. We cannot really say per allele per generation. Where there is one spot, you cannot call them incidence of frequency.

Russell: Gary Sega has measured ethylations per nucleotide in germ cells following intraperitoneal ENU injections. I can't remember the actual figures, but ethylation levels are totally proportional to the injected amount over a fairly wide range of ENU exposures.

Glickman: But it provides a basis to compare data sets.

Dixon: I think the question that you were raising is: In your calculation are you assuming that new mutations are occurring during the growth of the cells, or that new mutations may only occur during the first generation after treatment?

Chu: Since the rate of spontaneous mutation to polypeptide changes as revealed by 2-D PAGE is very low, we assume that in ENU experiments all the electrophoretic mutants we observed at different sampling periods were all induced by the mutagen at one time and appeared as the mutant progeny. Therefore, we call it induced mutant frequency and not rate.

Glickman: I have to think about your answer, because I am still not used to thinking of discussing something like ENU and talking about mutations per generation afterwards. However, what you are suggesting is that mutations continue to be produced after the treatment, in the same way as the work of Nazim in yeast many years ago.

Chu: No, I don't imply that.

Glickman: But you wouldn't know, because your electromorphs are going to be the same or different from each culture. If you take a culture and divide it up into many pieces, you could ask: If you looked then at cultures 1, 2, 3, 4, and 5, that all were derived from A, are those the same or different, and do you have a persistent lesion? That's a major question. Is a lesion a persistent lesion? Do you have an unrepairable damage that continues to generate mutation, because that's what you are implying.

Chu: I don't know that.

Hutchinson: As a function of incubation time after treatment, you are seeing a constant number of mutant spots per million spots scanned. That implies that you are generating mutants at some time after treatment and that these are not selectively lost.

Chu: Right, that implication is there. Also, the second conclusion of that is there. We have to assume that, because in that experiment, each time we only cloned 120 out of 10^7 cells and the probability of finding the same mutant is very small. If two mutants of the same phenotype (charge and/or apparent molecular weight) appeared at different sampling periods, they may be descendants of one mutant or may represent mutations of independent origin.

O'Neill: That to me is the problem. Your chances of seeing the same mutant twice are very low, when you're taking 120 cells out of 10^7 even assuming that an induced thioguanine resistant mutant frequency of 10^{-4}. Your chances of seeing the same mutant twice are probably essentially zero. So, every time you look at it you are going to see a different mutant, unless a mutant has a selective advantage.

Chu: That's right.

Glickman: There is one point in your calculation that I would like to address, but I don't quite know how. That is, when you think about mutation, you think about the statistics, and you think about killing. You have to ask the question about lethal sectoring. I believe that there is a fair bit of evidence that lethal sectoring in yeast, which is the organism that has been best studied, is not a random process. The question to you is, "Is there enough killing in these cells that lethal sectoring is a component that needs to be worked into your equations?"

Chu: Probably. In this particular ENU experiment, cells were exposed to 10 mg/ml ENU for 40 minutes each day for five consecutive days. The cell survival as measured by colony-forming ability on days 3 and 5 was about 40%.

Glickman: But, this is a different question than killing. In microorganisms, we measure viability by colony-forming units. We ask, Does this form a colony?

Chu: We do that too. Cell survival was measured in terms of colony-forming ability.

Glickman: But that is very different than having a cell that divides in two, one of which dies and the other goes on to produce. So, your viability is 100%, but your lethal sectoring, if it's not random, is changing your numbers on how you view mutation. I wonder if you can work those parameters into your thinking.

Chu: The viability was measured on day 3 and day 5. ENU treatment reduces the cloning efficiency. Instead of a plating efficiency of 70% in the controls, we reduced it to, say, 20 or 30%. So, there is some loss in cloning ability after the mutagen treatment. It was not by dye exclusion test to measure how many cells were killed. After we wash the mutagen out, we immediately plated the cells, by limiting dilution method in the 96-well plates, and calculated the clonability of the treated population. The number of clones that we picked at day 3 was about 40 that were analyzed.

de Serres: Barry, it might be easier to think about the original treatment giving rise to cryptic mutations that code for a different amino acid, where you don't have any striking change in specific activity. Now you've got a different frame of reference for changes that may occur in succeeding cell generations and that will become expressed as mutants in that particular background that wouldn't be mutant in the original background. You not only have to think in terms of damage that is immediately recognized as mutation but in terms of damage that is producing the silent or cryptic mutations. In other words, they don't show immediate expression. I think, if you think of it in those terms, you can see how, with succeeding cell replications over a period of time, new damage is going to be introduced that may then have an opportunity for expression in this genetic background that wouldn't have an opportunity for expression in the original background. That goes back to years ago, when we studied reversion of specific locus mutations that occur at particular sites. Much of this work showed that you got the amino acid back with which you originally started, but in other cases, you got changes elsewhere in the same coding unit that gives rise to other amino acids, so these were intralocal suppressors.

Glickman: I don't think the numbers argue for you.

de Serres: You don't know about the relative frequencies of these expressed versus cryptic changes.

Maher: It has already been mentioned by Ernie Chu that the number of mutants he looked at is very small. This is a very important point since the question that is being batted back and forth concerns whether ENU is forming adducts at various sites on the DNA, whether some are going to fall out in a couple of days, whereas some will be stable, and whether the critical one is misreplicated, and so forth. However, his system is not designed to answer these questions, since he is only looking at about 100 or so.

Chu: Clones. A very small sample, and yet we still detected mutants.

Maher: Yes. But, you have indicated yourself that your experimental design does not allow you to know if the ones you saw later on came up later on. They might have been there from day one.

Chu: That's right.

Albertini: Just picking up on your implication of use for this in vivo, do you have any feeling for the generation of variants during the time that you have looked at a clone? Obviously, if you look at a clone as one cell, you have to grow it up to a population in order to do the protein on it.

Chu: During the cell expansion, mutations occur and how large the fraction would depend on how early the mutation occurs. If the new mutation occurred very early, the whole clone will be a mutant. If the mutation occurred late during the cell expansion, the clone will constitute a mosaic population. However, the sensitivity of the 2-D gel would not pick up those if the proportion of mutant cells is small among 10^6 cells for loading to the gel.

Meuth: Hamlin has reported that there is a DNA replication origin in the *dhfr* gene. Has that been mapped clearly yet? Do any of your deletions originate there?

Chasin: Yes. These ERFs (early replicating fragments) in a pulse are about 30 kb 3′ of the gene, and in the large deletions they are definitely eliminated, but there's no way to tell what that means in terms of any further consequences. If the DNA is not there, you can't check to see whether it is replicated or not. We're not seeing, as we had hoped to see, a lot of common deletion endpoints as yet.

Hutchinson: Larry (Chasin), your result that a lot of these point mutations both depress messenger RNA and also impose an additional loss of protein is a very interesting point. The implication is that very few base substitutions change the activity of the protein. The average base substitution would be expected to merely change an amino acid. It is difficult to see how the change of an amino acid could lead to changes in splicing. So, it's a very curious result.

Chasin: What is beginning to bothers us, and which is why I was asking about the *aprt* systems this morning, is that we don't know exactly for what we are selecting. We know, for instance, that it doesn't take very much DHFR activity to give prototrophy for these three folate-related end-products (glycine, hypoxanthine, and thymidine), because we have revertants and transfectants that have about 1 or 2% DHFR activity and these grow in the absence of the three end products. One of our tests, when we isolate a putative mutant, is to determine if it is in fact auxotrophic; if it is not, we actually throw it away and not just put it in the freezer. Clearly, we may be missing a large class of leaky mutants this way.

Hutchinson: So, you may be isolating frameshifts, for example.

Chasin: We have sequenced some of these mutations, which I didn't have a chance to get to in my talk. We know the sequence change in five of them. One is a deletion of a C, a benzo[a]pyrene-induced mutation in exon 3. The others are base substitutions based on sequencing or on the appearance or disappearance of restriction sites in certain combinations. Five out of six are single-base substitutions. However, even if we are somehow selecting for the low mRNA phenotype, I don't think it makes it less interesting. That is, if we're selecting for something that must have low RNA as well as low activity, we are therefore picking out the nucleotides that cause these phenotypes. We can still look at the phenotype, even though the subset may be nonrepresentative.

Glickman: In terms of the low RNA, I was quite fascinated by your result. The reason is that we screened something like 200 spontaneous mutants, but we only found either 5 or 7 that had reduced RNA levels. I can also tell you that there is a graduate student by the name of David Heard in the lab who has been looking at the RNase being produced by the splice junction mutants to see whether they conform with the model and so forth. There is no difference in amount of RNA that is obvious from any of the mutants at which we have looked. So, that does seem to be a little bit different in *aprt* than what you have been finding.

Chasin: The splice-junction mutants, you say, are producing normal amounts of normal message?

Glickman: No, total message. The dot blots gave normal amounts. They are now doing Northerns. We are trying to ascertain whether the intron and exon arrangements are the way they are supposed to be, related to the sequences of the mutation.

Chasin: Our five spontaneous mutants all have high levels of mRNA, and four of them are missplicers.

Jensen: The missplices are high.

Chasin: That still doesn't really explain why out of 200 spontaneous mutants you shouldn't have seen a lot of these, unless it really does mean we are selecting for them.

Dixon: How sensitive is your method for determining multiple mutations?

Chasin: In most of the analyses, despite the background, we find one predominant change: I would say maybe 90% of the mutants that are informative (i.e., different from the wild-type pattern) are of this type.

In a few cases we see a complex pattern that could be due to multiple mismatches. Since the cutting at mismatched bases in the heteroduplex is often not complete, the presence of two mismatches would yield a complex pattern of partial digestion products.

Lieber: I just want to add that for *hgprt* in the TK-6 human lymphoblast, I have done 43 Northerns on so-called point mutations where we didn't see any change in the Southern. Ten of those also had much reduced RNA levels.

Gibbs: We have seen three cases in the human germ line HPRT mutations, of reduced HPRT mRNA.

Chasin: Three cases out of how many?

Gibbs: Fifty or sixty.

Chu: You may recall, Dr. Gibbs, that Bill Kelley's lab at Michigan did similar things with HPRT from Lesch-Nyhan patients. They did the amino acid sequencing, but at the same time they studied the messenger RNA. I don't recall their data. You might be more familiar with that.

Gibbs: They reported some cases where HPRT mRNA was absent, but no cases of reduced HPRT mRNA levels.

Chu: These are human in vivo mutations in Lesch-Nyhan syndrome.

Gibbs: Larry, have you found any RNase-sensitive sites outside the protein coding region in the 5′ or 3′ nontranslated sequence?

Chasin: No. The screen includes the full 5′ sequences that are untranslated leaders. We haven't found any there. It also includes a substantial amount of 3′ tail.

Gibbs: That's rather striking, isn't it, in terms of the number of cases that you see with reduced mRNA levels?

Chasin: In none of these mutants do we see zero RNA. If we push hard enough, we always see 2%. So, if there was a change that reduced the RNA down to 2%, but the protein coding region was left unsullied, we would not probably pick it up. It would come out as a wild-type.

Gibbs: I see!

Chasin: We don't have any relationship, because virtually all of our mutants have virtually no enzyme activity. Therefore, you can't make that kind of relationship between enzyme activity and anything else because we don't have enzyme activity. We have one mutant with a very small amount. We haven't tried to make a correlation.

Jensen: Your hypothesis about the degradation of the transcript, is there any evidence for that?

Chasin: In the case of the thalassemias, there are a couple of cases where there are nonsense mutations that result in a very much reduced cytoplasmic RNA. In those cases also, nuclear RNA levels had been checked, and they were surprisingly also found to be way down. So, there is that one precedent. It is hard to understand why a single-base substitution in those two cases in the beta-globin mRNA resulted in low nuclear levels. We might explain low cytoplasmic levels. Other than that, I'm not aware of any support.

Jensen: Are you in the midst of any interesting experiments that try to prove this?

Chasin: We checked transcription, and we checked message stability in order to look at the half-life of mature mRNA. We have not seen a transient primary transcript, where so far in our Northerns we are not even picking it up in the wild-type. That's a pretty good challenge, but I think we can do it. Again, we are impressed with the sensitivity of the RNase protection experiments. We have never really probed with intron sequences, so we are gearing up to make large intron probes and to look at the normal splicing events that take place in the amplified line where it should be easy to see. In fact, that is a project that is interesting in its own right.

We also have the ability, once we have cloned our mutant genes, to put them back into a double-deletion background, to coamplify them with a minigene other than *dhfr*, and then look at cells that have 50 or 100 copies of the mutant gene. This should get us over the technical problem of dealing with small amounts of RNA. That's the direction we are heading to look for aberrant primary transcripts in a few of these mutants.

Analysis of the *gpt* Gene as a Chromosomally Integrated Structure

Deletion Mutations Are Associated with the Differential-induced Mutant Frequency Response of the AS52 and CHO-K1-BH4 Cell Lines

KENNETH R. TINDALL* AND LEON F. STANKOWSKI, JR.[†]
*Laboratory of Genetics, National Institute of Environmental Health Sciences, Research Triangle Park, North Carolina 27709
[†] Pharmakon Research International, Inc., Waverly, Pennsylvania 18471

OVERVIEW

AS52 cells contain a single, functional copy of the bacterial guanine phosphoribosyl transferase (*gpt*) gene transfected and stably integrated into the Chinese hamster ovary (CHO) cell genome. These cells can be used to screen for mutations by selecting for resistance to 6-thioguanine (TG^r). Similar induced mutant frequency responses are observed following treatment of the AS52 (GPT) and the CHO-K1-BH4 hypoxanthine-guanine phosphoribosyl transferase (HPRT) cell lines with standard point mutagens. However, the induced mutant frequency response following treatment with various radiomimetic drugs or clastogens has been shown to be greater in AS52 cells than in CHO-K1-BH4 cells. We have demonstrated by Southern blotting that the differences observed are related to the types of mutations induced by these mutagens. Treatment with point mutagens yields mostly unaltered gene sequences, whereas treatment with radiomimetic drugs or clastogens yields mostly deletions. The increased mutant frequency response in AS52 cells appears to reflect an ability to recover large deletions as viable mutants due to the unique site of integration of the transfected *gpt* gene. Other loci, notably *hprt*, show a relatively weak mutagenic response to these deletion-inducing agents. We suspect that deletions affecting adjacent genes or essential DNA sequences (multilocus deletions) at the hemizygous *hprt* locus will be lethal. Multilocus deletions at the *gpt* locus presumably do not affect viability either because the site of integration is autosomal, and there are homologous chromosomal sequences that can complement function if an adjacent essential gene is deleted, or because the *gpt* gene has integrated into a site that can withstand extremely large deletions without affecting viability.

INTRODUCTION

Deletions and chromosome rearrangements occur in mammalian cells during specialized cellular processes such as immunoglobulin gene maturation, as well as during aberrant cellular processes such as the generation of chromosome aberrations or the activation of some oncogenes. Many studies of spontaneous or induced deletions in mammalian cells have used well defined cell lines or peripheral blood lymphocytes and have analyzed either the autosomal adenine phosphoribosyl transferase (*aprt*) or dehydrofolate reductase (*dhfr*) loci (Adair et al. 1983 and this volume; Nalbantoglu et al. 1983, 1986; Urlab et al. 1986; Meuth et al., this volume), or the hemizygous X-linked *hprt* (Albertini et al. 1985; Vrieling et al. 1985; Brown et al. 1986; Stankowski and Hsie 1986; Stankowski et al. 1986a; Thacker 1986) locus.

The *aprt* and *dhfr* loci used in these studies are hemizygous due to deletions of the corresponding *aprt* or *dhfr* regions of the homologous autosome. Such an allelic configuration allows the convenient selection of mutants. However, if *essential* sequences exist adjacent to the target gene, and if these sequences have been deleted from the homologous chromosome during the isolation of the hemizygous cell line, deletions extending from the target gene into these adjacent essential sequences will not be viable. Thus, the size of the deletion that can be recovered in such a system will be limited. Similarly, the sequences on the inactive X-chromosome cannot complement any essential sequences adjacent to the *hprt* locus that may have been deleted. Therefore, some deletion mutations will not be represented in isolated clones due to the lethal effect of deletions involving essential hemizygous sequences. In this presentation, we define the deletion of all or part of the target gene and any adjacent *essential* DNA sequences as "multilocus deletions."

In our studies of deletion mutations in mammalian cells, we have taken advantage of a CHO cell line, AS52, that carries a single, functional copy of the bacterial xanthine-guanine phosphoribosyl transferase (XPRT) gene (*gpt*) stably integrated into the CHO genome (Tindall et al. 1984, 1986). The *gpt* gene is analogous to the mammalian hypoxanthine-guanine phosphoribosyl transferase (*hprt*) gene, and either locus can be readily screened for mutations by selecting for resistance to the purine analog, 6-thioguanine (TG^r). In addition, the small size of the *gpt* gene (456 bp) allows for the easy manipulation of mutant *gpt* sequences for molecular analyses. AS52 cells have been demonstrated to be effective in the quantitative detection of both point and deletion mutations at the integrated *gpt* locus (Stankowski and Hsie 1986; Stankowski et al. 1986a). The induced mutant frequency at the *gpt* locus in AS52 cells following treatment with various radiomimetic drugs and clastogens has been shown to be greater than the induced mutant frequencies observed at the *hprt* locus in CHO-K1-BH4 cells (Stankowski and Tindall, this volume). No significant differences, however, are observed following treatment with standard point mutagens (e.g., UV-light, EMS, ICR-191) (Stankowski and Hsie 1986; Stankowski et al. 1986a). Survival curves of the two cell lines are comparable regardless of the mutagen treatment. This study presents molecular analyses of independently isolated spontaneous and induced TG^r mutants derived from the AS52 and CHO-K1-BH4 cell lines.

RESULTS

Southern blot analyses (Table 1) were performed on 231 XPRT and 110 HPRT mutants. For these studies, spontaneous mutants were isolated from an independent culture grown for at least one week in selective media to eliminate any preexisting XPRT or HPRT mutants. Induced mutants were isolated from independently treated cultures as previously described (Tindall et al. 1984; Stankowski and Hsie 1986; Stankowski et al. 1986a). To insure that mutants were independent, only one mutant per treatment flask was cloned for further analysis. A single dose (selected from a previously generated dose response curve) was used for each agent to preclude possible dose-effects on the mutant spectrum.

Table 1
Results of Southern Blot Analysis of TG^r Mutants Derived from the AS52 (*gpt*) and CHO-K1-BH4 (*hprt*) Cell Lines

Cell line mutagen	Number analyzed	No detectable alteration	Deletions partial	Deletions total[a]
(a) *AS52*				
None (spontaneous)	23	9	10	4
EMS	22	20	0	2
ICR-191	24	20	1	3
UV-light	26	21	0	5
Actinomycin D	23	5	6	12
Adriamycin	13	0	0	13
Bleomycin	28	2	6[b]	20
Mitomycin C	20	14	1	5
Formaldehyde	26	7	6	13
X-rays	26	0	1	25
(b) *CHO-K1-BH4*				
None (spontaneous)	23	18	4	1
EMS	21	21	0	0
ICR-191	22	22	0	0
UV-light	23	23	0	0
X-rays	21	6	3	12

Some of these data have been previously reported (Stankowski and Hsie 1986; Stankowski et al. 1986a).
[a]No sequences hybridized to the probes used (pSV2*gpt* for AS52 and pHPT12 for CHO-K1-BH4).
[b]Includes one complex rearrangement.

These experiments were performed to evaluate the molecular basis of the differential-induced mutant frequency response that had been observed in experiments using AS52 and CHO-K1-BH4 cells. Initial results demonstrated that EMS, UV-light, and ICR-191 were approximately equally cytotoxic to both cell lines and equally mutagenic at the *gpt* and *hprt* loci (Stankowski and Hsie 1986; Stankowski et al. 1986a). In addition, these three agents yielded primarily no change in the pattern of the Southern blot (Table 1) suggesting that these mutagens induce predominantly point mutations or deletions too small (< 100 bp) to be resolved by Southern blotting. However, significant differences were observed for mutant induction at the *gpt* and *hprt* loci by X-rays (Tindall and Hsie 1983; Tindall et al. 1984; Stankowski and Hsie 1986). For example, 700–800 versus 75–100 TG^r mutants/10^6 clonable cells were observed at a dose of 800 rads in AS52 or CHO-K1-BH4 cells, respectively. This treatment was equally cytotoxic in the two cell lines and yielded primarily deletions at their respective target loci (Table 1).

If the differential mutant frequency response is associated with deletions, then one might expect radiomimetic or clastogenic agents also to induce similar differences in the mutant frequency response and predominantly deletions. Further analysis of the mutational response of the AS52 and CHO-K1-BH4 cell lines following treatment with such agents has been reported (Stankowski et al. 1986b, 1987; Stankowski and Tindall, this volume). In these studies, seven antitumor drugs (actinomycin D [ACD], adriamycin [ADR], bleomycin [BLM], *cis*-diammine-dichloroplatinum, cyclophosphamide, daunomycin, and mitomycin C [MMC]) were chosen for analysis because they were expected to demonstrate the differential mutant frequency response based on reported radiomimetic or clastogenic activities. All seven agents were assayed in both the AS52 and CHO-K1-BH4 cell lines, and all exhibited the differential mutation response. That is, all seven agents induced more mutants per unit dose at the *gpt* locus in AS52 cells than at the *hprt* locus in CHO-K1-BH4 cells. As with all other agents examined to date, survival curves were similar in these two cell lines.

XPRT mutants were isolated following ACD, ADR, BLM, and MMC treatment and were evaluated by Southern blotting (Table 1). A representative Southern blot is shown in Figure 1. All ADR-induced mutants characterized (13/13; at 100 ng/ml) exhibit total deletions. ACD mutants (18/23; at 250 ng/ml) or BLM mutants (26/28; at 50 μg/ml) are also primarily either partial or total deletions. These treatments produce at least a tenfold increase in mutant frequencies in AS52 cells at relative survivals of 10–30%. Although the data are limited, drug-specific differences in mutational spectra are apparent. ACD- and BLM-induced putative point mutations (no detectable alterations, Table 1) and deletions, whereas ADR induced only total deletions. These results may reflect multiple drug-specific mutational mechanisms, and more detailed molecular analyses are in progress.

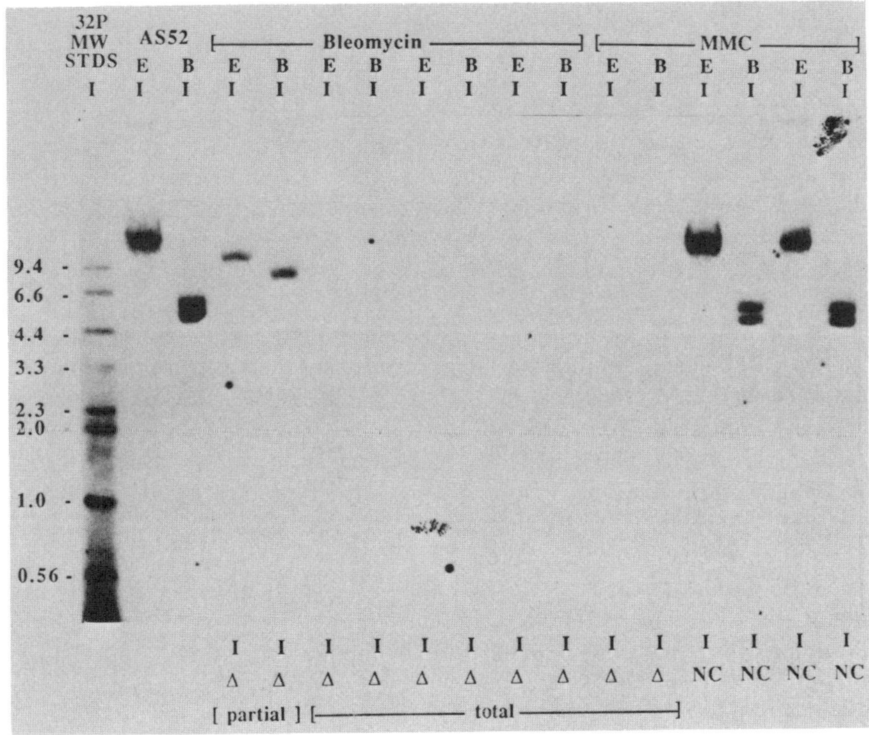

Figure 1

Southern blot of bleomycin- or mitomycin C-induced mutant and parental (AS52) *gpt* sequences. ^{32}P MW standards (Oncor, Inc.) are end-labeled λ*Hin*dIII fragments that include two additional fragments (3.3 and 1.0 kb) derived from pBR322. Each DNA (10 μg) was digested with either *Eco*RI (*E*) or *Bam*HI (*B*) and separated by electrophoresis in a 0.8% agarose gel. DNAs were transferred to Zeta-probe membrane, hybridized to a primer-labeled pSV2*gpt* probe, washed under stringent conditions, and exposed overnight with screens to Kodak XAR-5 X-ray film. *Eco*RI generates a single band migrating at 11.0 kb, and *Bam*HI generates two bands migrating at 4.9 and 4.3 kb in the parental AS52 DNA. Partial and total deletions (Δ) as well as mutants with no change (N.C.) in the pattern of band migration are presented.

In contrast, the majority of XPRT mutants induced by MMC (14/21; at 300 ng/ml) were putative point mutations. These results appear contrary to the prediction that MMC, a potent clastogen that exhibits the differential-induced mutant frequency response, should induce primarily deletions. Two points regarding the MMC data are worthy of comment. First, the concentration of MMC used (300 ng/ml) was relatively low, due to survival considerations, and is in the region of the dose-response curve where large differences between the XPRT and HPRT mutant frequencies are not observed. Second, the MMC-induced HPRT mutant frequency increases to a maximum (with some scatter) of ~100 TG^{r} mutants/10^6 clonable cells. The MMC-induced XPRT mutant frequency increases to ~700 6-TG^r mutants/10^6 clonable cells at 900 ng/ml (Stankowski et al. 1987). Thus, MMC may produce point mutations at low doses; however, at higher doses it may produce predominantly deletions. Such a dose-effect would explain the clear induction of HPRT mutations in CHO-K1-BH4 cells as well as the putative point mutations observed in the XPRT mutants analyzed. This hypothesis predicts that AS52 mutants isolated following treatment with 900 ng/ml MMC will be mostly deletions. Such experiments are in progress. Alternatively, the differences observed may reflect differences in the presence of sequence specific MMC-sensitive sites within the two loci (Ueda et al. 1984).

Also reported in Table 1 is the first molecular analysis of formaldehyde (FORM)-induced mutations in mammalian cells. Most of the FORM-induced mutants examined (19/26; at 50 μg/ml of 37% FORM) exhibited total or partial deletions of *gpt*. This observation is consistent with the observed clastogenic activity of FORM in AS52 and CHO-K1-BH4 cells (J.R. San Sebastian and L.F. Stankowski, unpubl.). In addition, some putative point mutations were also observed among the FORM mutants. However, based upon the induced mutant frequency and relative survival (~100 TG^r mutants/10^6 clonable cells at 1–5% survival), it is highly probable that all of the mutations analyzed were FORM induced. These data are consistent with a mechanism in which most of the FORM-induced mutations are deletions, thereby accounting for the lack of a mutagenic response at the *hprt* locus in CHO-K1-BH4 cells.

The relatively poor or negative-induced mutant frequency response of the CHO-K1-BH4 cell line to such agents limits the usefulness of analyzing HPRT mutants derived from these cells, since there is a high probability that any individual mutant isolated may be spontaneous in origin. Thus, further molecular analyses of HPRT mutants have not been performed.

Finally, differences were observed in the spectrum of spontaneous mutations arising at the *gpt* or *hprt* loci (Table 1). Approximately 60% (14/23) of the spontaneous XPRT mutants exhibited deletions, whereas only 23% (5/23) of the spontaneous HPRT mutants exhibited deletions. Since the AS52 and CHO-K1-BH4 cell lines are similar with regard to (1) spontaneous mutation rates, (2) the induced mutant frequency response following treatment with classical point mutagens, (3) the lack of a mutational response following treatment with cytotoxic nonmutagens, and (4) mutant spectra for EMS, ICR-191, UV-light and X-rays (Table 1; Stankowski and Tindall, this volume), we do not believe that the differences in the spontaneous mutant spectra reflect an inherent instability of the *gpt* integration site. Rather, it is likely that the *gpt* integration site may allow the recovery of viable multilocus deletions, therefore accounting for the observed differences in the spontaneous spectrum as well as for the minor differences in spontaneous mutation rates.

DISCUSSION

We have demonstrated by Southern blotting that most agents exhibiting the differential mutant frequency response yield predominantly deletion mutations at *gpt* (the same may be true for MMC at higher doses). Previous studies have demonstrated that treatment with the classical point mutagens, EMS, UV-light, and ICR-191, yields similar induced mutant frequencies at either *hprt* or *gpt* and mostly unaltered gene sequences as detected by Southern blotting (Stankowski et al. 1986a). These data are consistent with reports of point mutational changes characterized in mammalian cells following treatment with EMS (Lebkowski et al. 1986), UV-light (Lebkowski et al. 1985; Hauser et al. 1986) and ICR-191 (Fuscoe et al. 1982).

Earlier observations that treatment with X-rays yields mostly deletions at both loci and significantly more mutants per unit dose at *gpt* than at *hprt* (Stankowski and Hsie 1986) suggested that radiomimetic drugs or clastogens might be expected to exhibit the differential mutant frequency response. If the differential mutant frequency response reflects the recovery of viable multilocus deletions, then agents exhibiting this response should induce mostly deletions. Indeed, results to date indicate that most radiomimetic and clastogenic agents exhibit the differential mutant frequency response (Stankowski et al. 1987; Stankowski and Tindall, this volume) and also induce primarily deletion mutations at the *gpt* locus (Table 1).

Formaldehyde is clearly clastogenic in AS52 and CHO-K1-BH4 cells (J.R. San Sebastian and L.F. Stankowski, unpubl.) and induces a differential mutant frequency response in the two cell lines. The reported mutagenic activity of FORM at the thymidine kinase (*tk*) locus in human TK6 cells (Goldmacher and Thilly 1983) is consistent with our observations that formaldehyde induces predominantly deletion mutations. The *tk* alleles in TK6 cells are heterozygous (Yandell et al. 1986), and this allelic configuration should allow the recovery of FORM-induced *tk* mutations. As predicted, FORM is not (or only very weakly) mutagenic at *hprt*.

The data presented (Table 1; see also Stankowski and Tindall, this volume) also suggest that these agents may induce mutations by multiple mutagenic pathways, producing both point mutations or small deletions as well as the larger putative multilocus deletions. In addition, the MMC results suggest that the mutational spectrum generated by this drug may change as a function of dose. Such dose-dependent changes in mutant spectra are well established with a variety of mutagens in *E. coli* (Glickman et al. 1980) and have been implicated in mammalian cells (Chu 1971).

We propose that the differential-induced mutant frequency response is the result of lethal deletions at the *hprt* locus but not at the *gpt* locus. Deletions of the target gene that include adjacent essential sequences (multilocus deletions) will not be viable because of the hemizygous nature of *hprt* and these adjacent sequences. Multilocus deletions at *gpt*, however, may be viable as a result of the chromosomal site of *gpt* integration. The two most likely possibilities are that (1) the *gpt* gene has integrated into an autosome, and the deletion of an adjacent essential sequence is not lethal due to complementary function of a second copy of these essential sequences on the homologous chromosome, or (2) the *gpt* gene has integrated into a site that can withstand extremely large deletions without affecting cell viability (e.g., the telomeric region of a chromosome).

Certainly, there are numerous reports of deletions isolated at *hprt* and other hemizygous loci. Some of these deletions are at least 40–45 kb since the entire *hprt* gene has been deleted. Deletions of at least 200 kb have been observed at the *dhfr* locus (Urlab et al. 1986). Deletion mutations at the hemizygous *aprt* locus in CHO-AT3-2 cells appear to extend only 5′ to the *aprt* gene. No deletions have been isolated that extend beyond the *aprt* structural gene in the 3′ direction (Adair, this volume). These data suggest a selection bias against deletions in the 3′ direction of the *aprt* locus in these hamster cells. Thus, whereas deletion mutations occur at hemizygous loci, the recovery of mutants will represent only a subset of the deletions produced since multilocus deletions in these systems will be inviable. The precise definition of a multilocus deletion is, therefore, dependent upon the system and upon the locus of interest, since the juxtaposition of adjacent essential sequences will be both gene and species specific. In CHO-K1-BH4 cells there appears to be a limitation in the recovery of mutants induced by agents that induce primarily deletions of *gpt* in AS52 cells. Thus, the AS52 cell line should be useful in the detection and characterization of deletion mutations because a greater proportion of the deletions will be recovered for subsequent molecular analysis. We are in the process of characterizing the *gpt* integration site, defining the size of the deletions occurring at this locus, and cloning and defining deletion endpoints.

ACKNOWLEDGMENTS

The expert technical assistance of William G. Tuman is greatly appreciated. The technical assistance of Donald L. Halderman is also gratefully acknowledged. The authors also thank Connie Stankowski for her help in the preparation of this manuscript. This research was sponsored by Pharmakon Research International, Inc., and the National Institute of Environmental Health Sciences.

REFERENCES

Adair, G.M., R.L. Stallings, K.K. Friend, and M.J. Siciliano. 1983. High frequency structural deletion as the basis for functional hemizygosity of the *aprt* locus in CHO cells. *Proc. Natl. Acad. Sci.* **80:** 5961.

Albertini, R.J., J.P. O'Neill, J.A. Nicklas, N.H. Heintz, and P.C. Kelleher. 1985. Alteration of the *hprt* gene in human *in vivo*-derived 6-thioguanine-resistant T lymphocytes. *Nature* **316:** 369.

Brown, R., A. Stretch, and J. Thacker. 1986. The nature of mutants induced by ionising radiation in cultured hamster cells. II. Antigenic response and reverse mutation of HPRT-deficient mutants induced by gamma rays or ethyl methanesulfonate. *Mutat. Res.* **160:** 111.

Chu, E.H.Y. 1971. Mammalian cell genetics. III. Characterization of X-ray-induced forward mutations in Chinese hamster cell cultures. *Mutat. Res.* **11:** 23.

Fuscoe, J.C., J.P. O'Neill, R. Machanoff, and A.W. Hsie. 1982. Quantification and analysis of reverse mutations at the *hgprt* locus in Chinese hamster ovary cells. *Mutat. Res.* **96:** 15.

Glickman, B.W., K. Rietveld, and C.S. Aaron. 1980. Gamma-ray induced mutational spectrum in the *lacI* gene of *Escherichia coli*. Comparison of induced and spontaneous spectra at the molecular level. *Mutat. Res.* **69:** 1.

Goldmacher, V.S. and W.G. Thilly. 1983. Formaldehyde is mutagenic for cultured human cells. *Mutat. Res.* **116:** 417.

Hauser J., M.M. Seidman, K. Sidur, and K. Dixon. 1986. Sequence specificity of point mutations induced during passage of a UV-irradiated shuttle vector plasmid in monkey cells. *Mol. Cell. Biol.* **6:** 277.

Lebkowski, J.S., J.H. Miller, and M.P. Calos. 1986. Determination of DNA sequence changes induced by ethyl methanesulfonate in human cells, using a shuttle vector system. *Mol. Cell. Biol.* **6:** 1838.

Lebkowski, J.S., S. Clancy, J.H. Miller, and M.P. Calos. 1985. The *lacI* shuttle: Rapid analysis of the mutagenic specificity of ultraviolet light in human cells. *Proc. Natl. Acad. Sci.* **82:** 8606.

Nalbantoglu, J.O., O. Goncalves, and M. Meuth. 1983. Structure of mutant alleles at the *aprt* locus of Chinese hamster ovary cells. *J. Mol. Biol.* **167:** 575.

Nalbantoglu, J.O., D. Hartley, G. Phear, G. Tear, and M. Meuth. 1986. Spontaneous deletion formation at the *aprt* locus of hamster cells: The presence of short sequence homologies and dyad symmetries at deletion termini. *EMBO J.* **5:** 1199.

Stankowski, L.F., Jr. and A.W. Hsie. 1986. Quantitative and molecular analyses of radiation-induced mutation in AS52 cells. *Radiat. Res.* **105:** 37.

Stankowski, L.F., Jr., K.R. Tindall, and A.W. Hsie. 1986a. Quantitative and molecular analysis of ethyl methanesulfonate- and ICR 191-induced mutation in AS52 cells. *Mutat. Res.* **160:** 133.

Stankowski, L.F., Jr., K.R. Tindall, E.G. Godek, W.G. Tuman, and G.J. Kasper. 1986b. Quantitative and molecular analyses of mutations induced by anticancer drugs at the *gpt* and *hprt* loci in AS52 and CHO-K1-BH4 cells. *Environ. Mutagen.* (suppl. 6) **8:** 80.

Stankowski, L.F., Jr., K.R. Tindall, E.G. Godek, R.J. Matthews, and R.W. Naismith. 1987. Comparison of the cytotoxicity and mutagenicity of seven anticancer drugs in AS52 and CHO-K1-BH4 cells. *Mutat. Res.* (in press).

Thacker, J. 1986. The nature of mutants induced by ionising radiation in cultured hamster cells. III. Molecular characterization of HPRT-deficient mutants induced by gamma-rays or alpha-particles showing that the majority have deletions of all or part of the *hprt* gene. *Mutat. Res.* **160:** 267.

Tindall, K.R. and A.W. Hsie. 1983. Detection of deletion mutations at the *gpt* locus in pSV2*gpt* transformed CHO cells. *UCLA Symp. Mol. Cell. Biol. New. Ser.* **11:** 625.

Tindall, K.R., L.F. Stankowski, Jr., R. Machanoff, and A.W. Hsie. 1984. Detection of deletion mutations in pSV2*gpt* transformed cells. *Mol. Cell. Biol.* **4:** 1411.

———. 1986. Analysis of mutation in pSV2*gpt* transformed Chinese hamster ovary cells. *Mutat. Res.* **160:** 121.

Ueda, K., J. Morita, and T. Komano. 1984. Sequence specificity of heat-labile sites in DNA induced by mitomycin C. *Biochemistry* **23:** 1634.

Urlab, G., P.J. Mitchell, E. Kas, L.A. Chasin, V.L. Funanage, T.T. Myoda, and J. Hamlin. 1986. Effect of gamma rays at the dihydrofolate reductase locus: Deletions and inversions. *Somatic Cell Mol. Genet.* **12:** 555.

Vrieling, H., J.W.I.M. Simons, F. Awert, A.T. Natarajan, and A.A. van Zeeland. 1985. Mutations induced by X-rays at the HPRT locus in cultured Chinese hamster cells are mostly large deletions. *Mutat. Res.* **144:** 281.

Yandell, D.W., T.P. Dryja, and J.B. Little. 1986. Somatic mutations at a heterozygous autosomal locus in human cells occur more frequently by allele loss than by intragenic structural alteration. *Somatic Cell Mol. Genet.* **12:** 255.

Analysis of Spontaneous and Mutagen-induced Mutations in a Chromosomally Integrated Gene

CHARLES R. ASHMAN* AND RICHARD L. DAVIDSON
Center for Genetics
University of Illinois College of Medicine
Chicago, Illinois 60612

OVERVIEW

We have determined the DNA base sequence changes in 43 spontaneous and 28 ethyl methanesulfonate (EMS)-induced mutant genes recovered from a mouse fibroblast cell line. The target gene was the *Escherichia coli* xanthine (guanine) phosphoribosyl transferase (*gpt*) gene that is contained within a retroviral shuttle vector integrated into mouse chromosomal DNA. The mutant *gpt* genes were selected in the mouse cells and then rescued from the mouse cells and introduced into *E. coli* as part of a bacterial plasmid. Most of the EMS-induced mutations were found to be G:C→A:T transitions. A variety of mutational events were observed among the spontaneous mutant genes that were sequenced. Many of the spontaneous mutants had deletions of less than 20 bp and a very strong deletion "hot spot" was observed.

INTRODUCTION

The determination of the actual base changes that occur when a gene is mutated is an important step in elucidating the detailed mechanisms of mutagenesis. For example, knowledge of the type of base substitutions produced by a particular mutagen, coupled with a knowledge of the DNA lesions that the mutagen produces, can give important clues as to which lesions are biologically relevant and how these lesions are processed. In addition, the observation of a preference for certain mutational sites (hot spots) can suggest a role for the DNA sequence surrounding the mutational site in the mutational process. Finally, determination of mutagenic specificity in the presence or absence of a particular repair system can suggest mechanisms by which the repair system modulates the mutagenic process.

Recently, several systems have been described that utilize shuttle-vector plasmids to study mutational specificity in mammalian cells (Lebkowski et al.

*Present address: Department of Radiation Oncology, University of Chicago, 5841 S. Maryland Avenue, Box 442, Chicago, Illinois 60637.

Banbury Report 28: Mammalian Cell Mutagenesis

1985; Seidman et al. 1985; Ashman et al. 1986; Drinkwater and Klinedinst 1986; Glazer et al. 1986). In these systems, a small bacterial target gene is introduced into the mammalian cells as part of a shuttle vector. Following exposure of the mammalian cells to a mutagen, the shuttle-vector sequences are recovered from the mammalian cells and introduced into bacteria. In bacteria, large amounts of the target-gene sequence can be produced for DNA sequence analysis. The major advantage of these shuttle-vector systems is that the target-gene sequence can be rapidly recovered and analyzed.

We recently described the development of a shuttle-vector system that utilizes a retroviral shuttle vector containing the *E. coli gpt* gene as a target (Ashman et al. 1986). This vector was introduced into the mouse fibroblast A9 cell line (Littlefield 1964) that is deficient in hypoxanthine-guanine phosphoribosyl transferase (HPRT). A cell line (A9I2) that stably expresses the *gpt* gene was isolated, and it was shown that this cell line contained a single copy of the vector integrated into chromosomal DNA in a proviral form. Shuttle-vector sequences were recovered from A9I2 cells by fusing the mouse cells to monkey COS cells (Gluzman 1981), extracting low-molecular-weight DNA from the fused cells, and using this DNA to transform *E. coli.* In this paper, we describe the isolation and characterization of both spontaneous and EMS-induced *gpt* gene mutations isolated from A9I2 cells. Among the spontaneous mutants sequenced, the predominate mutational event was a small deletion, whereas G:C→A:T transition mutations predominated among the EMS-induced mutants.

RESULTS

The construction of the vector pZipGptNeo and the A9I2 cell line has been described in detail (Ashman et al. 1986). Briefly, pZipGptNeo was constructed by the introduction of a 930-bp fragment containing the *gpt* gene into the retroviral shuttle vector pZipNeoSV(X)1 (Cepko et al. 1984). The latter vector has the properties of a shuttle vector due to the presence of the SV40 origin of replication, the pBR322 origin of replication, and the neomycin (*neo*) gene from Tn*5*. The presence of the SV40 and pBR322 origins of replication enables the vector to replicate in mammalian cells and bacteria, whereas the *neo* gene provides a selectable marker in bacteria (kanamycin resistance) and mammalian cells (G418 resistance). The A9I2 cell line was constructed by the infection of A9 cells with pZipGptNeo virus. The expression of both the *neo* and *gpt* genes enables the cells to grow in both hypoxanthine aminopterin thymidine (HAT) and G418 media.

For the isolation of spontaneous mutants, a series of A9I2 cultures were established and grown in HAT medium supplemented with G418. After 10 days of growth in this medium, the cells from each culture were harvested and inoculated into dishes containing medium supplemented with G418, hypoxanthine, and thymidine. After an additional 10 days of growth to allow for the expression of mutations, the cells from each culture were harvested, and mutants were selected by plating the cells in medium containing 6-thioguanine (TG). After 10–14 days, TG resistant (TG^r) colonies were isolated and expanded for further study. In nine independent experiments, TG cells were observed at a frequency of approximately 2×10^{-5}.

For the isolation of EMS-induced mutants, cultures of A9I2 cells were exposed to 500 μg/ml of EMS for 18 hours. After a 10-day expression period in G418 medium, TG^r mutants were selected. Generally, a five- to tenfold increase in TG^r mutant frequency was observed in EMS-treated cultures relative to untreated control cultures.

A total of 77 spontaneous and 28 EMS-induced mutants have been isolated. Only one mutant was isolated from each separate culture of A9I2 cells in order to ensure that the mutants were independent in origin. Following isolation of the mutants, attempts were made to rescue vector sequences from every mutant by fusion with monkey COS cells. Following fusion, the SV40 origin of replication within the retroviral vector is apparently activated, and the vector sequences can excise from chromosomal DNA by intrastrand recombination between the viral long-terminal repeats (Botchan et al. 1980). Due to the presence of the *neo* gene and the pBR322 origin of replication, the resulting circular molecules have the properties of a bacterial plasmid and can be recovered by transformation into *E. coli*.

Vector sequences were recovered from each of the 28 EMS-induced mutants following COS cell fusion. However, it was found that vector sequences could be recovered from only 43 of the 77 spontaneous mutants. The loss of ability to recover vector sequences from a mutant clone could be the result of an event in which these sequences were lost or rearranged. It was found that almost all (33 of 34) of the mutant cell lines from which vector sequences could not be recovered had lost their resistance to G418, whereas G418 resistance was retained in almost all (42 of 43) of the remaining spontaneous mutants. These results are consistent with the idea that the nonrecoverable mutants are the result of a "large" event such as deletion, inversion, or translocation. High-frequency deletion events involving retroviral sequences have been observed in other studies (Shtivelman et al. 1984; Jolly et al. 1986).

In all cases in which shuttle-vector sequences could be recovered from mutant cells, the alterations in the *gpt* gene were determined by DNA sequencing. Dideoxy-sequencing reactions were carried out directly on double-stranded plasmid DNA (Zagursky et al. 1985) using a series of oligonucleotide primers located 5′ to and within the *gpt* coding region. Among the EMS-induced mutants, it was found that 25 of the 28 mutant genes sequenced have a single-base-pair substitution, and in every case this substitution was a G:C→A:T transition. Each of the base substitutions identified would be expected to produce a change in an amino-acid codon. The locations within the *gpt* gene of these mutations are presented in Table 1. The 25 G:C→A:T transitions occurred at 19 different sites within the gene. The three mutants in the EMS-induced collection that did not have a single-base substitution had deletions of 3, 34, and 893 bp. On the basis of the considerations presented below, these deletion mutants probably represent spontaneous rather than EMS-induced mutations.

A variety of mutational events have been observed among the 43 mutant *gpt* genes that were recovered from the spontaneous mutants. In contrast to the EMS-induced mutations, a relatively small fraction (11 of 43) are single-base substitutions. In addition, among the spontaneous base substitutions, a greater variety of mutations was observed compared to the EMS-induced mutants (Table 2). Transition and transversion mutations were observed in roughly equal numbers, and there was no evidence of any base-substitution hot spots.

The largest class of mutational events among the spontaneous mutations (29 of 43) was deletion. The sizes and locations of these deletions are presented in Table 3. The most striking feature of the spontaneous-deletion

Table 1
EMS-induced Base Substitution Mutants

Base change	No. of isolates	Position
G:C→A:T	1	3
G:C→A:T	1	23
G:C→A:T	1	27
G:C→A:T	1	37
G:C→A:T	1	49
G:C→A:T	1	55
G:C→A:T	3	86
G:C→A:T	1	140
G:C→A:T	1	145
G:C→A:T	2	170
G:C→A:T	2	176
G:C→A:T	1	241
G:C→A:T	1	242
G:C→A:T	1	262
G:C→A:T	3	332
G:C→A:T	1	391
G:C→A:T	1	401
G:C→A:T	1	402
G:C→A:T	1	418

The first base of the *gpt* coding sequence is designated as base no. 1.

Table 2
Spontaneous Base Substitution Mutations

Base change	No. of isolates	Position
G:C→A:T	1	92
G:C→A:T	1	139
G:C→A:T	1	242
A:T→G:C	1	119
A:T→G:C	1	169
G:C→C:G	1	46
G:C→C:G	1	139
G:C→C:G	1	417
A:T→T:A	1	1
A:T→T:A	1	52
A:T→C:G	1	257

The first base of the *gpt* coding sequence is designated as base no. 1.

collection is the existence of a very strong deletion hot spot. The same deletion was isolated from 16 independent cultures. Another interesting feature of these deletions is their small size. Despite the fact that deletions greater than 1200 bp have been sequenced using this system, 11 of 13 different deletions were less than 20 bp in length. A third feature of the spontaneous deletions is that they appear to be nonrandomly distributed throughout the

Table 3
Spontaneous Deletion Mutations

No. of isolates	No. of bases deleted	5′ Base deleted
1	1256	−38, −37, −36
1	1	24
1	2	107
1	7	107
1	7	120–123
1	7	149, 150
1	8	150, 151
1	17	210–212
1	8	213, 214
1	59	228, 229
16	3	366–369
1	3	370
2	3	381, 382

The positions of the most 5′ base deleted in the *gpt* sequence are indicated. In several cases, the exact location of the deletion breakpoint could not be determined due to the presence of a repeated sequence at the breakpoints.

gpt gene. As shown in Table 3, in most cases, a deletion either shares a breakpoint with a second deletion or there is a deletion with a breakpoint less than 15 bp away.

In addition to the deletions and base substitutions, three other spontaneous mutations were sequenced. One mutant has a +1 frameshift that consists of the insertion of an A:T base pair within a run of five A:T base pairs. Two other mutants had multiple changes. In one, the sequence 5′-CGTAAA-3′ is changed to 5′-TTTAA-3′ and the other has five single-base substitutions at sites throughout the *gpt* gene.

DISCUSSION

The great advantage of using shuttle vectors to study mutational specificity in mammalian cells is the ease with which mutant genes can be recovered following mutagenesis in the mammalian cell environment. In contrast to most shuttle-vector systems (Lebkowski et al. 1985; Seidman et al. 1985; Drinkwater and Klinedinst 1986), our system utilizes a single-copy, chromosomally integrated shuttle vector, rather than a multicopy, autonomously replicating vector. Also, mutant genes are selected in mammalian cells rather than following transfer to *E. coli*. Although recovery of mutant genes is somewhat more difficult in our system, it does have certain other advantages. One advantage is that it is possible to quantitate and partially characterize large mutational events that are lost when nonintegrated plasmids are used. A second advantage is that one can be certain that spontaneous mutations have been generated in the mammalian cell rather than in the bacteria subsequent to transfer. Finally, it is possible that integrated genes may more closely resemble endogenous genes with respect to DNA damage and repair.

A point of concern during the development of this system was that mutations would be generated during the recovery process and that it would not be possible to distinguish these mutations from mutations generated in A9I2 cells prior to the rescue. Two lines of evidence presented in this study suggest very strongly that this is not a serious problem. First, despite the fact that the same recovery procedure was used for all of the mutations, the mutational spectra of the spontaneous and EMS-induced mutants are quite different. Second, if mutations were being generated during recovery, one would expect to observe mutants with base-sequence alterations at two different sites. One mutation would have been generated prior to recovery and the other during recovery. However, of 71 mutant genes that were recovered and sequenced in this study, only 1 has base-sequence alterations at more than one site. We have concluded that mutations generated during the recovery process have not made a significant contribution to the mutational spectra we have determined.

The observation that the EMS-induced mutations were largely G:C→A:T transitions is in agreement with other studies utilizing the *lacI* gene of *E. coli* as a target in both bacterial and human cells (Coulondre and Miller 1977; Lebkowski et al. 1986). These mutations are thought to arise as a result of O^6-alkylguanine-mispairing during DNA replication. The study in human cells utilized an autonomously replicating shuttle vector, demonstrating, at least in the case of EMS, that similar types of mutations can be induced in both extrachromosomal and chromosomally integrated genes.

The most interesting feature of the spontaneous mutant collection was the observation of a very strong 3-bp deletion hot spot. The mutations at the deletion hot spot occur in the sequence 5′-TGATGA-3′ and result in the deletion of one TGA. The DNA sequence in the vicinity of this hot spot resembles that of a very strong spontaneous mutation hot spot in the *lacI* gene of *E. coli* (Farabaugh et al. 1978). The *lacI* hot spot occurs at the tandem repeating sequence 5′-CTGGCTGGCTGG-3′ and the mutations consist of either the addition or deletion of one unit of four nucleotides (CTGG). Since other deletions have been observed in the immediate vicinity of the hot spot, it seems likely that sequences other than the 5′-TGATGA-3′ are involved in the generation of these deletions. As mentioned earlier, the spontaneous deletions appear to be clustered rather than randomly distributed throughout the *gpt* gene. It appears that there are regions within the gene where deletions are more likely to occur, although the breakpoints of these deletions may not be exactly the same in each event.

ACKNOWLEDGMENT

We thank Pamela Broeker for expert technical assistance. This work was supported by U.S. Public Health Service Grant CA-31781 from the National Institutes of Health.

REFERENCES

Ashman, C.R., P. Jagadeeswaran, and R.L. Davidson. 1986. Efficient recovery and sequencing of mutant genes from mammalian chromosomal DNA. *Proc. Natl. Acad. Sci.* **83:** 3356.

Botchan, M., W. Topp, and J. Sambrook. 1980. Studies on simian virus 40 excision from cellular chromosomes. *Cold Spring Harbor Symp. Quant. Biol.* **44:** 709.

Cepko, C., B.E. Roberts, and R.C. Mulligan. 1984. Construction and applications of a highly transmissible murine retrovirus shuttle vector. *Cell* **37:** 1053.

Coulondre, C. and J.H. Miller. 1977. Genetic studies of the *lac* repressor. *J. Mol. Biol.* **117:** 577.

Drinkwater, N.R. and D.K. Klinedinst. 1986. Chemically induced mutagenesis in a shuttle vector with a low background mutant frequency. *Proc. Natl. Acad. Sci.* **83:** 3402.

Farabaugh, P.J., U. Schmeissner, M. Hofer, and J.H. Miller. 1978. Genetic studies of the *lac* repressor. VII. On the molecular nature of spontaneous hotspots in the *lac I* gene of *Escherichia coli*. *J. Mol. Biol.* **126:** 847.

Glazer, P.M., S.N. Sarkar, and W.C. Summers. 1986. Detection and analysis of UV-induced mutations in mammalian cell DNA using a lambda phage shuttle vector. *Proc. Natl. Acad. Sci.* **83:** 1041.

Gluzman, Y. 1981. SV40-transformed simian cells support the replication of early SV40 mutants. *Cell* **23:** 175.

Jolly, D.J., R.C. Willis, and T. Friedmann. 1986. Variable stability of a selectable provirus after retroviral vector gene transfer into human cells. *Mol. Cell. Biol.* **6:** 1141.

Lebkowski, J., J.H. Miller, and M.P. Calos. 1986. Determination of DNA sequence changes induced by ethyl methanesulfonate in human cells using a shuttle vector system. *Mol. Cell. Biol.* **6:** 1838.

Lebkowski, J.S, S. Clancy, J.H. Miller, and M.P. Calos. 1985. The *lac I* shuttle: Rapid analysis of the mutagenic specificity of ultraviolet light in human cells. *Proc. Natl. Acad. Sci.* **82:** 8606.

Littlefield, J.W. 1964. Three degrees of guanylic acid-inosinic acid pyrophosphorylase deficiency in mouse fibroblasts. *Nature* **203:** 1142.

Seidman, M.M., K. Dixon, A. Razzaque, R.J. Zagursky, and M.L. Berman. 1985. A shuttle vector plasmid for studying carcinogen-induced point mutations in mammalian cells. *Gene* **38:** 233.

Shtivelman, E., R. Zakut, and E. Canaani. 1984. Frequent generation of nonrescuable reorganized Moloney murine sarcoma viral genomes. *Proc. Natl. Acad. Sci.* **81:** 294.

Zagursky, R.J., K. Baumeister, N. Lomax, and M.L. Berman. 1985. Rapid and easy sequencing of large linear double-stranded DNA and supercoiled plasmid DNA. *Gene Anal. Tech.* **2:** 89.

Probing Mammalian Clones: A New Method with Application for Mutagenesis Research

TOBY G. ROSSMAN, LARRY M. RUBIN, AND JIH-HENG LI
Institute of Environmental Medicine
New York University Medical Center
New York, New York 10016

OVERVIEW

The ability to perform colony hybridization in bacteria has enabled the screening of colonies containing DNA sequences of interest in the absence of selection. We have recently developed a method to screen mammalian cell clones for desired sequences in the absence of selection. Entire plates of colonies are transferred to a nitrocellulose filter and processed for hybridization with the desired probe. By a slight change in protocol, the filters can be processed to reveal either DNA or RNA sequences. By combining this method with replica-plating, it is possible to isolate clones containing alterations in copy number or level of expression of either endogenous or transfected sequences. End-labeled oligonucleotide probes can also be used for probing clones, and we are now exploring the use of these probes to isolate specific mutations.

INTRODUCTION

The identification of variant colonies of mammalian cells has depended upon the availability of conditions that allow the growth of the rare variant clones but select against the original population. Mammalian genetic markers that have proved to be useful in mammalian cell mutagenesis studies select on the basis of resistance to toxic agents such as 6-thioguanine or trifluorothymidine. The resulting mutant colonies must then be isolated, expanded, and the DNA extracted and analyzed, in order to obtain information about the type of mutation involved.

A new technique for probing clones, developed in our laboratory originally to study gene amplification under nonselective conditions (Rossman et al. 1987), has a number of potential applications for mutagenesis research. The technique involves the replica-plating and in situ hybridization of entire dishes of mammalian clones. With this technique, it is possible to directly select for clones with variations in copy number of any DNA or RNA sequence (endogenous or transfected) for which a probe exists.

Banbury Report 28: Mammalian Cell Mutagenesis

RESULTS AND DISCUSSION

The general method used for probing clones of mammalian cells is outlined in Figure 1. If the clones are to be scored for a particular sequence, but not isolated, the replica-plating steps can be eliminated. The method is easy to perform and results can be obtained in a few days (T. Rossman and L. Rubin 1987).

Using nick-translated, ^{32}P-labeled simian virus 40 (SV40) DNA as a probe, Chinese hamster CO60 cells (a line transformed by SV40) were cloned and hybridized (Fig. 2). The X-ray film shows clonal heterogeneity with regard to the number of SV40 sequences contained in each clone. This technique was used to isolate clones containing low copy numbers of SV40 sequences for gene amplification studies (Rossman et al. 1987).

Figure 3 demonstrates the applicability of this technique for studying single-copy genes and their transcripts. Clones of Chinese hamster V79 cells were hybridized with a ^{32}P-labeled hypoxanthine-guanine phosphoribosyl transferase (*hprt*) cDNA probe. When the filter is processed for DNA hybridization (NaOH-treated), faint hybridization signals can be detected from each clone, due to the single copy of the X-linked *hprt* gene in this male cell line. (When the same cells are probed with SV40 DNA, which they do not

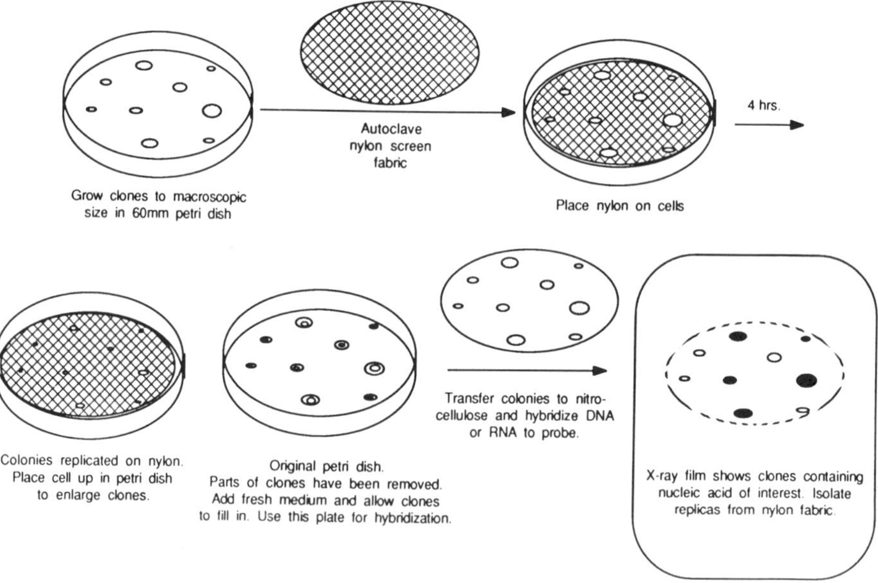

Figure 1

Isolation of mammalian cell clones containing amplified, transfected, or expressed sequences of interest.

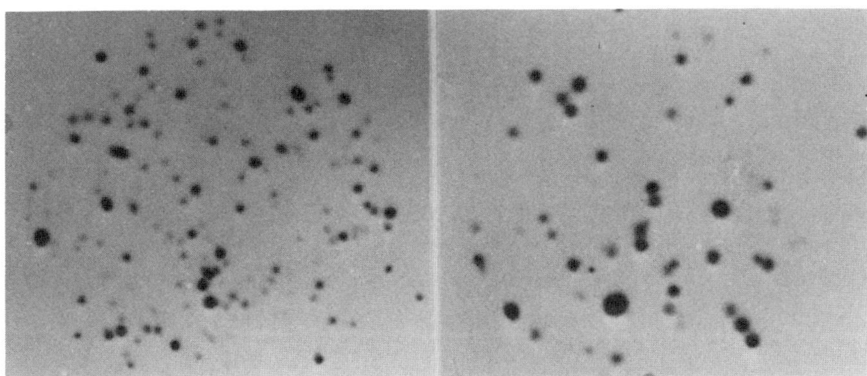

Figure 2
Clones of Chinese hamster CO60 cells probed for SV40 sequences. Heterogeneity in copy numbers is evident.

contain, no signal can be detected). By omitting the NaOH treatment, the cellular RNA remains intact and can be hybridized (Fig. 3, No NaOH), giving a much stronger hybridization signal due to the multiple mRNA transcripts of this expressed gene. We have also demonstrated that (unexpressed) metallothionine sequences cannot be detected unless the filters are processed for DNA hybridization (NaOH-treated) and that influenza RNA sequences can

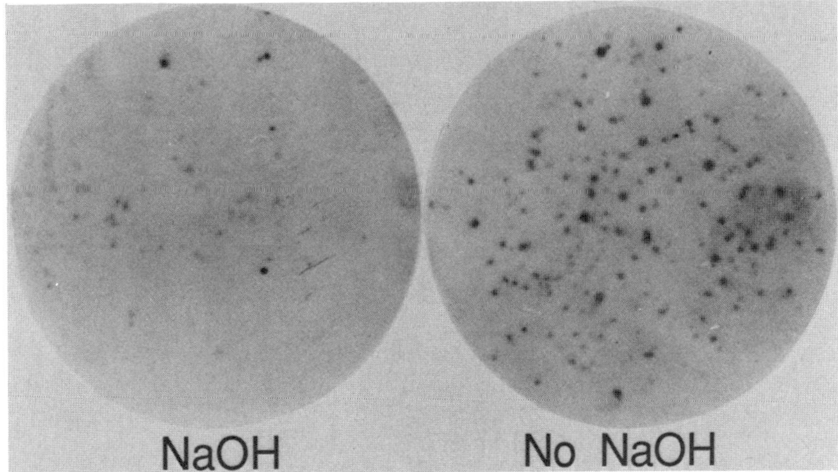

Figure 3
Clones of Chinese hamster V79 cells probed with *hprt* cDNA. NaOH: DNA sequences; No NaOH: RNA sequences.

only be detected in influenza-infected cells if the filters are processed for RNA hybridization (No NaOH) (T. Rossman and L. Rubin 1987).

Recently, we have begun to explore the use of end-labeled oligonucleotide probes for colony hybridization. When a 19-mer corresponding to a portion of wild-type mouse Harvey *ras* including the 12th codon, was end-labeled and hybridized to colonies of NIH 3T3 cells, all colonies showed a hybridization signal (J.-H. Li, unpubl.). We are currently determining the optimal conditions for differential hybridization using oligonucleotides containing a single base change. It may be possible to perform oncogene mutagenesis experiments in the absence of selection for tumorigenicity.

SUMMARY

Direct probing of mammalian clones can be of use in evaluating mutational spectra. One obvious application would be the identification of mutant clones containing deletions of any region desired. Also, mutant clones that fail to express a particular sequence could be identified by probing for RNA sequences. The construction of new mammalian cells for mutational studies, via transfection of bacterial genes or minigenes, might also be simplified by a technique allowing the selection of transfectants with low copy numbers.

ACKNOWLEDGMENTS

This work was supported by grants CA-35631 (National Cancer Institute) and ES-03847 (National Institute of Environmental Health Sciences) and is part of NYU Institute of Environmental Medicine Center programs supported by grants ES-00260 from the National Institute of Environmental Health Sciences, grant CA-13343 from the National Cancer Institute and by the American Cancer Society grant SIG-9.

REFERENCES

Rossman, T.G. and L.M. Rubin. 1987. Isolation of mammalian cell containing non-selectable amplified, transfected or expressed sequences. *Somatic Cell Mol. Genet.* (in press).

Rossman, T.G., L.M. Rubin, T.M. Gilmer, and M.L. Steinberg. 1987. Heterogeneity in the response of inserted viral sequences to carcinogen treatment: A clonal approach. In *The role of DNA amplification in tumor initiation and promotion* (ed. H. zur Hausen et al.). J.B. Lippincott, New York. (In press.)

Molecular Analysis of Shuttle Vectors and SV40

EBV-*lacI* Shuttle Vectors: Addition of a Selection for LacI Mutants

DAVID J. VAN DEN BERG, ROBERT B. DUBRIDGE, AND MICHELE P. CALOS
Department of Genetics
Stanford University School of Medicine
Stanford, California 94305

OVERVIEW

Mutation can be conveniently studied in human tissue-culture cells using shuttle vectors. Vectors based on Epstein-Barr virus (EBV) replicate in human cells with a very low background mutation frequency. We have used the *lacI* gene of *Escherichia coli* on such vectors as a target for mutation. After mutagenesis, the vectors are returned to *E. coli*, where mutations are rapidly scored and analyzed. Previously, we screened for LacI mutants using indicator plates on which I^+ colonies appeared white and I^- colonies were blue. We report here a selective system in which only the I^- colonies appear. This system preserves all the advantages of using *lacI* in a forward mutation, shuttle-vector system, but makes detection of I^- mutants even simpler, faster, and less expensive.

INTRODUCTION

The complexity of studying mutation in human cells at the molecular level can be reduced by focusing on a well-defined target gene that can be easily retrieved from human cells and analyzed in bacteria. Our approach to carrying out such a strategy has focused on shuttle vectors, small plasmids that can replicate in both mammalian and bacterial cells. We have placed the bacterial *lacI* gene on shuttle vectors as the target for mutation and introduced the vectors into mammalian cells. After mutagenesis takes place, vector DNA is recovered and introduced into *E. coli*, where mutations can be rapidly detected, localized, and analyzed at the DNA sequence level. The *lacI* gene offers a 1000-bp target for the isolation of forward mutations of all types. A sophisticated genetic system for *lacI* (Miller 1980) allows characterization of mutations using genetic techniques. Furthermore, a sizeable body of information exists for mutation of *lacI* in *E. coli* (Miller 1983), so that the behavior of the same target gene in bacteria and mammalian cells can be compared.

The first shuttle-vector experiments utilized vectors based on simian virus

Banbury Report 28: Mammalian Cell Mutagenesis

40 (SV40) (Calos et al. 1983; Razzaque et al. 1983). However, these experiments were limited by a high-background frequency of mutation associated with transfection (Lebkowski et al. 1984). The experiments were also necessarily short-term in nature, because replication of SV40 vectors was not compatible with long-term survival of the transfected mammalian cells.

Both of these limitations were overcome by using shuttle vectors based on EBV (Drinkwater and Klinedinst 1986; DuBridge et al. 1987). These vectors are carried indefinitely in the nuclei of human cells. Clonal lines carrying unmutated vectors can therefore be established and propagated. The vectors in such cell lines do not accumulate mutations at an appreciable frequency during passage (DuBridge et al. 1987). Thus, these cell lines are suitable for studies of mutation.

The great sensitivity provided by the low background of spontaneous mutations in EBV-*lacI* vectors invites studies of rare mutations (i.e., spontaneous mutations or mutations induced by low doses of mutagens). These studies necessarily involve analysis of large numbers of vector molecules. If every bacterial colony generated by a shuttle-vector molecule must be counted and scored for mutation on an indicator plate, a significant amount of work is involved. The amount of work and materials could be reduced if only the rare vectors carrying mutations in *lacI* gave rise to colonies when introduced into bacteria. This situation can be achieved by creating a selection for LacI mutants, in which only colonies carrying I^- vectors can grow.

Our strategy for creating such a selection has been to place the gene for chloramphenicol acetyltransferase (*cat*) under the control of the *lac* repressor, the product of the *lacI* gene. In the presence of the wild-type *lacI* gene, transcription of *cat* will be repressed, and growth in medium containing chloramphenicol will be precluded. Bacteria receiving I^+ shuttle vectors will not grow on plates containing chloramphenicol; however, bacteria receiving I^- vectors will not repress *cat* expression, and such bacteria will form colonies on plates containing chloramphenicol. A simple selection for I^- mutants is thus achieved. The construction and operation of this selective system are described below.

RESULTS

The screening system we have previously used for detection of LacI mutants is depicted in Figure 1A. This screen takes advantage of the normal role of *lacI* in transcriptional regulation of *lacZ*. The wild-type repressor binds to the *lac* operator 5′ to *lacZ*, blocking transcription. The *lacZ* gene product, β-galactosidase, is therefore not expressed in an I^+ situation. A mutation in *lacI* that inactivates the repressor will permit β-galactosidase to be made. The

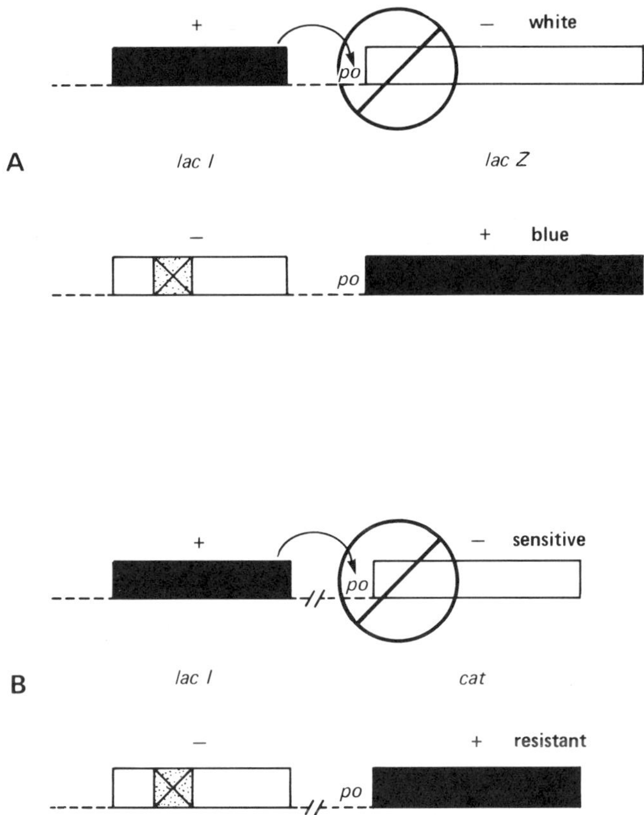

Figure 1

Screening and selective systems for detection of LacI mutants. (*A*) Screen; (*B*) selection. The upper part of each section depicts the I^+ situation, and the lower part diagrams the I^- situation.

presence of this enzyme can be detected on indicator plates containing X-Gal (5-bromo-4-chloro-3-indolyl-β-D-galactoside), since β-galactosidase cleaves X-Gal to form indigo, rendering the I^- colonies blue.

Selections for I^- mutants, in which only I^- colonies can grow, also exist (Miller 1972). These selections involve growth on minimal plates containing galactoside derivatives. These plates are more complex and expensive than rich plates, and it takes longer for colonies to grow on them. We preferred a selective system that was based on rich plates, since it would be simpler, faster, and less expensive.

The selective system we devised is shown in Figure 1B. The *cat* gene, conferring resistance to chloramphenicol, has been placed under the transcriptional control of *lacI* by fusing the *cat* coding sequence to the promoter and operator of *lacZ*. Thus, *cat* will not be transcribed in I^+ cells. In I^- cells, *cat* will be expressed, and such cells will form colonies when grown on plates containing chloramphenicol.

The optimal way to arrange the selective system would be to place the *cat* indicator gene in the recipient bacterial cell rather than on the shuttle vector. This arrangement ensures that the *cat* gene will not then be subject to mutation along with the target gene. This arrangement also permits us to use the chloramphenicol selection on any *lacI* shuttle vector, whether or not it contains the *cat* gene.

We established the *cat* gene in our recipient MC1061*recA* F′150Hg strain (Casadaban and Cohen 1980; Calos et al. 1983; Miller et al. 1984; see below) by placing *cat* on a plasmid. For this purpose, we used a plasmid based on pSC101. The pSC101 replicon was used because it is compatible with the pML replicon present on our shuttle vectors. The pML plasmid is derived from pBR322, which has the replication region from ColE1 and a gene conferring resistance to ampicillin.

The pSC101*cat* plasmid we constructed is shown in Figure 2. The first step in this construction was to create a translational fusion between the beginning of *lacZ* and *cat*. This fusion was created by placing a 0.82-kb *Bam*HI fragment containing the *cat* gene from pCM-4 (Pharmacia) into the unique *Bam*HI site in pUC18. This construct contains the first 14 codons of *lacZ* and the entire coding region of *cat*, with *cat* expression now under control of the *lac* promoter/operator. A 0.95-kb *Pst*I fragment containing the *lac* promoter/operator-*cat* fusion was excised from a pUC18 derivative and inserted into the *Pst*I site 250 bp downstream from the kanamycin resistance gene of pSLB20-8 to create pCAT. The pSLB20-8 plasmid is a kanamycin-resistant derivative of pSC101 obtained from D. Biek and S. Beaucage. It was formed from pPM30 (Meacock and Cohen 1980) as follows: The 0.1-kb *Bam*HI-*Eco*RI fragment containing the *lac* operator was removed from pPM30, and the *Eco*RI site

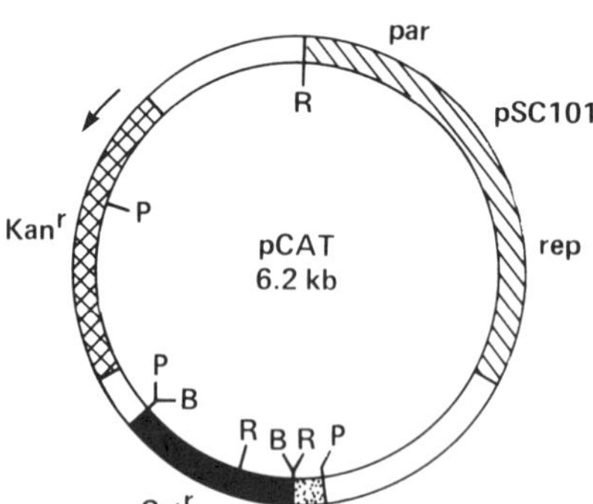

Figure 2
Structure of pCAT. The replication (*rep*) and partition (*par*) regions of pSC101 are indicated, as are the *lac* operator (*o*) and promoter (*p*). (*B*) *Bam*HI, (*E*) *Eco*RI, (*P*) *Pst*I.

was recreated by blunting the *Bam*HI end and adding an *Eco*RI linker. A 1.3-kb fragment containing the kanamycin resistance gene from Tn*5* was inserted into the *Sca*I site in the ampicillin resistance gene to create pSLB20-8. The salient features of pCAT are that it contains the *cat* gene under *lac* promoter/operator control, it contains the replication region from pSC101, and it confers resistance to kanamycin.

The pCAT plasmid was transformed into MC1061*recA* F′150Hg bacteria and kanamycin-resistant colonies were selected. This pCAT-containing MC1061 recipient strain, called McCAT, has the following four features relevant to shuttle-vector experiments: (1) The strain transforms well, giving 10^6 to 10^8 colonies per μg of plasmid DNA. (2) The RecA phenotype inactivates the bacterial SOS system, making it more likely that mutations detected are formed in the human cell and not the bacterial cell. (The *recA56* allele was introduced by cotransduction with a sorbitol gene containing a Tn*10* insertion. Therefore, the strain is resistant to tetracycline.) (3) The strain carries an F′*lac pro* episome with the Δ*150lacI-Z* deletion. This deletion removes most of the *lacI* coding sequence and the first part of *lacZ* (Gho and Miller 1974). It constitutively expresses the latter part of *lacZ*, which can act as an α-acceptor in *lacZ* α-complementation (Miller 1972). Therefore, when our *lacI* shuttle vectors are introduced, the product of the α portion of *lacZ* can associate with the carboxy portion of the gene product to give β-galactosidase activity. This activity will be detected as a blue color on plates containing X-Gal if the shuttle vector harbors an I^- mutation. The F′150 episome was marked by transposing Tn*501* to it (Weiss and Falkow 1983). Tn*501* confers resistance to mercury. (4) Finally, McCAT carries pCAT so that colonies containing I^- vectors can be selected on plates carrying chloramphenicol. Therefore, the McCAT strain can be used to select I^- mutants as well as to screen for them.

After extraction from human cells, shuttle-vector DNA is transformed into frozen competent McCAT cells and the cells are plated on HACK medium. HACK medium is LB agar (Luria broth; Miller 1972) containing mercury (9 μg/ml $HgCl_2$), ampicillin (50 μg/ml), chloramphenicol (15 μg/ml), and kanamycin (10 μg/ml). We include 50 μg/ml of methicillin when ampicillin is used to prevent the growth of satellite colonies. These plates select for the presence of the F′150 episome, pCAT, and the shuttle vector. The shuttle vector must contain an I^- mutation to give a colony on HACK plates. Colonies appearing on HACK plates are streaked onto HACK-X plates (HACK plus 40 μg/ml X-Gal) for purification and for verification of I^- phenotype by appearance of blue color and chloramphenicol resistance.

To obtain the mutation frequency, the total number of colonies (I^+ and I^-) must be determined. To obtain this number 5% of the sample is spread onto plates lacking chloramphenicol (MAX plates). Multiplying this colony num-

Table 1
Comparison of the Detection of I^- Mutants by Selection (HACK) or Screening (MAX)

Plate	Total colonies	I^-	Percentage of I^-
HACK	53,680	261	0.49
MAX	24,634	118	0.48

ber by 20 gives an indication of the total colony number. These MAX plates contain LB agar, mercury, ampicillin/methicillin, and X-Gal. The rest of the sample is spread on a HACK plate to obtain the I mutants.

We have tested the ability of this *cat* selection to detect I^- mutants. The system appears to have a negligible background of false positives, since no chloramphenicol-resistant colonies appear after 20 hours incubation when I^+ plasmid DNA is used for transformation or when McCAT cells are plated on chloramphenicol medium without ampicillin/methicillin. Therefore, the occurrence of spontaneous mutants resistant to 15 μg/ml of chloramphenicol is very rare. Any such mutants would be detectable in our experiments, since they would yield white instead of blue colonies upon streaking on HACK-X plates.

We have also tested to ensure that the selection system quantitatively detects all I^- mutants of interest. As shown in Table 1, when a mixture of I^+ and I^- vector DNA is transformed into McCAT cells and assayed on HACK and MAX plates, the same number of I^- colonies are detected. In this experiment, a 200:1 mixture of I^+ pSVi2 (Calos et al. 1983) and I^- pSVi2 DNA (carrying the *T63* I^- missense mutation) was transformed into McCAT cells. Part of the transformation was assayed for colony number on MAX plates and I^- mutants on HACK-selective plates, and the remainder was assayed on MAX plates, where I^- mutants were identified as blue colonies. The percentage of I^- mutants detected by either system is virtually identical. Therefore, the chloramphenicol selection system detects all I^- colonies that would have been identified as blue colonies in the screening system. We have tested I^- vector DNA containing mutations that give various shades of blue colonies, from light to dark, on X-Gal plates. All of these classes of I^- mutants appear to be quantitatively detected by the selection system.

DISCUSSION

We have developed a simple selection system to make the detection of I^- mutations in shuttle vectors easier. In this system, only vectors harboring I^- mutations can give rise to bacterial colonies on plates containing chloramphenicol. This selection system offers several advantages over the screening systems we have previously used. Since only I^- colonies will grow on the plates, large numbers of cells can be spread onto a small number of plates, saving a great deal of time and plates. More effort is saved, since large numbers of colonies do not need to be counted. Furthermore, the I^- mutant colonies are easier to identify when they appear as isolated colonies than when they appear as rare blue colonies among large numbers of white colonies. The reduced number of plates also produces a significant reduction in expenses, particularly since costly X-Gal is not used in the selection plates.

This selection system is particularly useful in situations in which large numbers of vector molecules must be screened for rare mutations. The EBV-*lacI* shuttle vectors we have developed have low enough background mutation frequencies to make the selection system a distinct advantage (DuBridge et al. 1987). The amount of shuttle-vector DNA that can be extracted from a plate of human cells in the current systems is sufficient to produce several thousand bacterial colonies. However, new advances are making it possible to increase vector copy number in the human cells so that much larger bacterial colony numbers can be obtained from the same number of human cells (R.B. DuBridge et al., in prep.). In the future, it will be possible to recover at least 10^5 to 10^6 vector molecules from a single plate of human cells. In this situation, a selective system for the detection of I^- shuttle vectors is essential.

ACKNOWLEDGMENTS

D.J.V. and R.B.D. were supported by National Institutes of Health training grants. This work was supported by grants to M.P.C. from the National Cancer Institute and the American Cancer Society.

REFERENCES

Calos, M.P., J.S. Lebkowski, and M.R. Botchan. 1983. High mutation frequency in DNA transfected into mammalian cells. *Proc. Natl. Acad. Sci.* **80:** 3015.

Casadaban, M.J. and S.N. Cohen. 1980. Analysis of gene control signals by DNA fusion and cloning in *Escherichia coli. J. Mol. Biol.* **138:** 179.

Drinkwater, N.R. and D.K. Klinedinst. 1986. Chemically induced mutagenesis in a shuttle vector with a low-background mutant frequency. *Proc. Natl. Acad. Sci.* **83:** 3402.

DuBridge, R.B., P. Tang, H.C. Hsia, P.-M. Leong, J.H. Miller, and M.P. Calos. 1987. Analysis of mutation in human cells by using an Epstein-Barr virus shuttle system. *Mol. Cell. Biol.* **7:** 379.

Gho, D. and J.H. Miller. 1974. Deletions fusing the *i* and *lac* regions of the chromosome in *E. coli*: Isolation and mapping. *Mol. Gen. Genet.* **131:** 137.

Lebkowski, J.S., R.B. DuBridge, E.A. Antell, K.S. Greisen, and M.P. Calos. 1984. Transfected DNA is mutated in monkey, mouse, and human cells. *Mol. Cell. Biol.* **4:** 1951.

Meacock, P.A. and S.N. Cohen. 1980. Partitioning of bacterial plasmids during cell division: A *cis*-acting locus that accomplishes stable plasmid inheritance. *Cell* **20:** 529.

Miller, J.H. 1972. *Experiments in molecular genetics.* Cold Spring Harbor Laboratory, Cold Spring Harbor, New York.

———. 1980. Genetic analysis of the *lac* repressor. *Curr. Top. Microbiol. Immunol.* **90:** 1.

———. 1983. Mutational specificity in bacteria. *Annu. Rev. Genet.* **17:** 215.

Miller, J.H., J.S. Lebkowski, K.S. Greisen, and M.P. Calos. 1984. Specificity of mutations induced by transfected DNA by mammalian cells. *EMBO J.* **13:** 3117.

Razzaque, A., H. Mizusawa, and M.M. Seidman. 1983. Rearrangement and mutagenesis of a shuttle vector plasmid after passage in mammalian cells. *Proc. Natl. Acad. Sci.* **80:** 3010.

Weiss, A.A. and S. Falkow. 1983. Transposon insertion and subsequent donor formation promoted by Tn*501* in *Bordetella pertussis. J. Bacteriol.* **153:** 304.

Analysis of Mutation Induction in Mammalian Cells with a Simian Virus 40-based Shuttle Vector

KATHLEEN DIXON,* JANET HAUSER,* NARENDRA TUTEJA,*† MIROSLAVA PROTIĆ-SABLJIĆ,*‡ EMMANUEL ROILIDES,* PETER J. MUNSON, AND ARTHUR S. LEVINE***

*Section on Viruses and Cellular Biology and
**Laboratory of Theoretical and Physical Biology
National Institutes of Child Health and Human Development
National Institutes of Health, Bethesda, Maryland 20892

OVERVIEW

We have used a simian virus 40 (SV40)-based shuttle vector (pZ189) to analyze the sequence specificity of point mutations induced in mammalian cells. This vector can replicate in both primate cells and bacterial cells, and it carries the *Escherichia coli-supF* suppressor tRNA gene that serves as the mutagenesis marker. The small size of the mutagenesis target and its placement within the vector serve to minimize background mutant frequencies, often observed with SV40-based vectors. To characterize the mutational specificity of UV-radiation and benzo[a]pyrene dihydrodiol-epoxide (B[a]PDE) in this system, the vector DNA was treated with either agent in vitro and then introduced into mammalian cells by transfection. After replication, the vector DNA was isolated and screened for mutations in bacterial cells. Analysis of the *supF* gene in mutant vectors by DNA sequencing reveals that the spectrum of base-substitution mutations is very different for the two mutagens. Furthermore, there appears to be a correlation between the type of mutational change observed and the type of DNA damage induced by the mutagen. Thus, the mutations appear to be targeted to sites of DNA damage. However, mutational hot spots do not necessarily correlate with sites of excessive DNA damage. Indirect effects on mutagenesis in this system have been observed when the mammalian cells are treated with a DNA-damaging agent before transfection with either untreated or mutagen-treated vector DNA. We believe that this vector system provides a versatile tool for studying mechanisms of mutagenesis in mammalian cells.

†Present address: Jules Stein Eye Institute, University of California, Los Angeles, California 90024; ‡Permanent address: Laboratory of Marine Molecular Biology, Center for Marine Research Zagreb, Rudjer Boskovic Institute, Zagreb, Yugoslavia.

INTRODUCTION

To understand the underlying mechanisms of mutagenesis, it is necessary to investigate mutation frequencies, the distribution of mutations within single genes and among different genes, the specific types of mutational changes that occur (i.e., deletions, additions, DNA rearrangements and/or base substitutions), and the characteristics of these types of mutations in terms of site specificity and sequence changes. In mammalian cells, selective tests have been developed that allow the investigation of mutation frequencies at a number of different genetic loci; however, characterization of specific mutational changes at the DNA sequence level is not possible in some cases and in others it is technically very laborious. Recently, a new approach to the analysis of the sequence specificity of mutagenesis in mammalian cells has been developed that involves the use of shuttle-vector plasmids. These plasmids can replicate in mammalian cells, and they have selectable mutagenesis markers that can be screened in bacteria. Mutations induced during passage of the vector through mammalian cells can be analyzed by genetic techniques in bacteria or by direct DNA sequencing. The vectors currently in use are of three general types: The first type is based on SV40 (Lebkowski et al. 1985; Seidman et al. 1985) or bovine papillomavirus (Ashman and Davidson 1985); these vectors replicate autonomously in the nucleus of transfected cells. The second type is based on the Epstein-Barr virus (Drinkwater and Klinedinst 1986; DuBridge et al. 1987); in stable transformants, these vectors replicate as free plasmids in the nucleus, but the replication frequency is under cellular control. The third type is integrated in the cellular genome and replicates along with other cellular genes (Ashman et al. 1986; Glazer et al. 1986); these vectors can only be rescued by treatments that allow the DNA to be released from the cellular genome or by in vitro DNA-packaging techniques.

Our analysis of the sequence specificity of mutagenesis in mammalian cells has been carried out using a shuttle vector of the first type. There are several advantages to this type of vector system: (1) Experiments can be performed readily in most primate cells without the necessity of first establishing stable transformants, required with the other two types of vector systems. (2) The vector DNA can be treated with a mutagen in vitro before transfection into mammalian cells, or the vector-containing cells can be exposed to the test agent after transfection. (3) Recovery of sufficient quantities of vector DNA from the mammalian cells is relatively quick and easy. (4) The positions at which mutagen treatment introduces damage to the DNA can be readily determined. The major disadvantage of this type of system is that there is a relatively high level of spontaneous mutation generated as a consequence of the mere transfection of the mammalian cells with the vector DNA. In the pZ189 system that we have used (Seidman et al. 1985), this problem has been alleviated by the choice of a very small mutagenesis target ($\sim$200 bp) placed between two functions required for plasmid growth and selection in bacteria. Although the resulting frequency of spontaneous mutants ($\sim 5 \times 10^{-4}$) is still well above that observed for cellular genes, this background is sufficiently low to allow detection of induced mutants at frequencies that are 10–20-fold higher. Furthermore, the spontaneous level is reduced five- to tenfold in certain human cell strains (i.e., adenovirus 293 cells and xeroderma pigmentosum cells [Bredberg et al. 1986]). It is not yet clear whether the episomal replication mode of this shuttle vector influences the mutational spectrum; comparison with results derived from other types of vector sytems, as well as from experiments with endogenous cellular genes, should resolve this question.

RESULTS

The pZ189 Shuttle-vector Plasmid

The pZ189 shuttle-vector plasmid (Fig. 1; Seidman et al. 1985) contains the early region of SV40, which permits replication of the vector in mammalian cells (monkey and human); pBR327 replicative functions and the β-lactamase gene, together permitting growth and selection in bacterial cells; and the *supF* suppressor tRNA gene, which serves as a mutagenesis marker. When pZ189 transforms the bacterial selector strain MBM7070 (Seidman et al. 1985), blue colonies are formed on appropriate selective medium due to suppression of an amber mutation in the β-galactosidase gene of the bacterial cells; SupF$^-$ mutants of pZ189 can be identified by the formation of white or light-blue colonies. To study mutagenesis in mammalian cells, the damaged (or undamaged) pZ189 DNA is introduced into mammalian cells by transfection and allowed to replicate. After about 48 hours, vector DNA is isolated from the cells, treated with *Dpn*I to remove nonreplicated input DNA, and then introduced into bacterial cells by transformation to screen for SupF$^-$ mutants. The mutational change in the SupF$^-$ mutants is then determined by direct DNA sequencing. The *supF* gene is a particularly good mutagenesis marker, since most base-substitution mutations within the tRNA coding sequence are likely to lead to inactivation of gene function; so far, single-base substitution mutations have been identified at more than half of the possible sites within the tRNA gene.

UV-induced Mutations

When pZ189 DNA is treated with UV-radiation (500 J/m^2; 254 nm) before transfection of monkey cells, *supF*$^-$ mutants are recovered at a frequency of about 20 times the spontaneous background. Replication of the plasmid in the monkey cells is about 10–20% that of the unirradiated control under these

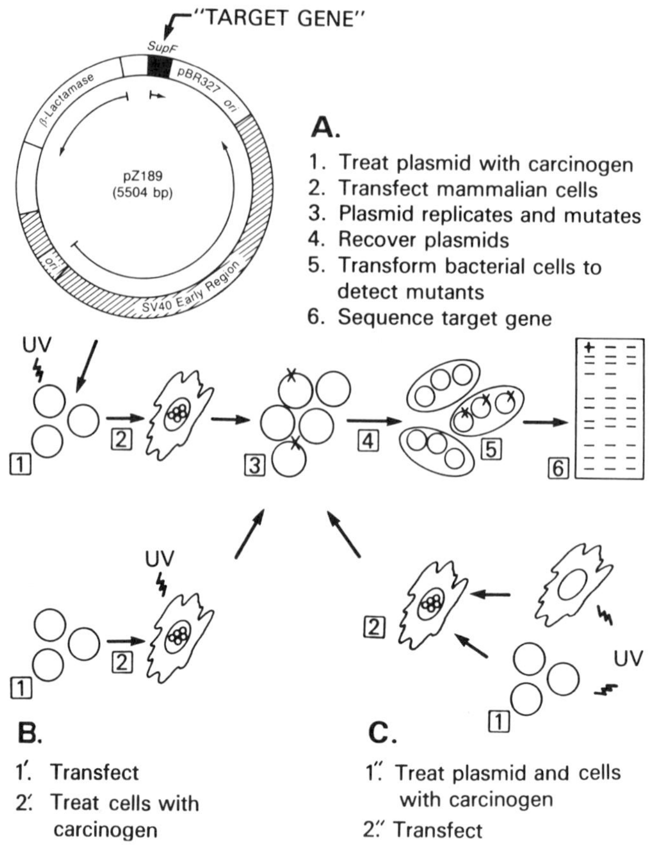

Figure 1

The pZ189 shuttle-vector system for studying mutagenesis in mammalian cells. Construction and characterization of the vector have been described previously (Seidman et al. 1985).

conditions. Most UV-induced mutations are either single- or tandem double-base substitutions, and the majority are G:C→A:T transitions (Table 1) (Hauser et al. 1986). A similar mutational spectrum was found with UV-irradiated pZ189 in normal human cells (Bredberg et al. 1986). We have used mathematical modeling and statistical analysis to compare the sites and frequencies of the UV-induced mutations (Fig. 2) with the sites and frequencies of UV-induced photoproducts in the *supF* gene of pZ189. We find (Munson et al. 1987) a good correlation between the two if we invoke the "A" insertion rule first proposed by I. Tessman (cited in Tessman 1985); i.e., the DNA polymerase preferentially inserts A opposite UV-induced photoproducts (CC, CT, TC, or TT dimers), so that the majority of mutations are

Table 1
Relative Frequencies of Base-substitution Mutations

	Treatment		
	none	UV[a]	B[a]PDE[b]
Transitions			
G:C→A:T	64	61	22
A:T→G:C	0	7	2
Transversions			
G:C→T:A	28	13	34
G:C→C:G	7	5	34
A:T→C:G	0	4	0
A:T→T:A	1	9	7

[a]UV influence, 500 J/m^2.
[b]B[a]PDE concentration, 19–48 μM.

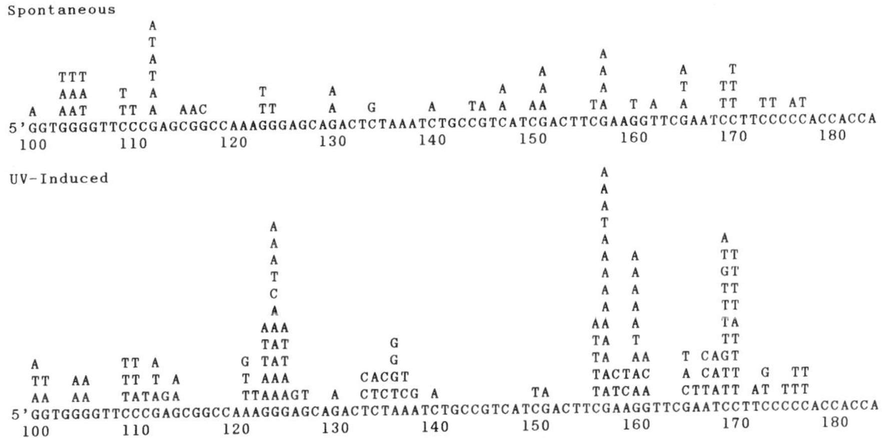

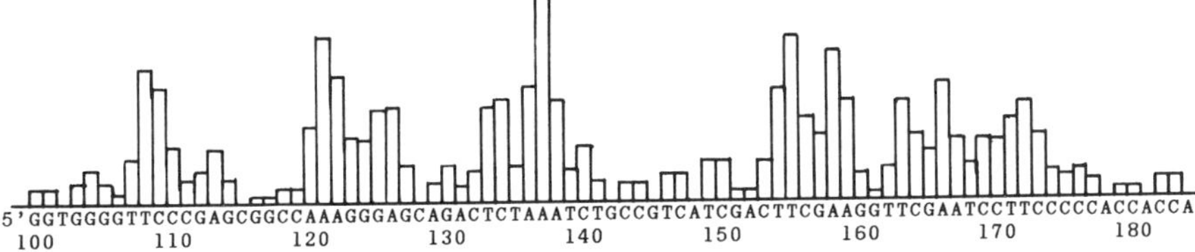

Figure 2
Base substitution mutations observed in the coding sequence of the *supF* gene. Both single and multiple mutations are included. To obtain UV-induced mutations, pZ189 was treated with 500 J/m^2 UV-radiation before transfection. Positions and frequencies of UV-induced photoproducts were determined by the DNA polymerase stop assay as described previously (Munson et al. 1987). Relative heights of bars indicate relative frequency with which the pyrimidine of each base pair participates in photoproduct formation (either cyclobutane pyrimidine dimer or [6-4]-photoproduct).

G:C→A:T transitions, and mutations at A:T base pairs occur rarely. Although this model adequately explains mutation frequencies at most sites in the *supF* gene, there are some notable hot spots (e.g., positions 156 and 169) that cannot be explained in this manner. The same hot spots were also observed in normal human cells and in cells from repair-deficient xeroderma pigmentosum patients (Bredberg et al. 1986), indicating that these changes are not dependent on the species nor the repair capacity of the cell. Perhaps unidentified DNA structural elements, or differences in the response of the DNA polymerase to DNA damage occurring within particular DNA sequence environments, are responsible for these mutational hot spots.

To determine whether cyclobutane pyrimidine dimers are important in mutation induction by UV-radiation, we treated UV-irradiated (500 J/m^2) pZ189 DNA with *E. coli* DNA photolyase (Sancar et al. 1984), which selectively resolves cyclobutane pyrimidine dimers restoring the bases to normal. This DNA was then used to transfect monkey cells. Removal of at least 90% of the dimers restored 80% of the infectivity of the UV-irradiated DNA and reduced the mutation frequency by a factor of 5 (Protić-Sabljić et al. 1986). The spectrum of mutations was somewhat altered so that there were relatively fewer tandem double-base changes and G:C→A:T transitions. Thus, the majority of mutations caused by UV-irradiation of the vector DNA are due to cyclobutane pyrimidine dimers.

Benzo[a]pyrene 7,8-dihydrodiol-9,10-epoxide-induced Mutations

B[a]PDE is thought to be the metabolic intermediate of benzo[a]pyrene (B[a]P) that is ultimately responsible for mutagenesis and carcinogenesis by this aromatic hydrocarbon (for review, see Conney 1982; Pelkonen and Nebert 1982). To determine the mutational spectrum of B[a]PDE, pZ189 was treated with B[a]PDE before transfection of monkey cells. An increase in mutant frequencies of as much as 40-fold was observed at the highest concentrations of B[a]PDE used. Sequence analysis of 54 B[a]PDE-induced mutations in the *supF* gene revealed that almost all mutations occur at G:C base pairs and about 30% are single-base deletions (Fig. 3; Table 1). Preferential mutagenesis at G:C base pairs is consistent with the mode of binding of B[a]PDE to DNA; B[a]PDE forms adducts mostly with the 2-amino group of G residues in DNA (Weinstein et al. 1976). We used the DNA polymerase stop assay to determine the sites and frequencies of B[a]PDE adduct formation in the *supF* gene of pZ189; we detected binding exclusively to G. There does not appear to be a strong selectivity in base misincorporation on templates carrying B[a]PDE:G adducts; all three base substitutions occur at about equal frequency (Table 1). We have initiated experiments to compare the mutational spectrum observed when DNA is treated in vitro with B[a]PDE with the mutational spectrum observed when cells that contain

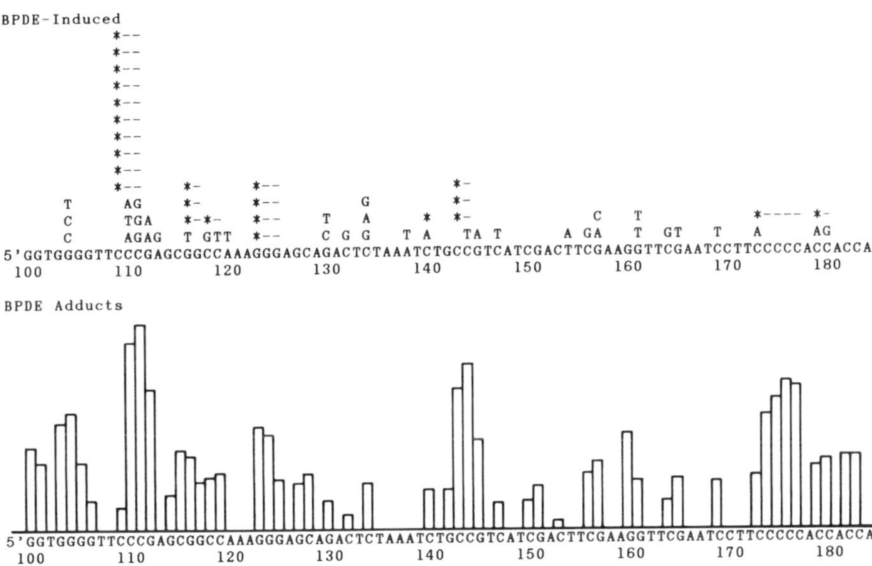

Figure 3

Base-substitution mutations and single-base deletions (*) observed in the coding sequence of the *supF* gene. Both single and multiple mutations are included. To obtain B[a]PDE-induced mutations, pZ189 was treated with 19–48 μM B[a]PDE before transfection. Positions and frequencies of B[a]PDE adducts were determined by the DNA polymerase stop assay as described previously (Munson et al. 1987). Relative heights of bars indicate relative frequency with which the G of each G:C base pair participates in adduct formation.

untreated plasmid DNA are treated with B[a]P or B[a]PDE. In this way, we should be able to determine what role other possible metabolic products of B[a]P may play in mutagenesis and whether B[a]PDE interacts with chromatinized DNA in the cell differently than it interacts with naked DNA in vitro.

Indirect Effects of DNA-damaging Agents on Mutant Frequencies

One advantage of pZ189 and other shuttle-vector systems of the first type described above is that they offer the possibility of studying indirect effects of DNA-damaging agents on mutagenesis in mammalian cells. The whole cell and the vector DNA can be treated independently with the same or different agents to assess the relative contributions of direct template damage and indirect cell-wide effects on the frequency and sequence characteristics of induced mutations. Our preliminary results with this type of experiment indicate that an indirect, inducible response to UV-radiation or mitomycin-C treatment of cells can be observed; the frequency of mutants increases when UV-treated pZ189 is replicated in pretreated monkey cells (Table 2). We are

Table 2
Effect of Mitomycin-C Pretreatment on Frequency (%) of supF⁻ Mutants

Treatment of pZ189	Treatment of cells	
	none	mitomycin C[a]
None	<0.04	<0.03
500 J/m^2	1.2	1.8

[a]Mitomycin C, 1 μg/ml for 2.5 hr.

eager to determine whether this inducible enhancement of mutagenesis is accompanied by a change in the spectrum of mutational changes that can be observed.

DISCUSSION

The pZ189 shuttle-vector plasmid has proven to be a very useful tool for investigating the mechanisms of mutagenesis in mammalian cells. Our investigations of induced mutagenesis have allowed the development of models to describe the response of the mammalian DNA polymerase(s) to UV-induced DNA damage and the identification of the primary premutagenic lesion as the cyclobutane pyrimidine dimer. Since B[a]PDE induces a very different spectrum of mutations in the same shuttle-vector system, the mutations observed must be a consequence of the particular type of damage introduced into the template DNA and not some unique characteristic of the mutagenesis-test system itself. By treating the vector DNA and the mammalian cells independently with mutagens, we have begun to analyze the relative contributions of template damage and cell-wide effects to mutagenesis. Further validation of this vector system as a mutagenesis assay will come from comparison of the mutational spectra observed here with those observed with the use of other types of shuttle-vector systems and viruses, as well as endogenous cellular genes. We believe that, at least for some applications, the advantages of the simplicity and versatility of this shuttle-vector system outweigh the disadvantage of the relatively high spontaneous background mutagenesis observed with this type of system.

REFERENCES

Ashman, C.R. and R.L. Davidson. 1985. High spontaneous mutation frequency of BPV shuttle vector. *Somatic Cell Mol. Genet.* **11:** 499.

Ashman, C.R., P. Jagadeeswaran, and R.L. Davidson. 1986. Efficient recovery and sequencing of mutant genes from mammalian chromosomal DNA. *Proc. Natl. Acad. Sci.* **83:** 3356.

Bredberg, A., K.H. Kraemer, and M.M. Seidman. 1986. Restricted ultraviolet mutational spectrum in a shuttle vector propagated in xeroderma pigmentosum cells. *Proc. Natl. Acad. Sci.* **83:** 8273.

Conney, A.H. 1982. Induction of microsomal enzymes by foreign chemicals and carcinogenesis by polycyclic aromatic hydrocarbons: G.H.A. Clowes Memorial Lecture. *Cancer Res.* **42:** 4875.

Drinkwater, N.R. and D.K. Klinedinst. 1986. Chemically induced mutagenesis in a shuttle vector with a low-background mutant frequency. *Proc. Natl. Acad. Sci.* **83:** 3402.

DuBridge, R.B., P. Tang, H.C. Hsia, P.-M. Leong, J.H. Miller, and M.P. Calos. 1987. Analysis of mutation in human cells by using an Epstein-Barr virus shuttle system. *Mol. Cell. Biol.* **7:** 379.

Glazer, P.M., S.N. Sarkar, and W.C. Summers. 1986. Detection and analysis of UV-induced mutations in mammalian cell DNA using a lambda phage shuttle vector. *Proc. Natl. Acad. Sci.* **83:** 1041.

Hauser, J., M.M. Seidman, K. Sidur, and K. Dixon. 1986. Sequence specificity of point mutations induced during passage of a UV-irradiated shuttle vector plasmid in monkey cells. *Mol. Cell. Biol.* **6:** 277.

Lebkowski, J.S., S. Clancy, J.H. Miller, and M.P. Calos. 1985. The *lac*I shuttle: Rapid analysis of the mutagenic specificity of ultraviolet light in human cells. *Proc. Natl. Acad. Sci.* **82:** 8606.

Munson, P.J., J. Hauser, A.S. Levine, and K. Dixon. 1987. Test of models for the sequence specificity of UV-induced mutations in mammalian cells. *Mutat. Res.* **179:** 103.

Pelkonen, O. and D.W. Nebert. 1982. Metabolism of polycyclic aromatic hydrocarbons: Etiologic role in carcinogenesis. *Pharmacol. Rev.* **34:** 189.

Protić-Sabljić, M., N. Tuteja, P.J. Munson, J. Hauser, K.H. Kraemer, and K. Dixon. 1986. UV-induced cyclobutane dimers are mutagenic in mammalian cells. *Mol. Cell. Biol.* **6:** 3349.

Sancar, A., F.W. Smith, and G.B. Sancar. 1984. Purification of *Escherichia coli* DNA photolyase. *J. Biol. Chem.* **259:** 6028.

Seidman, M.M., K. Dixon, A. Razzaque, R. Zagursky, and M.L. Berman. 1985. A shuttle vector plasmid for studying carcinogen-induced point mutations in mammalian cells. *Gene* **38:** 233.

Tessman, I. 1985. UV-induced mutagenesis of phage S13 can occur in the absence of the recA and umuC proteins of *Escherichia coli. Proc. Natl. Acad. Sci.* **82:** 6614.

Weinstein, I.B., A.M. Jeffrey, K.W. Jennette, S.H. Blobstein, R.G. Harvey, C. Harris, H. Autrup, H. Kasai, and K. Nakanishi. 1976. Benzo[a]pyrene diol epoxides as intermediates in nucleic acid binding in vitro and in vivo. *Science* **193:** 592.

Analysis of Carcinogen-induced Point Mutations in a Simian Virus 40 Genetic Assay

ALAIN SARASIN, FRANÇOIS BOURRE, AND ALAIN GENTIL
Laboratory of Molecular Mutagenesis
Institut de Recherches Scientifiques sur le Cancer
B.P. n° 8 94802—Villejuif Cedex, France

OVERVIEW

We have analyzed induced point mutation in mammalian cells, using a simian virus 40 (SV40) genetic assay based upon reversion of temperature-sensitive mutants to a wild-type phenotype. Several types of DNA-damaging treatments are shown to be mutagenic with this assay. The spectrum of UV light- and acetoxy-acetylaminofluorene (AAAF)-induced mutations are quite different. Indeed, UV-induced mutations appear to be targeted opposite potential DNA lesions, whereas AAAF-induced mutagenesis does not seem to be directly targeted, but involves the stabilization of a specific quasi-palindromic sequence on SV40 DNA. This clearly shows that specificity of mutagenesis is due both to the classes of DNA lesions and to the nature of sequences inside the target gene.

INTRODUCTION

Mutations are involved in the initiation of the carcinogenesis process, as recently demonstrated by the activation of some specific cellular oncogenes (Sukumar et al. 1983). The relationship between unrepaired genomic lesions and hypermutability leading to tumorigenesis is evident in the cancer-prone disease xeroderma pigmentosum (Hanawalt and Sarasin 1986). The analysis of the molecular mechanisms giving rise to gene alterations, therefore, appears as a major goal in modern cancer research.

SV40 is a double-stranded DNA papovavirus composed of 5243 bp. After viral infection, the early transcription gives rise to two early proteins, the small t antigen and the large T antigen (Fig. 1). Large T antigen binds near the viral origin of replication and is needed to initiate DNA replication. This large T antigen is the only virus-encoded protein necessary for viral replication. The late transcription gives rise to four proteins, the VP_1, VP_2, and VP_3 capsid proteins and the agnoprotein, that are involved in viral encapsidation (Griffin 1981). SV40 is thus a useful probe for studying mutagenesis in mammalian cells, since it is repaired and replicated by the host-cell machi-

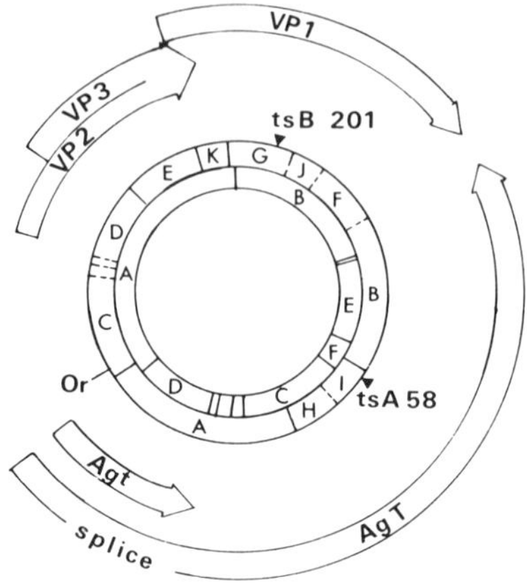

Figure 1

Genetic map of SV40. Internal circle refers to *Hin*f1 restriction sites and external circle refers to *Hin*cII (- -) + *Hin*dIII (—) restriction sites. Each fragment is labeled by a letter corresponding to decreasing relative size. The tsA58 and tsB201 mutations are mapped as solid triangles. (Or) is the origin of replication. The coding sequence of the small t (Agt), the large T (AgT), and the capsid proteins (VP_1, VP_2, VP_3) are shown by arrows.

nery. Moreover, SV40 DNA exists as a minichromosome resembling the eukaryotic chromatin.

Our mutation assay is based on the reversion from a temperature-sensitive phenotype growth to a wild-type phenotype growth at 41°C. Two temperature-sensitive mutants have been used: (1) the early tsA58 that does not allow the initiation of viral DNA replication at the restrictive temperature and (2) the late tsB201 mutant that does not produce virions at 41°C (Tegtmeyer and Ozer 1971; Lai and Nathans 1975). Permissive monkey cells were infected or transfected with in-vitro-treated virus or DNA. After one lytic cycle at 33°C, the viral progeny was rescued. Titers of the viral progeny were determined at 33°C and 41°C, and the mutation frequency was defined as the ratio of the titer at 41°C to that at 33°C (Sarasin and Benoit 1980). Revertants able to grow at 41°C were isolated and plaque-purified, and their DNA was prepared. Mutations able to complement the original temperature-sensitive mutation were located by the marker-rescue technique (Lai and Nathans 1975) and then sequenced by the Maxam and Gilbert method (1980) or by the chain-terminating inhibitor method (Sanger et al. 1977).

Two DNA damaging agents were used in our experiments: UV light at 254 nm and the active metabolite of the chemical carcinogen acetylaminofluorene, AAAF. These two treatments induced DNA lesions, chiefly pyrimidine dimers and pyrimidine (6-4)pyrimidone, Py(6-4)Py, for UV (Lipke et al. 1981), and bulky adducts on guanine residues for the AAAF. Intercalation of this compound between the nucleic bases leads, after covalent binding to the guanine, to a strong local denaturation (Fuchs et al. 1976). DNA damages induced by these two treatments were therefore very different, and it was of interest to determine their relative effects on the classes of mutations induced in mammalian cells. Our SV40 genetic assay revealed a striking difference between mutation spectra produced by these two types of DNA lesions on the same target gene.

RESULTS

One of the advantages of using exogenous probes for analyzing mutation induction is the possibility of treating the genome in vitro, even with compounds very toxic for cells, and of quantifying precisely the number and type of lesions before transfection into permissive cells. We were thus able to show that apurinic sites, produced by heat under acidic conditions, were particularly mutagenic when processed in monkey cells (Gentil et al. 1984). Using the same SV40 genetic target, we showed that a great difference existed in lethal potency and mutation efficiency between UV-lesions, apurinic sites, and AAAF adducts. If one hypothesized that these various lesions were repaired with similar efficiency in the host cells, apurinic sites were four to eight times more mutagenic as compared to these other DNA lesions (Fig. 2). The high potency for lethality and mutagenesis of apurinic sites has also been found in prokaryotes (Shaaper and Loeb 1981).

UV-irradiation (254 nm) of the tsA58 or tsB201 SV40 mutants strongly induced mutagenesis as measured in a reversion assay (Gentil et al. 1982; Sarasin et al. 1982). The mutation frequency was even higher when the host cell was treated with a DNA-damaging agent prior to infection with UV-irradiated tsA58 (Sarasin and Benoit 1986). This biological assay was among those used to analyze induced SOS functions in mammalian cells (Sarasin and Hanawalt 1978). Phenotypical revertants were isolated and purified in order to allow their molecular characterization. The marker rescue was carried out to locate the reversion sites of the tsA58 revertants, and the complementing restriction fragments were found to be the H, I, and B fragments isolated after the *Hin*dIII-*Hin*cII digestion of the revertant DNAs. The reversion sites of tsA58 mutation were all located in the carboxy-terminal half of the T-antigen gene, and no complementation with the A fragment was found (Fig. 3). This suggests that several functional domains are present in this protein. DNA sequencing of the complementing restriction fragments of the 17 analyzed UV-induced revertants revealed that one single-base substitution was at least present on every complementing fragment. All of these point mutations resulted in an amino acid substitution in the T-antigen protein, leading to the recovery of a functional T antigen at 41°C. All revertants conserved the original tsA58 mutation (a G:C→ A:T transition) (Bourre and Sarasin 1983).

The observed mutations were mostly located opposite potential UV-induced DNA lesions. This result was also true for a UV-induced mutant for which we found, without selective pressure, a new *Hin*dIII restriction site, in

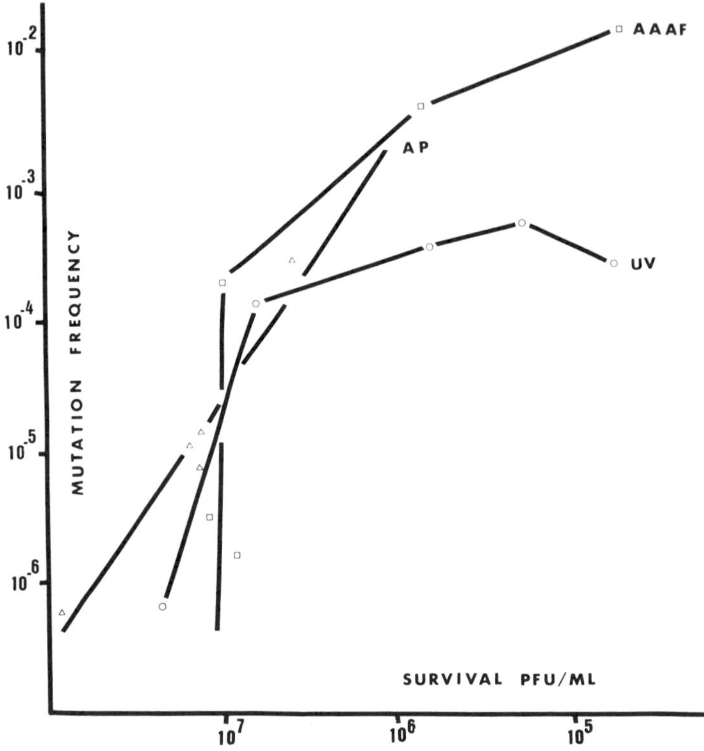

Figure 2
Mutation frequency, as measured by reversion of the tsB201 mutant, as a function of the survival of SV40 containing various DNA damages:apurinic sites (AP). (△) AP; (○) UV-induced lesions; (□) AAAF-induced lesions.

the late region of the SV40 genome, due to a single-base change. The repartition of all SV40 UV-induced mutations is almost identical between transitions and transversions (Table 1), but it seems that the transitions occur preferentially at the 3′ side of the putative lesion and transversions at the 5′ side. The two major lesions induced by UV light are the pyrimidine dimer and Py(6-4)Py. Since these two lesions are produced at the same sequence site, we cannot determine which of them, if not both, is the mutagenic lesion. With the 2 kJ/m^2 UV dose we have used in our experiments, the frequency of pyrimidine dimer formation is about ten times greater than for the Py(6-4)Py photoproduct (Bourre et al. 1985). However, preliminary experiments with the photoreactivating enzyme seem to indicate that both types of DNA lesions are mutagenic on the SV40 genome.

SV40 has also been used to study mutagenesis induced by AAAF in control cells and in cells pretreated with UV in order to induce a possible enhanced-

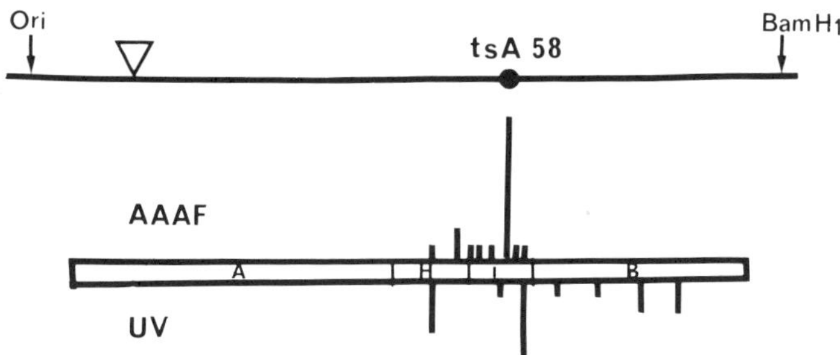

Figure 3
Distribution of reversion sites of the tsA58 mutation along the T-antigen gene. The upper line shows the T-antigen gene of SV40. (Ori) origin of replication; (▽) the early splice; (●) tsA58 mutation. The restriction site *Bam*HI is shown. A, H, I, B, correspond to the *Hin*cII + *Hin*dIII restriction enzyme fragments of the T-antigen gene of SV40. The vertical bars refer to the reversion sites of tsA58 obtained after UV-irradiation (under the open horizontal bar) or AAAF treatment (above the open horizontal bar). Their sizes are proportional to the number of occurrences. Seventeen mutations and twenty mutations have been determined for UV- and AAAF-induced revertants, respectively.

recovery pathway. DNA of a temperature-sensitive SV40 mutant was treated in vitro with increasing concentrations of AAAF in dimethyl sulfoxide. Cells were transfected with damaged DNA, and survival and mutation frequencies of the viral progeny were then measured as described above. Progeny survival decreased as a function of the number of AAAF adducts per SV40 genome but only after a threshold was reached, corresponding to 7 or 40 adducts, depending upon the concentration of the input DNA (Fig. 4A). The number of adducts was computed, without any presumption about their chemical structure, from the radioactivity associated with SV40 DNA that had been treated in vitro with tritiated AAAF. Since it is well known that recombinational processes may occur between damaged molecules (Gentil et al. 1983), the threshold that was observed may indeed reflect a multiplicity effect linked to the total number of molecules infecting the cells. The lethal hit, corresponding to the 37% survival and calculated from the exponential portion of the curves, is roughly equal to 85 adducts per SV40 genome and independent of the DNA concentration. Pretreatment of cells with 10 J/m^2 of UV 24 hours prior to transfection did not induce enhanced virus survival (Fig. 4A). The mutation frequency increased as a function of the number of adducts, even for low numbers of adducts (Fig. 4B). Consequently, at approximately 40 adducts per genome, we observed an increase in mutation frequency that was more than two orders of magnitude above a 10^{-6} background level, without any significant decrease in survival. No enhanced mutation frequency was observed in the viral progeny from UV-irradiated cells.

Table 1
Characterization of the UV-induced Reversion Sites of tsA58 and tsB201 Mutants on the SV40 Genome

	No. of occurrences	UV-photoproducts Py–Py 5′ position	UV-photoproducts Py–Py 3′ position	Undeterminable[a]	Untargeted[b]
Transition					
T→C	9	2	7	—	—
C→T	14	3	1	8	2
Subtotal	23	5	8	8	2
Transversion					
T→G	7	1	2	4	—
T→A	7	7	—	—	—
C→G	3	—	—	2	1
C→A	2	—	—	2	—
Subtotal	19	8	2	8	1
Total	42	13	10	16	3

[a]These mutation sites are included in a pyrimidine track. It is therefore impossible to determine the exact lesion involved.
[b]These mutations do not occur opposite a known potential UV lesion.

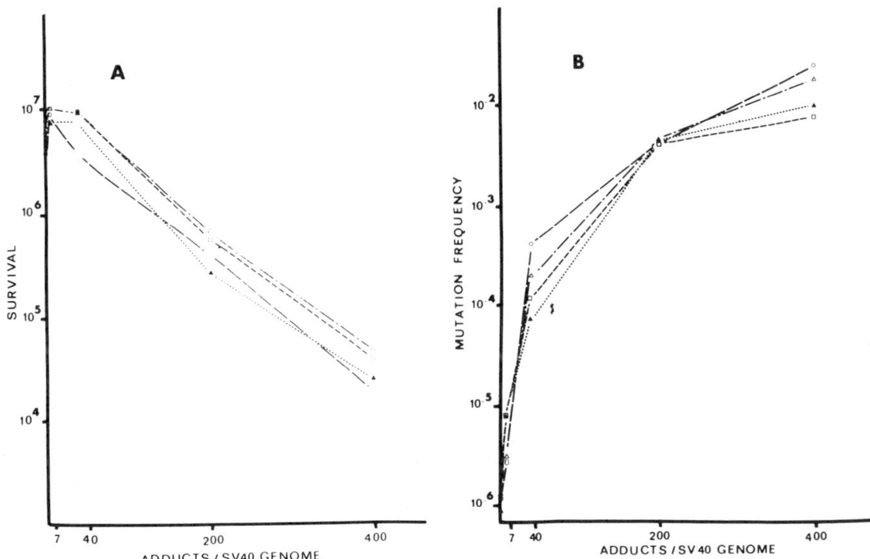

Figure 4

Survival in pfu/ml (*A*) and mutation frequency (*B*) in the viral progeny from tsB201 SV40 DNA treated in vitro with various concentrations of AAAF, as a function of the number of DNA adducts. (○) 0.1, (△) 0.2, or (□) 1 μg of treated DNA were transfected into untreated cells or (▲) 0.2 μg of DNA was transfected into cells, which were UV-irradiated 24 hr prior to transfection.

The mutation spectrum along the large-T-antigen gene of AAAF-induced revertants is completely different from the spectrum found with UV-induced revertants (Fig. 3). This striking difference indicates that our reversion assay is differently sensitive to various DNA adducts and, conversely, that UV lesions and AAAF adducts are subject to a different cellular processing, leading then to different types of mutations. Whereas the UV-induced mutations are targeted on UV-induced lesions, the AAAF-induced mutations are not located opposite potential AAAF-induced deoxyguanine adducts. Indeed, 19 mutations among 20 analyzed were due to base substitution at the A:T site. A hot spot of mutagenesis was found after AAAF treatment at the site responsible for the tsA58 phenotype. Of the AAAF-induced revertants, 60% were due to an A:T→G:C transition restoring the true wild-type genotype (Gentil et al. 1986). This hot spot, which was not targeted to a major AAAF adduct, was not found after UV-irradiation.

DISCUSSION

We have described a genetic assay, using SV40 as a molecular probe, to characterize cellular mechanisms leading to mutation in mammalian cells.

SV40 has been the first exogenous system giving molecular data and is still a very efficient and sensitive probe to detect the effect of DNA-damaging agents, particularly those that are toxic for the cell. It is difficult to conceive how to study the effect of apurinic sites, induced by heat under acidic conditions, directly on genomic DNA.

The results obtained on SV40 mutagenesis with UV light and AAAF treatment are interestingly different and clearly show the advantages and limitations of this genetic assay. UV-irradiation of SV40 DNA leads to base substitutions that are mostly located opposite potential UV-induced lesions. This result agrees with those found in bacteria (Miller 1985) and has been confirmed by using shuttle vectors (Lebkowski et al. 1985) as probes. AAAF treatment of SV40 DNA leads to base substitutions that are not located directly opposite potential AAAF-induced lesions. We found a particular hot spot of AAAF-induced mutations on the large-T-antigen gene, which was not a UV-induced mutation site, although the involved DNA sequence could have given rise, in theory, to UV-induced base substitution (Bourre and Sarasin 1983). We have proposed two models to explain this hot spot of mutation, involving the stabilization of specific quasi-palindromic structures by AAAF adducts (Gentil et al. 1986). If this hypothesis holds true, this implies that AAAF can reveal specific DNA structures and could therefore be used as a probe for analyzing specific secondary structures in mammalian chromatin. These sequences exhibiting particular secondary or more complex structures could be sites for both preferential defect in DNA repair (Hanawalt and Sarasin 1986) and preferential processing in mutation induction.

Limitations of our genetic assay are due to the fact that the integrity of the SV40 target genes is necessary for a normal virus cycle. Any mutation leading to the inactivation of the large T antigen or VP_1 protein will not allow virus survival and will therefore not be screened. In order to detect other classes of mutations, such as frameshift or deletion, we have inserted a series of synthetic oligonucleotides in the intron of the large-T-antigen gene. The appearance or the loss of restriction enzyme recognition sites on these inserted sequences constitutes our genetic screening assay and allows the detection of all mutation types without selective pressure. The chemical addition of modified bases or adducts on these inserted sequences allows us to characterize site-specific mutagenesis at a unique position in these SV40-hybrid genomes.

ACKNOWLEDGMENTS

We wish to thank Mrs. A. Benoit and A. Margot for their technical assistance. This work has been supported by grants from the Association pour la

Recherche sur le Cancer (Villejuif, France), the Fondation pour la Recherche Médicale (Paris, France), and the Commission of the European Communities (N° BI6-E-163-F, Brussels, Belgium).

REFERENCES

Bourre, F. and A. Sarasin. 1983. Targeted mutagenesis of SV40 DNA induced by UV-light. *Nature* **305:** 68.

Bourre, F., G. Renault, P.C. Seawell, and A. Sarasin. 1985. Location and quantification of UV-induced lesions in SV40 DNA. *Biochimie* **67:** 293.

Fuchs, R.P.P., J.F. Lefevre, J. Pouillet, and M.P. Daune. 1976. Comparative orientation of the fluorene residue in native DNA modified by N-acetoxy-N-2 acetylaminofluorene and two 7-halogeno-derivatives. *Biochemistry* **15:** 3347.

Gentil, A., A. Margot, and A. Sarasin. 1982. Enhanced reactivation and mutagenesis after transfection of carcinogen-treated monkey kidney cells with UV-irradiated simian virus 40 (SV40) DNA. *Biochimie* **64:** 693.

———. 1983. Effect of UV-irradiation on genetic recombination of simian virus 40 mutants. In *Cellular responses to DNA damage* (ed. E.C. Friedberg and B.A. Bridges), p. 385. A.R. Liss, New York.

———. 1984. Apurinic sites cause mutations in simian virus 40. *Mutat. Res.* **129:** 141.

———. 1986. 2-(N-acetoxy-N-acetylamino)fluorene mutagenesis in mammalian cells: Sequence-specific hot spot. *Proc. Natl. Acad. Sci.* **83:** 9556.

Griffin, B.E. 1981. Structure and genomic organization of SV40 and polyoma virus. In *Molecular biology of tumor viruses,* 2nd edition, revised: *DNA tumor viruses* (ed. J. Tooze), p. 61. Cold Spring Harbor Laboratory, Cold Spring Harbor, New York.

Hanawalt, P.C. and A. Sarasin. 1986. Cancer prone hereditary diseases with DNA processing abnormalities. *Trends Genet.* **2:** 124.

Lai, C.J. and D. Nathans. 1975. A map of temperature sensitive mutants of simian virus 40. *Virology* **66:** 70.

Lebkowski, J.S., S. Clancy, J.H. Miller, and M.P. Calos. 1985. The *lac*I shuttle: Rapid analysis of the mutagenic specificity of ultraviolet light in human cells. *Proc. Natl. Acad. Sci.* **82:** 8606.

Lipke, J.A., L.K. Gordon, D.E. Brash, and W.A. Haseltine. 1981. Distribution of UV-light induced damage in defined sequence of human DNA: Detection of alkaline-sensitive lesions at pyrimidine nucleoside-cytidine sequences. *Proc. Natl. Acad. Sci.* **78:** 3378.

Maxam, A.M. and W. Gilbert. 1980. Sequencing end-labelled DNA with base-specific chemical cleavages. *Methods Enzymol.* **65:** 499.

Miller, J.H. 1985. Mutagenic specificity of ultraviolet light. *J. Mol. Biol.* **182:** 45.

Sanger, F., S. Nicklen, and A.R. Coulson. 1977. DNA sequencing with chain-terminating inhibitors. *Proc. Natl. Acad. Sci.* **74:** 5463.

Sarasin, A. and A. Benoit. 1980. Induction of an error-prone mode of DNA repair in UV-irradiated monkey kidney cells. *Mutat. Res.* **70:** 71.

———. 1986. Enhanced mutagenesis of UV-irradiated simian virus 40 occurs in mitomycin C-treated host cells only at a low multiplicity of infection. *Mol. Cell. Biol.* **6:** 1103.

Sarasin, A. and P.C. Hanawalt. 1978. Carcinogens enhance survival of UV-irradiated simian virus 40 in treated monkey kidney cells: Induction of a recovery pathway? *Proc. Natl. Acad. Sci.* **75:** 346.

Sarasin, A., F. Bourre, and A. Benoit. 1982. Error-prone replication of ultraviolet-irradiated simian virus 40 in carcinogen treated monkey kidney cells. *Biochimie* **64:** 815.

Shaaper, R.M. and L.A. Loeb. 1981. Depurination causes mutations in SOS-induced cells. *Proc. Natl. Acad. Sci.* **78:** 1773.

Sukumar, S., V. Notario, D. Martin-Zanca, and M. Barbacid. 1983. Induction of mammary carcinomas in rats by nitroso-methyl urea involves malignant activation of H-*ras* 1 locus by single point mutations. *Nature* **306:** 658.

Tegtmeyer, P. and H.L. Ozer. 1971. Temperature sensitive mutants of simian virus 40: Infection of permissive cells. *J. Virol.* **8:** 56.

Comments

Meuth: What is the host range for the EBV-based shuttle vector?

Calos: It's human cells. It appears any kind of human cell works, and also, monkey cells work. Also, according to Sugden, it can replicate in dog cells, but it doesn't replicate in rodent cells.

Meuth: What is the story about the background here? Where does it come from?

Calos: It's at the level where we run up against the background in *E. coli*. So, either the mutations are actually formed in *E. coli*, or it's similar to the spontaneous background of a gene in a human cell. Therefore, I think it's essentially spontaneous mutation.

Meuth: However, there are some small deletions?

Calos: We have seen very few of these. So far, they have been point mutations. We also have a couple of insertions, which might be the kind of thing you have.

Evans: Michele, would large deletions also be lethal for your virus, since your origins are not too far away from your target?

Calos: There is a restriction on the size of the deletion you could get, definitely. These small vectors are geared toward picking up intragenic mutations. That means that basically, with these little tiny vectors, we do not have the ability to see these large deletions. We are trying to address that issue, basically by building much bigger vectors. Epstein-Barr virus itself is 180 kb, and so we know that the origin can carry that amount of DNA. Therefore, we are building cosmids now that are based on the EBV origin that will have, say, 50 or 100 kb that are open for deletion, and we're hoping that we might be able to pick up some of these larger events that way.

Glickman: A question directed to all of you, really: You have all done UV, and we have done UV; and we all have to, because of our systems, to use different doses. For example, we are at 5 J. I think, Kathleen, you're at 500 J. However, there is a problem here, and that problem has to do with comparing. You get saturation for the cyclobutane lesions, and you get the increased productions of the 6-4's photoproducts; so, we are not really looking at the same damage distribution. Two questions that need to be addressed are, "What dose are you using?" and "What are the lesions?"

Before I let you try to answer that, I wanted to ask one other question, and that is a very important one. I have heard contradictory rumors with the Strauss method. Some people are saying that the 6-4's photoproducts block replication and, therefore, are also picked up. Some people have said that's not true. I think that, in order to assess the answer, we need to know your feelings regarding that question.

Dixon: I think I would like to answer that. There were several questions. Question one concerns the change in spectrum of photoproducts as a function of the UV dose. There certainly is some change. On the other hand, the major saturation for pyrimidine dimers comes at fluences higher than 500 J. In any case, the spectrum of photoproducts can be measured at different UV doses. I measured the spectrum at the dose that I was using for mutagenesis, because that is what I wanted to find out. However, theoretically, by using other techniques, you could measure the spectrum of photoproducts at the lower doses and for each particular gene of interest.

The other point, is about the 6-4's photoproducts and whether they block DNA synthesis in the Strauss stop assay. Michael Seidman and Doug Brash have recently measured the positions of 6-4s and cyclobutane dimers in the *supF* gene, and they have very graciously allowed me to see their results before publication. I have compared them to the results that I have obtained with the stop assay, and there is extremely good agreement. It is pretty clear that in cases where there are about half-and-half 6-4s and cyclobutane dimers, I'm measuring both.

Glickman: So, you're not differentiating between dimers and 6-4, yet?

Dixon: No. I might be slightly underestimating the 6-4s, but in the positions where I can make that determination, I don't think it's significant.

Sarasin: We have carried out our experiments with SV40 at 2000 J/m^2, and we looked also at the distribution of both (6-4) photoproducts and pyrimidine dimers by using the Haseltine technique. We find at 200 J/m^2 that we have already a plateau for the amount of pyrimidine dimers, while the 6-4s start to increase a little bit after 2000 J/m^2. However, the interesting result, I don't think it was previously published by Haseltine, is that the frequency of 6-4 was measured for all possible sites along 340 bp. Not only the type of pyrimidine-pyrimidine sequence is important but also the adjacent sequence and the nucleotide environment. In fact, as Kathleen said, I think in each case we have to measure the UV-lesion distribution for the whole DNA sequence that is involved in the mutagenesis study, because the flanking sequence is very important.

Hutchinson: I just want to comment on the correlation between incidence of photoproducts and of mutations. This is a very important correlation. However, let me point out a complicating factor. Only 1% or less of lesions leads to mutations, at least in regards to UV and X-ray mutations, which are the ones I understand best. That means that 99% of the products you are observing aren't forming mutations.

Dixon: I think that our having seen that there were mutational hot spots that do not correspond with hot spots for UV photoproduct formation, certainly means that there are additional factors.

Davidson: This is a question for both *lacI* and *supF*. In those cases where there might be data on the mutational spectra of hot spots for those genes in bacteria, is there enough data on mutagenesis in mammalian cells to say whether they are the same or different, not in terms of base pairs or substitutions, when they are mutated either in cells or in bacteria? Those you would expect to be the same in terms of hot spots?

Calos: Yes, in our UV study. UV has been looked at extensively in *lacI* in *E. coli* by Jeffrey Miller, and he has three, strong hot spots. Two of those coincide identically to what we found in human cells in *lacI*. So, it looks like there is correspondence, even on the level of hot spots.

Maher: The data that I know about involve mutations induced by N-acetoxy-N-acetylaminofluorene.

In *Salmonella typhimurium*, N-AcO-AAF forms the same DNA adduct as it does in human cells. However, in the bacterial cells the predominant mutations are frameshifts, whereas in human cells the predominant mutations induced by N-AcO-AAF are base substitutions.

Hsie: I would like to ask Michele to comment on what I am going to ask. First of all, would you elaborate on the experiment you mentioned about XP cells and then comment on the possibility of using that system for genetic monitoring of the population at risk?

Calos: We have the vector in several XP complementation groups. It is, of course, more susceptible to UV light. It does have a certain parallel to the Ames system, in that they have used repair-crippled cells in *Salmonella* to sensitize the cells to mutation. Most of our data now are in the XP-D complementation group, which is not the one with the strongest defect. But, I guess what you are getting at is perhaps that one could detect the elevated mutation frequency readily with this kind of system and use it as a diagnostic tool. I think that has some potential.

Hutchinson: A characteristic of all the shuttle vector data I have seen is the existence of multiple mutations in the same mutant. I was interested to see that in the data for mutations induced in chromosomes, the *gpt* data and Barry Glickman's data for *aprt*, there are not these multiple mutations. The most obvious reason is that very high doses were used on the shuttle vectors, but this does not seem to account for the multiple mutations, making the assumption of random production of mutagenic lesions. I just wondered what do the shuttle vector experts think about this?

Calos: First of all, we haven't seen any of that so far with the EBV vectors.

Hutchinson: I haven't seen any EBV vector data except Norman Drinkwater's, but what about the SV40-based vectors?

Dixon: I thought Barry Glickman did, in fact, see multiple mutations with UV in the *aprt* gene.

Glickman: However, they tended to be separated by one base pair.

Dixon: That's what we see in the pZ189 vector as well.

Glickman: In that case they are very close, but the rumor that I keep hearing is that there are some mutations that are separated by 50 base pairs. We don't have anything like that. Do you have things separated by 50 base pairs?

Dixon: We have a few that could be accounted for by the spontaneous background. In general, with UV, the multiple mutations are very close to each other, within 15 base pairs of each other.

Hutchinson: There are really two types of mutiple mutations. One has two base changes within a base pair or two of each other. These are characteristic of UV, and has been found in several different systems. There is another type of spacing, from three base pairs and up, which seem to be reasonably common in shuttle vectors.

Sarasin: In SV40 we have two mutations that could be between 50 and 200 base pairs apart. We have some experiments where we tried to UV-irradiate one part of SV40, and we found that, if we UV-irradiated this part, we can get mutation that was 500 base pairs away, in a nonirradiated DNA sequence.

Hutchinson: In your case, calculations can't rule out the possibility of two independent events because you are using 2000 J. However, at lower doses, independent events cannot account for the occurrance of two mutations in the same genome. This seems to be a characteristic of the shuttle-vector system, and I wonder if anybody has some insight as to what's going on.

Maher: With the vector that we have shown to you and using B[a]PDE with the human 293 cells that Michele used in many of her studies, we don't find that at all. Of all the changes induced by B[a]PDE, 70% involve single-base changes all alone, and the number involving 2 base changes is approximately 13%; i.e., 5 out of 86 are tandem; 3/86 are less than 20 bases apart; and 3/86 are more than 20 bases apart. That's less than 35% for the latter.

Hutchinson: That's still very high, that's the point.

Maher: Seventy percent are single.

Hutchinson: Yes, but in other systems 99% are single.

Calos: In what other systems?

Hutchinson: The bacterial systems. It's very, very rare to get a double mutation, except for those in which the base changes are adjacent, or separated by only a base pair or two.

Calos: No. In our NMU study, which was published in January in *MCB*, 61 out of 61 of the NMUs were single-base changes. That has been typical of what we have seen with them, although in some cases we have used the nonsense system, and we haven't sequenced through. However, even with the direct sequencing, we haven't been seeing it.

Hutchinson: Now, that was an EB viral based vector.

Calos: Yes. I agree, that this has only shown up in the SV40 systems. At least, I have only seen it in the SV40.

Glickman: I have, I think, what is an important point that I would like to get in. Every system has its Achille's heel or, as we have heard here, the small-/large-colony problem. I think the problem with shuttle vectors is not the single/double question. I think that will resolve itself. Where I see the major problem is trying to understand the numbers. The reason for that is we don't understand how these shuttle vectors replicate. You damage them, and then they do whatever they're going to do. They repair or they start to replicate, and they get to a block, a lesion, and they just are lost. Others replicate, mutate, and contribute very much to the population. We have no control over what is contributing, in terms of numbers. This is the same as the clonal effect that we talked about yesterday. It is particularly perturbing when you do something like compare a normal cell line with an XP cell line, where you don't really understand that replication over the lesions, and your lesions are different. I think that's a major problem. I wondered if you could try to address that issue, since it is the one we are asking. How can you effectively quantitate the mutations that are coming out? How can you identify their clonal origins?

Calos: I agree. That's one reason we have tried to make vectors that are more and more like chromosomes. That's one thing that was disconcerting about the SV40 vectors because they would replicate continuously. You wouldn't have any control about what was lost or gained. With the EBV, we don't understand everything yet about their replication, but it appears that they replicate once per S period in the cell cycle and that their copy number is very stable. We are moving toward even wanting to substitute a chromosomal origin for that to make it more like a chromosome, and I think already with EBV one has a good approximation.

In fact is that these systems require continuous validation from studies that are done on real chromosomes. So far, they haven't been completely validated. Of the specificities that we have seen, your data corroborates very well, for example, the UV. As you said, it's pretty much superimposable, the spectra that you see on the chromosomes, what one has seen in various shuttle-vector systems, the SV40 and so on.

So, I don't think there is a big discrepancy, but I myself am trying to move toward systems that resemble the chromosome more and more, so that the vector will be just like the chromosome but just smaller.

Sarasin: In your question, you imply that we know exactly how mutation occurs in general, and even with a simple unique gene we don't know that.

Davidson: I always find it somewhat amusing, in listening to Michele or in reading her papers, because everything that we say is an advantage in our type of chromosomally integrated system is a problem to be avoided and resolved in Michele's system. There is obviously a somewhat different point of view. That is one thing.

I think, in terms of Barry's question, though, that one might be concerned about loss of a vector due to different types of mutational lesions, that would then result in an inability to recover a certain type of mutation, because the vector is just geared after a certain type of alteration. That is one case where there might be an advantage to using a chromosomally integrated vector, and one can then assess it in different ways. So, I think, with a vector that is integrated into the chromosomes, can replicate in the chromosomes, and does not really depend upon the vector, you can get around a lot of those problems that you were raising.

Dixon: It seems to me that we do not have to arrive at the best system or say that one system is good and all the other systems are bad. I think that each system can be used to look at different questions having to do with mutagenesis, and each system has its advantages, and each system has its disadvantages. I could list both for my own system, and probably each of us could for his own system.

I think that you have to ask the right questions with the right system. However, as long as you do that, I don't think that it makes sense to try to decide which one is best overall.

Davidson: I agree. I think right now we have all been trying to be salespeople for our particular system, and I think they all do have particular advantages and the best information is going to come from comparative studies with them.

Dixon: Absolutely.

Recombination and Repair as Modulators of Mutational Processes in Mammalian Cells

Isolating Human DNA Repair Genes Using Rodent-cell Mutants

LARRY H. THOMPSON,* CHRISTINE A. WEBER,* KERRY W. BROOKMAN,* EDMUND P. SALAZAR,* SHERI A. STEWART,* AND DAVID L. MITCHELL†

*Biomedical Sciences Division
Lawrence Livermore National Laboratory
Livermore, California 94550
†Science Park, Research Division
The University of Texas System Cancer Center
Smithville, Texas 78957

OVERVIEW

The DNA repair systems of rodent and human cells appear to be at least as complex genetically as those in lower eukaryotes and bacteria. The use of mutant lines of rodent cells as a means of identifying human repair genes by functional complementation offers a new approach toward studying the role of repair in mutagenesis and carcinogenesis. In each of six cases examined using hybrid cells, specific human chromosomes have been identified that correct Chinese hamster ovary (CHO) cell mutations affecting repair of damage from UV- or ionizing radiations. This finding suggests that both the repair genes and proteins may be virtually interchangeable between rodent and human cells. Using cosmid vectors, three human repair genes that map to chromosome 19 have been cloned as functional sequences: excision repair complementing defective repair in Chinese hamster (*ERCC1*) by D. Bootsma's group in Rotterdam and *ERCC2* and X-ray repair complementing defective repair in Chinese hamster (*XRCC1*) in our laboratory. *ERCC1* was found to have homology with the yeast excision repair gene *RAD10*. Transformants of repair-deficient cell lines carrying the corresponding human gene show efficient correction of repair capacity by all criteria examined. Future experiments will test whether these genes correct the defects in the cells from human repair syndromes such as xeroderma pigmentosum (XP) and Bloom's syndrome.

INTRODUCTION

Mutations induced by environmental agents arise when unrepaired DNA damage is followed by DNA replication or when damage is repaired inaccurately. The nucleotide excision repair (NER) system, which acts on bulky covalent adducts and pyrimidine dimers induced by UV-radiation, is a major line of defense against mutagenesis and carcinogenesis in human cells. Cell cultures from individuals having the genetic disorder XP are defective in NER and show hypersensitivity to induced mutation and transformation (Maher et al. 1979; Maher and McCormick 1984). XP patients have an elevated risk for neoplasms in sunlight-exposed tissues as well as some internal sites (Kraemer et al. 1984). The XP syndrome appears to be extremely complex genetically. Nine complementation groups have been reported using unscheduled DNA synthesis as the repair assay (Fischer et al. 1985), indicating at least nine genes may underlie the repair process. Cells from each of these complementation groups show varying degrees of both hypersensitivity to killing by UV and reduced activity at the incision step of the repair process (Cleaver 1983). To date, the efforts of several laboratories to transfect and isolate genes that will correct the XP defects have been unsuccessful (Lehmann 1985; Schultz et al. 1985), perhaps because of the poor integration of foreign DNA in most human cell lines (Hoeijmakers et al. 1987).

An alternative approach to isolating human repair genes has been developed using DNA-repair-deficient mutants isolated in rodent cell lines (for review of mutants, see Collins and Johnson 1987). The molecular cloning of the first human DNA repair gene was accomplished by Westerveld and co-workers (1984), who used a CHO cell mutant as the recipient for DNA-mediated gene transfer. This gene, designated *ERCC1*, has been analyzed in detail. The amino acid sequence deduced from the cDNA has regions of homology with the protein encoded by the yeast excision repair gene *RAD10* (van Duin et al. 1986). This observation suggests that there may be significant evolutionary conservation of repair proteins even between lower and higher eukaryotes. If so, the study of mammalian repair proteins could be facilitated by what is known about the NER system in the yeast *Saccharomyces cerevisiae*, in which at least five proteins are essential for the incision step (Friedberg 1987).

Our laboratory has recently isolated a second human gene that corrects a different complementation group of UV-sensitive CHO mutants. We have also isolated a human gene involved in sensitivity to ionizing radiation and DNA-strand-break repair. The efficient correction of mutant hamster cells with both these human genes suggests conservation of repair-gene expression as well as repair-protein structure and function. Therefore, the use of rodent cell mutants may be highly effective for dissecting human repair processes.

RESULTS

The UV-sensitive mutants of CHO cells that show markedly reduced incision after UV exposure have been assigned to UV complementation groups 1–5 as presented in Table 1. The mutants in each of these five groups all have a similar degree of hypersensitivity to killing by UV (~ sevenfold compared to

Table 1
Status of Mapping and Cloning Human DNA Repair Genes

Mutant[a]	UV group[b]	Chromosome[c]	Gene name	Cloned?
EM9	—	19	*XRCC1*	yes
UV5	1	19	*ERCC2*	yes
UV20	2	19	*ERCC1*	yes
UV24	3	2	*ERCC3*	no
UV41	4	16	*ERCC4*	no
UV135	5	13	*ERCC5*	no
UV61	6	n.d.[d]	*ERCC6*	no

[a]Mutants were isolated as described previously (Busch 1980; Busch et al. 1980; Thompson et al. 1980, 1982b).
[b]Complementation group assignments were reported as follows: UV5, UV20, UV24, and UV41 (Thompson et al. 1981); UV135 (Thompson and Carrano 1983); UV61 (Thompson et al. 1987).
[c]Chromosomal assignments were made as follows: EM9 (Siciliano et al. 1986); UV5 and UV41 (M.J. Siciliano and L.H. Thompson, in prep.); UV20 (Thompson et al. 1985); UV24 and UV135 (L.H. Thompson et al., in prep.).
[d]n.d. indicates not determined.

the parental line, AA8). However, with respect to one particular property, namely sensitivity to killing by DNA interstrand cross-linking agents, the mutants show phenotypic diversity. Mutants in complementation groups 2 and 4 are highly sensitive to this class of DNA-damaging agent, whereas mutants in groups 1, 3, and 5 show little or no hypersensitivity (Hoy et al. 1985). Thus, mutants in these latter three groups appear very similar to XP cells in being hypersensitive to bulky monoadducts. (XP cells are not generally noted for being hypersensitive to cross-linking agents, although this issue has not been systematically addressed.) The sensitivity to cross-linking agents seen with the other two UV complementation groups indicates that these cells are defective in components essential for efficiently repairing both UV damage and cross-links. Sensitivity to cross-linking agents is a characteristic of cells from patients with Fanconi's anemia (Ishida and Buchwald 1982). The mutant UV61, which constitutes the sixth group of UV sensitivity in Table 1, differs from the others. Not only is this mutant less UV sensitive than the others (~threefold), but it also appears to have normal incision as measured by alkaline elution (L.H. Thompson, unpubl.). Thus, these properties resemble those of cells from both the XP variant and from Cockayne's syndrome (Friedberg et al. 1979). The phenotype of line EM9 is totally different. This mutation confers a much reduced ability to rejoin radiation- or chemical-induced strand breaks, a tenfold elevated baseline frequency of sister-chromatid exchange (SCE), and hypersensitivity to killing by ionizing radiations and certain simple alkylating agents (Thompson et al. 1982b). High SCE is a well-known feature of Bloom's syndrome cells.

As shown in Table 1, in each case that we have examined, there has proved to be a gene (or chromosome) in normal human cells that will correct the repair defect in the CHO mutant. This result could not have been fully anticipated since various studies have pointed to significant differences between rodent and human cells in the NER system (see Discussion section). As a simple working hypothesis, we assume hamsters and humans share common genetic loci involved in repair and that these loci encode gene products that are functionally interchangeable between species. We see that the first five human NER genes identified by functional complementation are distributed among four different chromosomes, suggesting that the CHO complementation groups correspond to distinct genes. The two NER genes assigned to chromosome 19 must also differ from each other. The first NER gene to be cloned, *ERCC1*, was isolated using another CHO mutant (line 43-3B) belonging to complementation group 2 (Westerveld et al. 1984). We recently isolated a second NER gene, *ERCC2*, on the basis of its ability to correct mutant UV5 functionally (C.A. Weber et al., in prep.). Both genes have been shown to correct only the mutants of the complementation group that were used in their isolation (A. Westerveld et al.; C. Weber et al.; both in prep.). This finding of specificity provides strong evidence that these two complementation groups involve different genes.

We have addressed in some detail the question of how well the human genes function in the CHO mutants. Transformants made with genomic DNA probably contain single copies of the human genes (based on band intensity in Southern blots), whereas cosmid transformants may contain multiple copies. Both genomic and cosmid transformants containing *ERCC2* show a full restoration of resistance to both killing and mutation induction at the *aprt* locus by UV-radiation (C. Weber et al., in prep.). The rate of incision measured immediately after UV-irradiation (Thompson et al. 1982a) is also returned to the normal level by the presence of the human gene in UV5 cells (C. Weber et al., in prep.). By using radioimmunoassays specific for Pyr(6-4)Pyo photoproducts and for cyclobutane (5-6) dimers, the kinetics of repair of UV damage can be examined in greater detail (Mitchell et al. 1985a). As shown in Figure 1, during the first 6 hours after irradiation, UV5 cells appear grossly deficient compared with parental AA8 cells in removing (6-4) products. For (5-6) products, normal CHO cells are known to excise only low percentages (Meyn et al. 1974), and the kinetics of the antibody sites are similar for UV5 and AA8 cells. At times beyond 6 hours, it is apparent that both classes of lesions are somehow modified (Mitchell et al. 1987), even in UV5 cells, since most sites are no longer recognized by the antibodies.

The kinetics of removal for (6-4) photoproducts in the secondary transformant of UV5 (line 5T4-1), in which UV resistance has been restored by *ERCC2*, resembles that of AA8 cells (Fig. 1). In both human and rodent

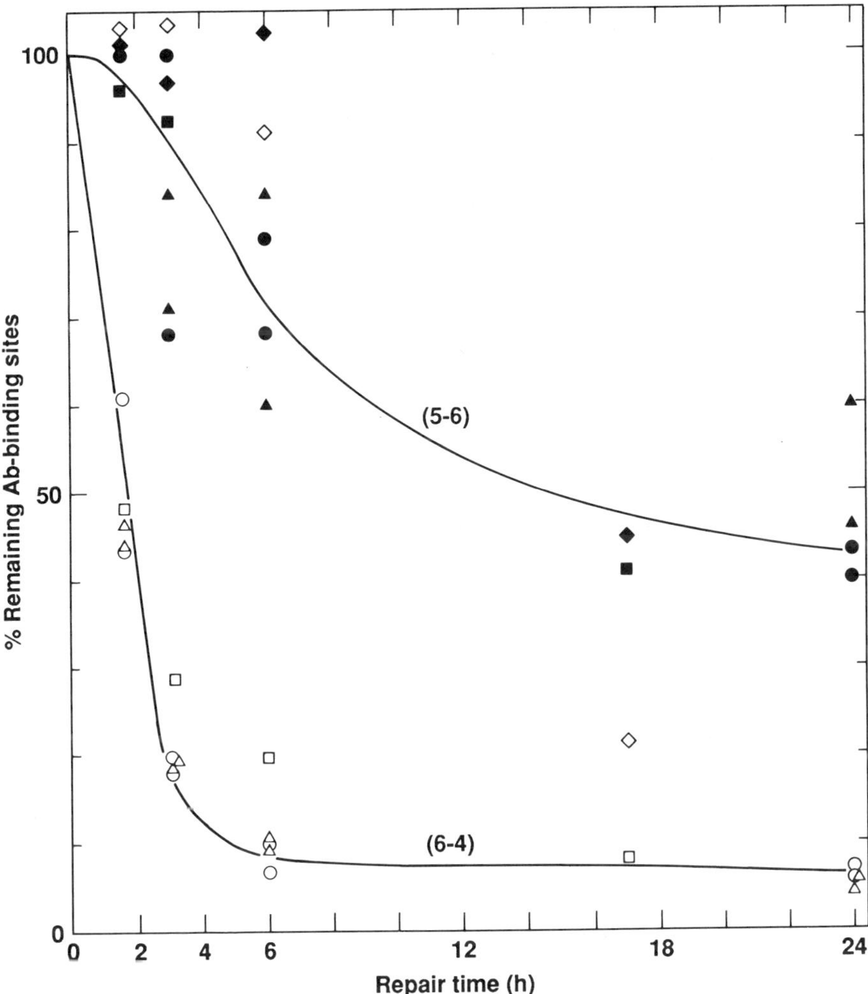

Figure 1

Kinetics of repair of UV-induced (6-4) photoproducts and cyclobutane dimers ([5-6] photoproducts). Photoproducts were measured using radioimmunoassays specific for each class of dimer (Mitchell et al. 1985a). Closed symbols represent (5-6) products and open symbols (6-4) products. (■,□) Wild-type CHO line AA8; (◆,◇) UV5 cells; (●,○,▲,△) *ERCC2* secondary transformant 5T4-1. Lines were fit to the data for 5T4-1. Data for UV5 and AA8 cells were taken from Mitchell et al. (1987).

cells, (6-4) products appear to be repaired much more rapidly than (5-6) products (Mitchell et al. 1985b). Moreover, in CHO cells the removal of (6-4) products appears to be more closely correlated with cell survival than does the repair of (5-6) products. The mutants in complementation groups 1–5 are all deficient at early times in repairing (6-4) products (Mitchell et al. 1987).

The CHO mutant EM9 is also corrected by a gene on human chromosome 19 (Table 1). This gene was cloned in our laboratory from a tertiary transformant using chlorodeoxyuridine (CldUrd), which is highly toxic to EM9 cells, as the selecting agent (Brookman et al. 1987; Thompson 1987). The gene is designated *XRCC1*. Cell survival and the kinetics of strand-break repair are restored to normal levels in *XRCC1* transformants, and the baseline frequency of SCE is also returned to the normal range (Brookman et al. 1987). Thus, *XRCC1*, which is the first isolated mammalian repair gene affecting cellular response to ionizing radiations, efficiently corrects the defect in EM9 cells.

DISCUSSION

From the results presented here, it is clear that using the repair mutants of CHO cells, and hopefully other rodent lines, provides a powerful way to identify and isolate human genes that are likely to be important in the repair processes of human cells. Both human genes we have studied function efficiently in correcting the CHO defects. Similar results were obtained by Zdzienicka et al. (1987), who found that the *ERCC1* gene fully reversed the UV-induced hypermutability to ouabain resistance and restored cell killing to a near-normal level (for UV-radiation) or normal levels (for several bulky mutagens). Several laboratories are currently using the other mutants listed in Table 1 in attempts to clone the complementing human genes using methods similar to ours (Thompson et al. 1987) and that of Westerveld et al. (1984). Within several years, a sizeable collection of human repair genes should be in hand, and efforts to characterize some of the encoded gene products should be under way.

A leading question is whether the human genes being isolated using rodent cell mutations will be capable of correcting the mutations responsible for any of the human repair syndromes such as XP, ataxia telangiectasia, or Bloom's syndrome. So far, the *ERCC1* gene does not correct any of the complementation groups of XP cells in which it has been tested (J. Hoeijmakers and A. Westerveld, pers. comm.). We have tested the *ERCC2* gene for correction only in the highly UV-sensitive XP-A cells (line XP12ROSV) and have seen no evidence for complementation (L. Thompson and C. Weber, unpubl.).

Further isolation and characterization of human (and rodent) repair genes should help resolve the dilemma of why rodent cells tend to have lower overall levels of UV-induced NER than human cells (Thompson et al. 1980;

Vijg et al. 1984; Yagi et al. 1984). This difference has been difficult to understand, since rodent and human cells tend to have very similar resistance to killing by UV-radiation (Takebe et al. 1974; Yagi et al. 1984). Evidence now points to there being preferential repair of active genes in rodent cells (Bohr et al. 1985; Madhani et al. 1986). Human cells also show this property with respect to the rate of repair (Mellon et al. 1986). This preferential repair of essential genes appears to correlate with UV resistance (Bohr et al. 1986).

Our results also show that the *ERCC2* secondary transformant of UV5 has an efficiency of cyclobutane dimer removal (Fig. 1) more like that of normal CHO cells than normal human cells (Mitchell et al. 1985b, 1987). This finding suggests that *ERCC2* does not control a hypothetical, rate-limiting component of repair that might be responsible for the overall difference in repair capacity between CHO and human cells.

ACKNOWLEDGMENT

This work was performed under the auspices of the U.S. Department of Energy by the Lawrence Livermore National Laboratory under contract W-7405-ENG-48.

REFERENCES

Bohr, V.A., D.S. Okumoto, and P.C. Hanawalt. 1986. Survival of UV-irradiated mammalian cells correlates with efficient DNA repair in an essential gene. *Proc. Natl. Acad. Sci.* **83:** 3830.

Bohr, V.A., C.A. Smith, D.S. Okumoto, and P.C. Hanawalt. 1985. DNA repair in an active gene: Removal of pyrimidine dimers from the DHFR gene of CHO cells is much more efficient than in the genome overall. *Cell* **40:** 359.

Brookman, K.W., L.H. Thompson, C.C. Collins, S.A. Stewart, J.L. Minker, and A.V. Carrano. 1987. Cloning of a human gene that restores normal SCE and strand-break repair to CHO mutant line EM9. *Environ. Mutagen.* (suppl. 8) **9:** 20.

Busch, D.B. 1980. "Large scale isolation of DNA repair mutants of Chinese hamster ovary cells." Ph.D. thesis, University of California, Berkeley.

Busch, D.B., J.E. Cleaver, and D.A. Glaser. 1980. Large scale isolation of UV-sensitive clones of CHO cells. *Somatic Cell Genet.* **6:** 407.

Cleaver, J.E. 1983. Xeroderma pigmentosum. In *The metabolic basis of inherited disease,* 5th edition (ed. J.B. Stanbury et al.), p. 1227. McGraw-Hill, New York.

Collins, A. and R.T. Johnson. 1987. DNA repair mutants in higher eukaryotes. *J. Cell Sci. Suppl.* **6:** 61.

Fischer, E., W. Keijzer, H.W. Thielmann, O. Popanda, E. Bohnert, L. Edler, E.G. Jung, and D. Bootsma. 1985. A ninth complementation group in xeroderma pigmentosum. *Mutat. Res.* **145:** 217.

Friedberg, E.C. 1987. The molecular biology of nucleotide excision repair of DNA: Recent progress. *J. Cell Sci. Suppl.* **6:** 1.

Friedberg, E.C., U.K. Ehmann, and J.J. Williams. 1979. Human diseases associated with defective DNA repair. *Adv. Radiat. Biol.* **8:** 85.

Hoeijmakers, J.H.J., H. Odijk, and A. Westerveld. 1987. Differences in the amount of transfected, integrated DNA between various rodent and human cell lines. *Exp. Cell Res.* **169**: 111.

Hoy, C.A., L.H. Thompson, C.L. Mooney, and E.P. Salazar. 1985. Defective DNA cross-link removal in Chinese hamster cell mutants hypersensitive to bifunctional alkylating agents. *Cancer Res.* **45:** 1737.

Ishida, R. and M. Buchwald. 1982. Susceptibility of Fanconi's anemia lymphoblasts to DNA-cross-linking and alkylating agents. *Cancer Res.* **42:** 4000.

Kraemer, K.H., M.M. Lee, and J. Scotto. 1984. DNA repair protects against cutaneous and internal neoplasia: Evidence from xeroderma pigmentosum. *Carcinogenesis* **5:** 511.

Lehmann, A.R. 1985. Use of recombinant DNA techniques in cloning DNA repair genes and in the study of mutagenesis in mammalian cells. *Mutat. Res.* **150:** 61.

Madhani, H.D., V.A. Bohr, and P.C. Hanawalt. 1986. Differential DNA repair in transcriptionally active and inactive proto-oncogenes: c-*abl* and c-*mos*. *Cell* **45:** 417.

Maher, V.M. and J.J. McCormick. 1984. Role of DNA lesions and excision repair in carcinogen-induced mutagenesis and transformation in human cells. In *Biochemical basis of chemical carcinogenesis* (ed. H. Geim et al.), p. 143. Raven Press, New York.

Maher, V.M., D.J. Dorney, A.L. Mendrala, B. Konze-Thomas, and J.J. McCormick. 1979. DNA excision-repair processes in human cells can eliminate the cytotoxic and mutagenic consequences of ultraviolet irradiation. *Mutat. Res.* **62:** 311.

Mellon, I., V.A. Bohr, C.A. Smith, and P.C. Hanawalt. 1986. Preferential DNA repair of an active gene in human cells. *Proc. Natl. Acad. Sci.* **83:** 8878.

Meyn, R.E., D.L. Vizard, R.R. Hewitt, and R.M. Humphrey. 1974. The fate of pyrimidine dimers in the DNA of ultraviolet-irradiated Chinese hamster cells. *Photochem. Photobiol.* **20:** 221.

Mitchell, D.L., C.A. Haipek, and J.M. Clarkson. 1985a. Further characterization of a polyclonal antiserum for DNA photoproducts: The use of different labelled antigens to control its specificity. *Mutat. Res.* **146:** 129.

———. 1985b. (6-4)Photoproducts are removed from the DNA of UV-irradiated mammalian cells more efficiently than cyclobutane pyrimidine dimers. *Mutat. Res.* **143:** 109.

Mitchell, D.L., R.M. Humphrey, G.M. Adair, L.H. Thompson, and J.M. Clarkson. 1987. The importance of (6-4) photoproducts and cyclobutane dimers for split-dose recovery in UV-irradiated normal and hypersensitive rodent cells. *Mutat. Res.* (in press.)

Schultz, R.A., D.P. Barbis, and E.C. Friedberg. 1985. Studies on gene transfer and on reversion to UV resistance in xeroderma pigmentosum cells. *Somatic Cell Mol. Genet.* **11:** 617.

Siciliano, M.J., A.V. Carrano, and L.H. Thompson. 1986. Assignment of a human DNA-repair gene associated with sister-chromatid exchange to chromosome 19. *Mutat. Res.* **174:** 303.

Takebe, H., S. Nii, M.I. Ishii, and H. Utsumi. 1974. Comparative studies of host-cell reactivation, colony forming ability and excision repair after UV irradiation of xeroderma pigmentosum, normal human and some other mammalian cells. *Mutat. Res.* **25:** 383.

Thompson, L.H. 1987. Using CHO cell mutants to study human repair genes. In *DNA repair* (ed. E. Friedberg and P. Hanawalt), vol. 3. Marcel Dekker, New York. (In press.)

Thompson, L.H. and A.V. Carrano. 1983. Analysis of mammalian cell mutagenesis and DNA repair using in vitro selected CHO cell mutants. *UCLA Symp. Mol. Cell. Biol. New Ser.* **11:** 125.

Thompson, L.H., K.W. Brookman, L.E. Dillehay, C.L. Mooney, and A.V. Carrano. 1982a. Hypersensitivity to mutation and sister-chromatid-exchange induction in CHO cell mutants defective in incising DNA containing UV lesions. *Somatic Cell Genet.* **6:** 759.

Thompson, L.H., D.B. Busch, K. Brookman, C.L. Mooney, and D.A. Glaser. 1981. Genetic diversity of UV-sensitive DNA repair mutants of Chinese hamster ovary cells. *Proc. Natl. Acad. Sci.* **78:** 3734.

Thompson, L.H., C.L. Mooney, K. Burkhart-Schultz, A.V. Carrano, and M.J. Siciliano. 1985. Correction of a nucleotide-excision-repair mutation by human chromosome 19 in hamster-human hybrid cells. *Somatic Cell Mol. Genet.* **11:** 87.

Thompson, L.H., J.S. Rubin, J.E. Cleaver, G.F. Whitmore, and K. Brookman. 1980. A screening method for isolating DNA repair-deficient mutants of CHO cells. *Somatic Cell Genet.* **6:** 391.

Thompson, L.H., K.W. Brookman, L.E. Dillehay, A.V. Carrano, J.A. Mazrimas, C.L. Mooney, and J.L. Minkler. 1982b. A CHO-cell strain having hypersensitivity to mutagens, a defect in DNA strand-break repair, and an extraordinary baseline frequency of sister chromatid exchange. *Mutat. Res.* **95:** 427.

Thompson, L.H., E.P. Salazar, K.W. Brookman, C.C. Collins, S.A. Stewart, D.B. Busch, and C.A. Weber. 1987. Recent progress with the DNA repair mutants of Chinese hamster ovary cells. *J. Cell Sci. Suppl.* **6:** 97.

van Duin, M., J. de Wit, H. Odijk, A. Westerveld, A. Yasui, M.H.M. Koken, J.H.J. Hoeijmakers, and D. Bootsma. 1986. Molecular characterization of the human excision repair gene *ERCC-1*: cDNA cloning and amino acid homology with the yeast DNA repair gene *RAD10*. *Cell* **44:** 913.

Vijg, J., E. Mullaart, G.P. van der Schans, P.H.M. Lohman, and D.L. Knook. 1984. Kinetics of ultraviolet induced DNA excision repair in rat and human fibroblasts. *Mutat. Res.* **132:** 129.

Westerveld, A., J.H.J. Hoeijmakers, M. van Duin, J. de Wit, H. Odijk, A. Pastink, R.D. Wood, and D. Bootsma. 1984. Molecular cloning of a human DNA repair gene. *Nature* **310:** 425.

Yagi, T., O. Nikaido, and H. Takebe. 1984. Excision repair of mouse and human fibroblast cells, and a factor affecting the amount of UV-induced unscheduled DNA synthesis. *Mutat. Res.* **132:** 101.

Zdzienicka, M.Z., L. Roza, A. Westerveld, D. Bootsma, and J.W.I.M. Simons. 1987. Biological and biochemical consequences of the human *ERCC-1* repair gene after transfection into a repair-deficient CHO cell line. *Mutat. Res.* **183:** 69.

Carcinogen-induced Homologous Recombination in Mammalian Cells

VERONICA M. MAHER,* YENYUN WANG,* NITAI P. BHATTACHARYYA,* J. JUSTIN MCCORMICK,* AND R. MICHAEL LISKAY[†]

*Carcinogenesis Laboratory–Fee Hall
Departments of Microbiology and Biochemistry
Michigan State University
East Lansing, Michigan 48824-1316
[†]Departments of Therapeutic Radiology and Human Genetics
Yale University School of Medicine
New Haven, Connecticut 06510

OVERVIEW

The ability of a series of mutagenic agents to induce homologous intrachromosomal recombination between duplicated genes in the chromosomes of mammalian cells was investigated. The agents tested included UV and ionizing radiation, *N*-methyl-*N'*-nitro-*N*-nitrosoguanidine (MNNG), mitomycin C (MC), (±)-7β,8α-dihydroxy-9α,10α-epoxy-7,8,9,10-tetrahydrobenzo[a] pyrene (B[a]PDE), *N*-acetoxy-2-acetylaminofluorene (*N*-AcO-AAF), and 4-nitroquinoline-1-oxide (4-NQO). The target cells were a thymidine kinase (TK)-deficient mouse L-cell strain containing a single integrated copy of a plasmid carrying two Herpes simplex virus *tk* (HSV *tk*) genes. Each HSV *tk* gene is inactivated by an 8-bp *Xho*I linker frameshift mutation at a unique site. In addition, there is a selectable marker gene in the intervening sequence between the two HSV *tk* genes. Following exposure to mutagens, the cells are selected for the ability to produce a functional HSV *tk* gene product that requires a productive recombinational event between the two nonfunctional genes. Cobalt-60 radiation did not induce a detectable increase in the frequency of recombination. MC induced a dose-dependent increase only at low doses. The other mutagens induced a dose-dependent increase in recombination over the entire range tested. When induced recombination frequencies were compared at equicytotoxic doses, the order of decreasing efficiency was MC > *N*-AcO-AAF=B[a]PDE=MNNG > 4-NQO > UV > > ^{60}Co. Analysis of the recombinants induced by MC, B[a]PDE, MNNG, and UV showed that the majority retained the HSV *tk* gene duplication, consistent with nonreciprocal transfer of wild-type genetic information (gene conversion). This is the event most frequently found in spontaneous recombinants with these target cells.

Banbury Report 28: Mammalian Cell Mutagenesis

INTRODUCTION

For several years, investigators have studied the mechanisms of meiotic and mitotic homologous recombination in lower eukaryotes and examined the effect of DNA-damaging agents on these processes. Such research has shown that UV and ionizing radiation and simple alkylating agents, such as methyl methanesulfonate (MMS), can stimulate recombination and that the frequency of recombination can depend on the DNA repair capacity of the cells. (For an extensive review of these kinds of studies in yeast, see Kunz and Haynes 1981.) However, less is known about the ability of such agents to induce homologous recombination in mammalian cells.

Homologous recombination is presumed to play an important role in the rearrangement of immunoglobulin genes (Brack et al. 1978; Sakano et al. 1981) and has been postulated as a mechanism for uncovering recessive genes involved in the development of certain childhood tumors (Cavenee et al. 1983; Leder et al. 1983). Interest in the mechanisms of homologous recombination in mammalian cells has led to the development of a number of systems for studying this process. Examples include methods for detecting extrachromosomal recombination (e.g., see Shapira et al. 1983; Brenner et al. 1984; Folger et al. 1984; Kucherlapati et al. 1984; Lin et al. 1984; Rubnitz and Subramani 1986), intrachromosomal recombination (e.g., see Lin and Sternberg 1984; Liskay et al. 1984; Smith and Berg 1984; and Stringer et al. 1985), as well as targeted recombination between exogenous DNA and a chromosomal sequence (e.g., see Smithies et al. 1984; Thomas and Capecchi 1986).

In an effort to understand the mechanisms of chemical and physical carcinogenesis, we have begun to investigate whether several well-studied carcinogenic (mutagenic) agents can induce homologous recombination between genes located within the mammalian cell chromosome and to characterize the kinds of recombination events induced (Wang et al. 1988). A comparative study of the efficiency of the various agents in inducing recombination could shed light on the mechanisms involved. Furthermore, it should be useful to compare the mutagenic efficiency of the agents with their recombinagenic efficiency.

The system we employ, designed by Liskay et al. (1984), involves TK-deficient mouse L cells containing a single, stably integrated copy of a plasmid. The plasmid carries duplicated copies of the HSV *tk* gene, each inactivated by an 8-bp *Xho*I linker mutation at a unique site. Approximately 18 hours after exposure to the mutagenic agents, the target cells are selected for expression of a functional HSV *tk* enzyme, which requires a productive recombinational event between the two nonfunctional genes.

The mutagens tested to date include two physical carcinogens (254 nm UV and ^{60}Co ionizing radiation), a simple alkylating agent (MNNG), an alkylating, cross-linking agent (MC), and three polycyclic aromatic carcinogens, (B[a]PDE, *N*-AcO-AAF, and 4-NQO). Using this system, we found that each agent, with the exception of ^{60}Co, induced a dose-dependent increase in the frequency of TK^+ recombinants. We are currently determining the kinds of recombinational events that occur and the mechanisms responsible for them and comparing the recombinagenic efficiency of the agents with their mutagenicity in mammalian cells.

RESULTS

Optimization of the Assay Conditions for Recombination

Before examining the effect of the various mutagens on the frequency of homologous recombination, we determined conditions for the assay, e.g., level of cytotoxicity, cell density at the time of selection for TK^+ cells, the length of the expression, and the selection period, and whether the TK^+ cells could form colonies with 100% efficiency under the conditions of the experiment, etc. (Wang et al. 1988). In view of the results, the following conditions were employed: Cells in exponential growth were trypsinized and resuspended in Eagle's minimal essential medium supplemented as described by Wang et al. (1988). For ionizing radiation, the cells were exposed in suspension and then plated at a density of 10^4 cells per cm^2 (5×10^5 cells per 100-mm-diameter dish). For all the other agents, the cells were plated at this density and allowed to attach for 6 hours before being exposed to mutagens. The number of dishes was adjusted to have at least 2×10^6 *surviving* target cells per dose. Selection for TK^+ cells, able to survive in selective medium containing *c*ytidine (2×10^{-5} M), *h*ypoxanthine (1×10^{-6} M), *a*minopterin (2×10^{-6} M), and *t*hymidine (3×10^{-5} M) (CHAT medium), was begun 18–20 hours later and continued for 14 days with refeeding every 2–3 days.

Unless otherwise noted, the doses used for inducing recombination were those that reduced cell survival to between 90% and 10% of the untreated controls. In each recombination experiment, cells treated at the same density as those to be assayed for recombination were assayed for survival by being plated at cloning densities immediately after the treatment and allowed 10–12 days to form macroscopic colonies. The cloning efficiency of the treated cells divided by that of the untreated cells (which ranged from 66% to 85%) was used to estimate the fraction of surviving cells in the original target population. This value was used to calculate the frequency of TK^+ recombinants per 10^6 surviving target cells.

Induction of Recombination by Mutagenic Agents

The ability of the various mutagens to cause cell killing and induce recombination between the two mutant HSV *tk* genes was determined for a range of doses. With the exception of ^{60}Co, each of the carcinogens tested increased the frequency of TK$^+$ cells two- to fivefold above the background frequencies, which averaged $18.6 \times 10^{-6} \pm 5 \times 10^{-6}$ in a total of 16 individual recombination experiments. A concentration of 0.25 μM B[a]PDE gave an observed frequency of recombination of 60.3×10^{-6} that was three times the background of that particular experiment, viz., 19.5×10^{-6}. The induced frequency was 40.8×10^{-6}. The dose of 8 μM MNNG gave an observed frequency of 95×10^{-6} that was five times the background of that experiment, viz., 19×10^{-6}. The induced frequency was 76×10^{-6}.

Liskay et al. (1984) reported that the spontaneous *rate* of recombination between the duplicated mutant HSV *tk* genes in cell line 333M is 3 per 10^6 cells per generation. Since the recombination event induced by the various mutagens we tested had to occur within one cell cycle, i.e., before CHAT selection began, the induced frequencies can be considered to represent the induced increase in *rate* of recombination. Considered in this way, the highest doses of B[a]PDE increased the *rate* of recombination up to 14-fold (i.e., up to 40 per 10^6 cells per generation) and those of MNNG increased the rate up to 25-fold (up to 76 per 10^6 cells per generation) (Wang et al. 1988).

Sensitivity of Recombinants to the Test Agents

If CHAT-resistant TK$^+$cells preexisting in the target population were more resistant than TK$^-$ cells to the cytotoxic effects of the test agents, this could cause an apparent increase in frequency of such recombinants. Evidence that this was not the case was obtained by isolating colonies of TK$^+$ recombinants that were found on plates receiving the highest dose of B[a]PDE and of MNNG and testing them for sensitivity to killing by B[a]PDE or by MNNG. The results showed that TK$^+$ recombinants were just as sensitive as the original parent target population and, therefore, did not represent mere survivors.

Dependence on the Presence of Duplicated HSV *tk* Genes

As a negative control, the ability of B[a]PDE and MNNG to induce CHAT-resistant cells in a population of mouse L cells containing only a single mutant copy of the HSV *tk* gene was investigated. No evidence of any TK$^+$ cells could be seen in the treated or untreated population (Wang et al. 1988). Letsou and Liskay (1986) showed previously that the spontaneous rate of formation of TK$^+$ cells in this particular cell line was less than 1 out of 3×10^8 per generation.

Induction of Recombination of MC and the Other Carcinogens

In our initial study of the cross-linking agent, MC, the concentrations used yielded survivals from 40% down to 15% and caused the cells to increase greatly in size by the time CHAT selection was begun. In those experiments, no increase in the frequency of TK^+ recombinants was observed. However, when lower doses were used (i.e., survival between 90% and 60% of the control) there was a dose-dependent increase in frequency of the TK^+ recombinants by MC (Wang et al. 1988). This requirement for high survival was not seen with UV, 4-NQO, or *N*-AcO-AAF. Just as with B[a]PDE or MNNG, doses that lowered the survival of the cells from 90% to 10% of the control induced a dose-dependent increase in the frequency of TK^+ recombinants.

Comparing the Agents for Recombination Efficiency

Since one cannot compare the frequency of recombination induced by radiation with that of chemicals on the basis of applied concentration, the agents were compared on the basis of equicytotoxic doses (recombination efficiency, i.e., the frequency of recombinants induced per mean lethal event). The results are shown in Figure 1. Analyzed in this way, MC was the most efficient agent and UV was the least efficient. The recombinagenic effects of the other mutagens were very similar to each other. (The negative ^{60}Co data have not been included in this comparison.)

Kinds of Recombination Events Induced by These Agents

The plasmid used to transfect the mouse L cells was designed to facilitate analysis of the kinds of recombination events that take place. In addition to duplicate HSV *tk* genes, each with an 8-bp *Xho*I linker frameshift mutation in a unique location, the plasmid contains the neomycin (*neo*) gene coding for Geneticin resistance located between the two HSV *tk* genes (Liskay et al. 1984). This selectable marker facilitates assaying the TK^+ recombinants for the retention or loss of the intervening DNA sequence. If the recombination event involves a single, reciprocal exchange within a chromatid or a single, unequal exchange between sister chromatids, only a single wild-type copy of the HSV *tk* cell gene will be present in the cell and the sequence containing the *neo* gene will be lost. In contrast, if the recombination event consists of a nonreciprocal transfer of wild-type genetic information, gene conversion, then the HSV *tk* gene duplication with the *neo* gene will be retained in the recombinant. One will be a wild type and lack the *Xho*I site; the other will still contain the original *Xho*I linker insertion frameshift mutation. Simply assaying TK^+ recombinants for resistance to Geneticin can distinguish which type of event occurred. Southern blot DNA analysis of such recombinants can

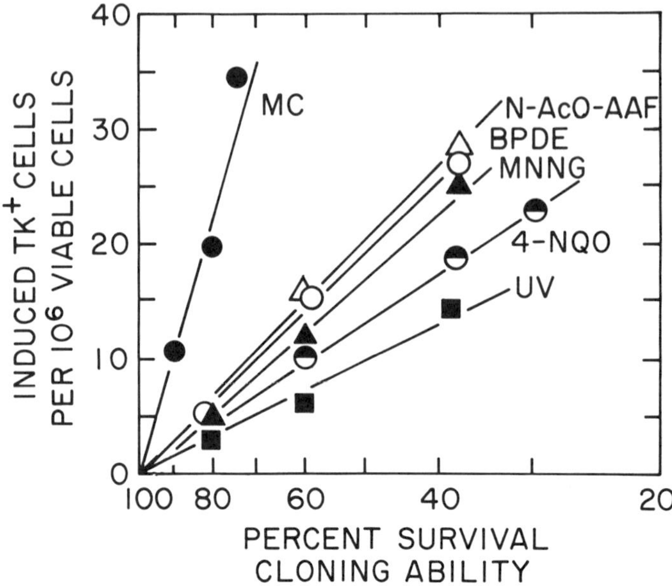

Figure 1

Comparison of the frequency of homologous recombination induced by the various carcinogens as a function of the extent of cell killing induced by each agent. (●) MC; (△) *N*-AcO-AAF; (○) B[a]PDE; (▲) MNNG; (◒) 4-NQO; (■) UV. These data were extrapolated from the respective survival and recombination curves obtained from a series of experiments.

readily be used to confirm these results, since each HSV *tk* gene was inserted at a unique restriction site locus in the original plasmid.

Liskay et al. (1984) showed that 85% of the spontaneous recombination events in the 333M cells are gene conversions. We isolated 13–22 recombinants from each of the populations receiving high doses of MC, B[a]PDE, MNNG, or UV and tested them for resistance to Geneticin, as well as continued ability to grow in a CHAT medium. The results showed that 85–90% of the recombinants induced by those mutagens resulted from type-A events, i.e., retained the intervening sequence. The results were confirmed by Southern blot DNA analysis (data not shown). Similar studies with recombinants induced by *N*-AcO-AAF, 4-NQO, and related agents are in progress.

DISCUSSION

Although Cavenee et al. (1983) provided strong evidence that homologous mitotic recombination could be causally involved in human retinoblastoma, and studies with procaryotes and lower eukaryotes have demonstrated that mutagenic/carcinogenic agents can induce such recombination, less is known about induction of homologous recombination in mammalian cells by such agents. The present studies were undertaken to determine if selected mutagens could increase the frequency and rate of homologous recombination of duplicated genes within the chromosome of mouse L cells. The results showed clearly that, with the exception of ^{60}Co, all the mutagens tested were capable of inducing a dose-dependent increase in such recombination. We (Konze-Thomas et al. 1982; Aust et al. 1984; Domoradzki et al. 1984; Howell et al. 1984; Wang et al. 1986) as well as other investigators have measured the frequency of mutations induced in mammalian cells by each of these agents except for MC. The present finding that these agents can also increase the rate of homologous recombination of duplicated genes within the chromosome of somatic cells in culture expands our understanding of the potentially harmful (mutagenic, carcinogenic) effects of such environmental agents.

Two kinds of recombination events occur spontaneously in these target cells: a single wild-type HSV *tk* gene occurs 15% of the time. It can be attributed to a single, reciprocal exchange between homologous genes on the same chromatid or an unequal, single exchange between sister chromatids and gene conversion between genes on the same chromatid or on sister chromatids. To date, we have tested four of the mutagens for the *kinds* of recombinant events induced (i.e., MC, B[a]PDE, MNNG, and UV). We find gene conversion to be four to five times more frequent than reciprocal exchange (Wang et al. 1988), just as is true for spontaneous recombination. This suggests that whatever mechanism is responsible for the observed carcinogen-induced increase in rate of recombination is affecting both types of events.

It is of interest to us to determine if the recombination event results from the presence of unexcised lesions present in the DNA during DNA replication or other times during the cell cycle or if the process is stimulated by cellular DNA excision-repair processes. Compared to human cells, mouse cell lines in culture are not very proficient in nucleotide excision repair (see, e.g., Yagi 1982; Elliott and Johnson 1985). However, Bohr et al. (1985) have suggested that such rodent cells in culture preferentially repair actively transcribed genes, which would include the HSV *tk* genes stably integrated into 333M cells. To examine this question more closely, we are currently integrating the plasmid with its duplicated HSV *tk* genes into human cells which differ in DNA excision-repair capacities.

ACKNOWLEDGMENTS

This research was supported in part by Department of Health and Human Services grants CA-21253 and GM-32741 from the National Institutes of Health and by a Leukemia Society of America Scholar Award to R.M.L.

REFERENCES

Aust, A.E., N.R. Drinkwater, K.C. Debien, V.M. Maher, and J.J. McCormick. 1984. Comparison of the frequency of diphtheria toxin and thioguanine resistance induced by a series of carcinogens to analyze their mutational specificities in diploid human fibroblasts. *Mutat. Res.* **125:** 95.

Bohr, V.A., C.A. Smith, D.S. Okumoto, and P.C. Hanawalt. 1985. DNA repair in an active gene: Removal of pyrimidine dimers from the DHFR gene of CHO cells is much more efficient than in the genome overall. *Cell* **40:** 359.

Brack, C., M. Hirama, R. Lenhard-Schuller, and S. Tonegawa. 1978. A complete immunoglobulin gene is created by somatic recombination. *Cell* **15:** 1.

Brenner, D.A., S. Kato, R.A. Anderson, A.C. Smigocki, and R.D. Camerini-Otero. 1984. The recombination and integration of DNAs introduced in mouse L cells. *Cold Spring Harbor Symp. Quant. Biol.* **49:** 151.

Cavenee, W.K., T.P. Dryja, R.A. Phillips, W.F. Benedict, R. Godbout, B.L. Gallie, A.L. Morphree, L.C. Strong, and R.L. White. 1983. Expression of recessive alleles by chromosomal mechanisms in retinoblastoma. *Nature* **305:** 779.

Domoradzki, J., A.E. Pegg, M.D. Dolan, V.M. Maher, and J.J. McCormick. 1984. Correlation between O^6methylguanine-DNA-methyltransferase activity and resistance of human cells to the cytotoxic and mutagenic effect of N-methyl-N′-nitro-N-nitrosoguanidine. *Carcinogenesis* **5:** 1641.

Elliott, G.C. and R.T. Johnson. 1985. DNA repair in mouse embryo fibroblasts. II. Responses of nontransformed, preneoplastic and tumorigenic cells to ultraviolet irradiation. *Mutat. Res.* **145:** 185.

Folger, K., K. Thomas, and M.R. Capecchi. 1984. Analysis of homologous recombination in cultured mammalian cells. *Cold Spring Harbor Symp. Quant. Biol.* **49:** 123.

Howell, J.N., M.H. Greene, R.C. Corner, V.M. Maher, and J.J. McCormick. 1984. Fibroblasts from patients with hereditary cutaneous malignant melanoma are abnormally sensitive to the mutagenic effect of simulated sunlight and 4-nitroquinoline-1-oxide. *Proc. Natl. Acad. Sci.* **81:** 1179.

Konze-Thomas, B., R.M. Hazard, V.M. Maher, and J.J. McCormick. 1982. Extent of excision repair before DNA synthesis determines the mutagenic but not the lethal effect of UV radiation. *Mutat. Res.* **94:** 421.

Kucherlapati, R.S., E.M. Eves, K.-Y. Song, B.S. Morse, and O. Smithies. 1984. Homologous recombination between plasmids in mammalian cells can be enhanced by treatment of input DNA. *Proc. Natl. Acad. Sci.* **81:** 3153.

Kunz, B.A. and R.H. Haynes. 1981. Phenomenology and genetic control of mitotic recombination in yeast. *Annu. Rev. Genet.* **15:** 57.

Leder, P., J. Battey, G. Lenoir, C. Moulding, W. Murphy, H. Potter, T. Stewart, and R. Taub. 1983. Translocations among antibody genes in human cancer. *Science* **222:** 765.

Letsou, A. and R.M. Liskay. 1986. Intrachromosomal recombination in mammalian cells. In *Gene transfer* (ed. R. Kucherlapati), p. 383. Plenum Press, New York.

Lin, F.-L. and N. Sternberg. 1984. Homologous recombination between overlapping thymidine kinase gene fragments stably inserted into a mouse cell genome. *Mol. Cell. Biol.* **4:** 852.

Lin, F.-L., K. Sperle, and N. Sternberg. 1984. Model for homologous recombination during transfer of DNA into mouse L cells: Role for DNA ends in the recombination process. *Mol. Cell. Biol.* **4:** 1020.

Liskay, R.M., J.L. Stachelek, and A. Letsou. 1984. Homologous recombination between repeated chromosomal sequences in mouse cells. *Cold Spring Harbor Symp. Quant. Biol.* **49:** 183.

Rubnitz, J. and S. Subramani. 1986. Extrachromosomal and chromosomal gene conversion in mammalian cells. *Mol. Cell. Biol.* **6:** 1608.

Sakano, H., Y. Kurosawa, M. Weigert, and S. Tonegawa. 1981. Identification and nucleotide sequence of a diversity DNA segment (D) of immunoglobulin heavy-chain genes. *Nature* **290:** 562.

Shapira, G., J.L. Stachelek, A. Letsou, L.K. Soodak, and R.M. Liskay. 1983. Novel use of synthetic oligonucleotide insertion mutants for the study of homologous recombination in mammalian cells. *Proc. Natl. Acad. Sci.* **80:** 1827.

Smith, A.J.H. and P. Berg. 1984. Homologous recombination between defective *neo* genes in mouse 3T6 cells. *Cold Spring Harbor Synp. Quant. Biol.* **49:**171.

Smithies, O., M.A. Koralewski, K.-Y. Song, and R.S. Kucherlapati. 1984. Homologous recombination with DNA introduced into mammalian cells. *Cold Spring Harbor Symp. Quant. Biol.* **49:**161.

Stringer, J.R., R.M. Kuhn, J.L. Newman, and J.C. Meade. 1985. Unequal homologous recombination between tandemly arranged sequences stably incorporated into cultured rat cells. *Mol. Cell. Biol.* **5:** 2613.

Thomas, K.R. and M.R. Capecchi. 1986. Introduction of homologous DNA sequences into mammalian cells induces mutations in the cognate gene. *Nature* **324:** 34.

Wang, Y., V.M. Maher, R.M. Liskay, and J.J. McCormick. 1988. Carcinogens can induce homologous recombination between duplicated chromosomal sequences in mouse L cells. *Mol. Cell. Biol.* **7:** (in press).

Wang, Y., W.C. Parks, J.C. Wigle, V.M. Maher, and J.J. McCormick. 1986. Fibroblasts from patients with inherited predisposition to retinoblastoma exhibit normal sensitivity to the mutagenic effects of ionizing radiation. *Mutat. Res.* **175:** 107.

Yagi, T. 1982. DNA repair ability of cultured cells derived from mouse embryos in comparison with human cells. *Mutat. Res.* **96:** 89.

Comments

Tindall: I have two questions for Veronica. First, in the gene conversion events that you have observed, is there a directional bias indicating the preferential conversion of either of the TK genes?

Maher: So far, for the carcinogen-induced events it's about 50/50, but we have only analyzed a small sample of such recombinants. However, Dr. Mike Liskay and his associates have examined the products of more than 150 spontaneous events and have found a 2 to 1 preference for mutating no. 8 over mutating no. 26. They do not have a mechanistic explanation for this preference.

Tindall: My second question concerns the type-B recombinants you described. It would be interesting to recover the product of the reciprocal recombinant event. Are you or anyone else making another construction that would include an origin of replication inserted between the two TK genes allowing you to use the Neo marker to recover the reciprocal recombinant product?

Maher: The beauty of the system I am using is that this recombination is occurring within the chromosome; not between two replicating plasmids that have been put in extrachromosomally. So, with direct repeats, one cannot recover the other product of the reciprocal exchange. Of course, it is possible to design a system such as you describe. But the piece that is cut out would have to be reincorporated to withstand G418 selection. However, with *inverted* repeated sequences, it is possible to recover the product of intrachromatid reciprocal exchanges since the section is not deleted. Instead, there is an inversion between the points of the crossover. Dr. Liskay and his group have demonstrated that this occurs readily and they recover the product.

Hsie: I just want to follow up on your last slide, the summation of several agents in which you showed a beautiful gene conversion event. Do you have any data or comparisons that have been made with the yeast system; how do they compare in terms of relative efficacy between the different agents?

Maher: I am not familiar with the data base in yeast. I do know that X-ray has been said to be a good recombinagen in yeast, but it was not very good at inducing recombination in our mammalian cell system.

Hsie: Having this recombination data, and having worked on mutagenesis for a long time, do you have any feeling what kind of proportion or what percentage of mutational events we have been seeing traditionally could have been attributed by recombinational events?

Maher: The recombination system that we are using here is a specially-designed set-up with two closely related, very closely linked *tk* genes. Therefore, it allows one to detect homologous recombination with a reasonable frequency. If the two genes were located far from one another, this would not be true. So this is an excellent system, but it's not comparable to what is occurring with mutations in the *hgprt* locus. The chance of mutations in *hgprt* resulting from recombination with some other *hgprt* gene is practically nil.

Evans: I was going to ask about X-ray induced recombination, but you have already said that you don't get too many.

Maher: We have not examined the question extensively, but what we have seen so far is that the frequency of recombinants is not high. However, the background in the X-ray experiments was a little high. So, perhaps we haven't given it a real chance. We still intend to study the question further.

Chasin: A question for Larry (Thompson): Almost all of your transfectants, by whatever criterion you looked at, were nearly identical to your wild type. Don't you find that a little surprising, considering that you might expect different amounts of expression in independent transfectants?

Thompson: It is a little surprising. In the one case where we saw intermediate X-ray sensitivity, that didn't correlate with the normal level of repair. In the case of the EM9 mutation, if the amount of gene product from the normal gene is usually in great excess, the level of expression really won't make that much difference.

One of the questions we have to consider is what the copy number of the gene might be. We think that, in the transformants made with genomic DNA, we have a single copy of the gene. That inference is based on the band intensity of the Southern blots; we don't think there are significant differences among the bands of different transformants.

In the case where we're transfecting the cosmids, we would expect that sometimes we would probably get multiple copies. In the case of the *ercc1* gene, which the Dutch group has studied, overexpressing the gene appears to be toxic to the cells. For example, they have tried to amplify the gene and get overexpression, and they were not able to do that. So, the level of the gene product appears to be extremely critical. It might be that there is an inherent counter selection against cells that have overexpression.

Chasin: How about hybrids? Did you get hybrids when you did your complementation tests? Are they the same as parental wild-type cells in terms of their sensitivity?

Thompson: In some crosses, the hybrids consistently had intermediate levels of sensitivity. In that situation, you would expect there to be only one copy of the normal gene and only one complementing chromosome present in the hybrids. I think in most instances the hybrids did have more sensitivity than the wild-type cells. So far, the transformants seem to be more resistant on the average than the hybrids.

Chu: May I add something to the same question Larry raised? In Phil Hanawalt's lab at Stanford, DNA repair at specific loci was studied. They found that the transcriptionally active genes repair more efficiently as compared to the nontranscribing flanking regions. I spent a year of sabbatical last year at Phil's lab and studied the repair of the dihydrofolate reductase gene in the wild-type CHO cells, a mutant that is UV-sensitive, and the mutant containing the human *ercc1* gene. The repair at the *dhfr* locus is efficient in the wild type, deficient in the UV-sensitive mutant, and in full repair capacity, but not exceeding the wild type, in the transfectant having the single copy of the *ercc1* gene.

Hutchinson: I just wanted to follow up on Larry's point about the sevenfold reduction. Much the same thing is seen in *E. coli*; that is, nonleaky *uvr* A, B, and C mutants all produce about the same sensitization to UV. If Larry's mutants are all nonleaky, and if the five proteins are all component parts of the same repair process, you would expect the same degree of sensitization in all five mutants.

One interesting question should be raised. From your table, it is implied that you have isolated many more mutants than just those five; presumably, you have taken the most solid one in each of the complementation groups for further analysis.

Thompson: During the process of doing the complementation groups, the mutants were definitely put through a filter, so to speak, that we could at that time do the test practically only with the cells that showed high degrees of sensitivity. The other mutants, which are many—an equal number—have not been examined and could represent alleles that might have different levels of sensitivity within the given complementation group. Also, there may be other complementation groups among that collection that couldn't be identified because of the lesser sensitivity.

Tindall: Have you looked at the possibility of transfecting into repair-deficient human cells to see if you can complement repair functions?

Thompson: We have started doing those experiments. The only result I can give you at the moment is that the *ercc2* gene does not correct XP-A cells, which are usually the most UV-sensitive of all the complementation groups.

Tindall: A few years ago a number of laboratories were attempting to isolate human repair genes by transfecting human repair-deficient cell lines. Obviously, the greatest advances have come using repair-deficient hamster cells as transfection hosts. Can you comment on the current status of using human repair-deficient cell lines in the isolation of DNA repair genes?

Thompson: A number of labs put a lot of effort into transfecting into human XP cells, for example, and couldn't get any genomic transformants. At the meeting I was at recently in the Netherlands, another set of laboratories is really still working on this problem quite diligently. Also, there was a nice report from Tanaka at Osaka University, showing quite convincing data for correction of XP-A cells by a mouse gene, which he is also tracking by repetitive sequences the way we did. He appears to have that gene now in a cosmid library, and he may be able to pull it out, but I don't have any idea how big it is. We might be quite close to seeing the XP-A gene, at least the mouse counterpart, being cloned.

Glickman: Veronica, I like the fact that you present your data in terms of adducts per nucleotides, because it gives you a way of doing valuable comparisons between different chemical agents. However, for us who are not very knowledgeable in what that means, could you give us an example of the B[a]PDE. For example, what molar exposure do you use, and what do you do?

Maher: For B[a]PDE, the dose that we would use for all our experiments is this: 0.1 μM, 0.2 μM. A dose of 0.15 μM will give you about 37% survival and will induce about 14 B[a]PDE adducts per million nucleotides.

Glickman: Let me then go one step further. You are comparing micromolar and survival. How do you determine the frequency of the adducts? Are you taking the DNA back out of the cells; and how quickly—how much time is involved in your experiment?

Maher: The exposure time is one hour with B[a]PDE in serum-free medium. The cells are harvested at the end of that hour. We use radioactively labeled compounds for the binding studies and extract the DNA, purify it, and determine the number of labeled residues bound per nucleotide.

Glickman: You're not worried that during the one hour you've had a certain amount of repair? Is that something that you know anything about?

Maher: It is not much of a problem with B[a]PDE since these adducts are removed relatively slowly, even in human fibroblasts. Furthermore, there is evidence that in mouse L cells, the removal of such adducts from DNA may be even slower than in human cells.

Glickman: The amount of the first hour is negligible, and this is going to be a good dosimeter for that case at least.

Evans: Larry, have you found the molecular basis for the inability of EM9 to repair DNA damage?

Thompson: No, we haven't. A number of assays were done on the EM9 mutant looking for a biochemical defect. Things that were looked at were the DNA ligases, AP endonucleases, and the integrity of the poly(ADP)-ribose polymerase system. In all of those studies, the results showed that the EM9 cells were just like the wild-type CHO, so we sort of gave up on the approach of looking for the defect in the gene product at that time, a couple of years ago. We are hoping that maybe, by getting the protein, we can get a better idea of what the defect is.

Meuth: Did you fractionate the ligases?

Thompson: Yes. That was done by Chan and Becker. They looked pretty carefully at both forms of DNA ligase, looked at heat sensitivity and things like that, as well as activity, and nothing turned up.

Meuth: Recent work with Bloom's by Tomas Lindahl's group shows consistent alterations of DNA ligase I.

Thompson: Right. I could add that Jim German, who has done a lot of work with Bloom's syndrome, which has these high SCEs like EM9, has now shown that our EM9 mutant is complemented by a chromosome in Bloom's cells. I guess it's not known yet whether that's chromosome 19. The fact that the two mutations complement suggests that they are probably in different loci. Tom Lindahl's group has now gotten fairly convincing data showing that, in a number of cultures from Bloom's patients, there is a defect in the DNA ligase I. We didn't see any such defect in EM9 cells.

Calos: Veronica, this stimulation of recombination, how do you view that mechanistically, the relationship between these adducts and recombination?

Maher: I haven't gotten the answer to exactly which mechanism(s) are being used. I can imagine that unrepaired lesions are stimulating recombination, but it could be that it is stimulated by the beginning of the repair process. Perhaps, excision repair makes a single-strand break, and then a strand invades and begins the recombination process, but we aren't sure yet. I am currently studying the ability of these cells to carry out excision repair of DNA damage caused by these particular agents. It's not known. But, if the cells are not opening up any strands, making nicks, then the unexcised adducts will interfere with DNA replication. The recombination mechanisms that are usually hypothesized involve single-strand gaps or double-strand gaps, but I can't make a distinction yet, at this point.

Chasin: Your second response to Abe's question raises an interesting question. There are scattered throughout the natural genes in the natural places all kinds of repeated sequences that should provide substrates for recombination. It's always somewhat of a mystery that we don't see more. Your numbers show at this dose of B[a]PDE you get about 10 times more *hgprt* mutations than TK recombinants, but this isn't a fair comparison because we don't know the state of the integrated *tk* genes. There is always the question of position effect with integrated genes. I wonder if you could get some information on this point—or maybe Mike (Liskay) has already—in particular, as to how stable that *tk* gene is. That is, after you have a TK^+, you could now measure BrdU resistance and see what the spontaneous mutation rate is for the integrated gene.

Maher: I did not test this but Dr. Mike Liskay has done so. He has found using spontaneous TK^+ recombinants formed by gene conversion that their rate of reversion to resistance to trifluorothymidine depends on the particular parental line. But in the two parental lines he examined, the most frequent event (10^{-2}–10^{-4}) was loss of the entire gene locus, probably by loss of the whole chromosome. In other cases the cells underwent a recombination event, a gene conversion, which produced a cell with two identical TK^- mutations. This occurred at a frequency of 10^{-5}. At a still lower frequency ($>10^{-6}$) he observed single reciprocal exchanges, with loss of the intervening sequence containing the neomycin gene.

Chasin: I don't doubt that you're inducing recombination. However, the question is why, when we look at APRT mutants or DHFR mutants, why don't we see deletions due to recombination between repeated sequences that are stuck within the gene? We don't really see that very much.

Maher: I agree that there is all kinds of DNA around in the cell that they could use.

Chasin: In fact, the *dhfr* gene contains several repeated sequences in various introns.

Maher: Ernie (Chu) has a good system to look for them. I don't.

Gibbs: I have one question for Larry (Thompson). Do you have DNA double-strand break repair deficient, ionizing radiation-sensitive cells in your library?

Thompson: EM9 does show some defect in double-strand break repair, but it's not pronounced. There are several CHO mutants that do have a gross defect in double-strand break repair, the XR-1 that Tom Stamato isolated and the set of mutants within one complementation group isolated by Penny Jeggo.

Gibbs: Are efforts underway to clone these genes?

Thompson: The Los Alamos group I think is quite close to cloning the human gene that corrects the XRS mutants of Penny Jeggo.

Hutchinson: I would like to make a plea that at the moment this be thought of as cells defective in the repair of damage as shown by neutral elution. It is becoming clear that several lesions are detected by neutral elution. Double-strand breaks are detected but so are other lesions.

Gibbs: What are they?

Hutchinson: Mildly alkali-labile (pH 9.6) structures of some kind and DNA protein cross-links. At the moment, results obtained by the neutral elution technique must be treated with care. Alkaline elution has some problems, but it is much more reliable than neutral elution.

Concluding Remarks

ERNEST H.Y. CHU
University of Michigan Medical School
Ann Arbor, Michigan 48107-0618

In the less than two decades since the first demonstration of experimental mutagenesis in cultured mammalian cells, and eight years since the publication of *Banbury Report 2* on the same topic, the progress made by numerous investigators is impressive and significant indeed. Great strides have been made, both in concept and methodology, in studies of mutation in human and rodent cells. The field shows vigor and every sign for continued growth and development, and we still have a great deal more to learn and to do.

Whereas mutagenic processes have been well worked out and understood in prokaryotic systems, why should we study mutagenesis in mammalian cells? Because a DNA is a DNA is a DNA, isn't the sex of *Escherichia coli* everything one needs to know to explain the genetics of the elephant? Nonetheless, in view of the organizational and functional complexities of the mammalian genome, we can expect that only through direct study of the genetic material of mammals can we fully understand the mutagenic mechanisms and consequences of genetic alterations in mammalian cells and organisms. The idea is to apply the knowledge and tools of molecular and cell biology to tackle biological problems that are unique in higher eukaryotes, including humans. It is precisely this strategy that highlights the theme of this conference, and perhaps indicates the trends of mutation research on mammalian cells in the years to come.

Mutation can be identified and measured at the phenotype, polypeptide, and nucleotide levels. Extensive but unbalanced accounts of all three approaches are presented in this volume. With the exception of a short report by my co-workers and me, there is little discussion on either quantitative mutagenesis, multilocus assay, or the use of protein variation as indicators of mutation. Furthermore, even though mutations have been successfully identified by amino acid sequence analysis in human hereditary disorders involving hemoglobin, HPRT, APRT, and others, there is a paucity of parallel studies on somatic cells.

Three sections of this volume are devoted to a review of selectable single-locus genetic markers and differential recovery of such mutants both in vitro and in vivo. This certainly reflects the intense interest, activity, and problems in this research area.

Following the pioneering work of D. Pious and R. DeMars, Nicklas et al. have been studying mutation at the human major histocompatibility complex

(HLA) in a human lymphoblastoid cell line. This system is particularly attractive because of (1) the existence of multilocus polymorphism in normal diploid cells, (2) the availability of specific antisera and DNA probes, (3) the possibility of discovering mitotic recombination and gene conversion, and (4) the potential to measure human somatic mutations in vivo in relation to health and disease.

Mutation leading to a deficiency in one of the salvage pathway enzymes (e.g., HPRT, APRT, TK) of purine and pyrimidine metabolism is not essential for cell survival unless the de novo biosynthetic pathways are genetically or chemically blocked. The gene for HPRT is X linked, whereas those for APRT and TK are autosomal. All three genes from human, mouse, and Chinese hamster have been cloned and characterized.

In several rodent cell lines, recovery of forward mutations from base-analog sensitivity to resistance at these loci seems to depend on (1) the heterozygous versus the hemizygous state, (2) the existence of putative vital genes adjacent to the locus under selection, (3) the proximity of the marker gene to the breakpoint in a chromosome rearrangement, (4) the kind of mutagen used, (5) the stringency of selection, and (6) other factors such as the cell-culture environment. These problems may be real or imagined, and the mechanism for the apparent discrepancies is not always clear. It is true that certain classes of mutants, such as large deletions encompassing the test locus in a hemizygous chromosome or region, are nonviable. It was pointed out by Glickman that genetic processes other than mutation, such as mitotic recombination and gene conversion, may contribute to the genetic variability at a locus in diploid state.

Hsie and Stankowski demonstrated equivalent cell survival, but about a tenfold higher mutagenicity to ionizing radiation and several radiomimetic compounds when the frequencies of the induced mutants to 6-thioguanine resistance (HPRT$^-$) were compared between those occurring at the single-copy bacterial *gpt* gene, introduced and integrated into an autosome of the host cell, and those at the endogenous *hprt* gene on the X chromosome in the parental cell. The explanation offered for the hypermutability of the bacterial gene was the tolerance and thereby higher probability for recovery of multilocus deletion mutants induced in the transfected as compared with the cellular gene. It would be interesting to determine whether the *gpt* sequence integrated into other chromosomal sites will be similarly hypermutable. However, the existing genetic system may be employed for the detection of environmental mutagens at an enhanced sensitivity.

De Serres reviewed the extensive results of fine structure analysis of the adenine-3 (*ad-3*) region in *Neurospora crassa.* The data offer a prediction of differential recovery of mutants in mammalian cells. The data are beautiful, the parallelism is appropriate, and the utility of the *Neurospora* test system is unquestioned. However, his suggested classification of mutants according to the size of the lesion, in an operational sense, did not enjoy wide-spread acceptance.

Russell and Rinchik summarized the results of genetic and molecular analysis of chromosomal regions surrounding the seven specific loci of the mouse. The large collection at Oak Ridge of radiation-induced deletion mutants of the mouse, especially those at the dilute short ear (*d se*) region on chromosome 9, offers an excellent opportunity for investigating a variety of important questions, including (1) the effect of heterozygous deletions, (2) the structure/function correlation of genes opposite the deleted regions, (3) targeted mutagenesis at the region corresponding to the largest deletion, and (4) risk assessment of the genome as a whole, based on the heterozygous effect of deletions and the estimate of the total number of essential genes.

Albertini et al. measured the mutant frequency at the *hprt* locus in human nontransformed T-lymphocyte clones. Using specific T-cell surface receptor markers, they were able to distinguish between unique and clonal progeny of 6-thioguanine resistant mutants that occur in vivo and are propagated in vitro. Jensen et al. used fluorescent antibodies against MN markers (glycophorin A, GPA) on human erythrocytes to quantify somatic mutations. The test system appears promising for monitoring human somatic mutation in vivo. The attractive features of this system are (1) the demonstrated genetics of the MN blood groups, (2) the known chemistry of glycophorins, and (3) the efficiency of laser-directed cell discrimination technology. There are, however, a number of limitations. First, only one locus is monitored. Second, it is impossible to clone mutants for further analysis or to distinguish between mutation and phenocopy. Perhaps the most serious difficulty is the variable time of occurrence of mutation during erythropoiesis leading to a differential clonal expansion of mutant red cell progenitors and a huge disparity in apparent mutant, but not mutation, frequency in normal and mutagen-exposed human subjects. These problems were shared by an earlier study on the ABO locus in erythrocytes by K.C. Atwood and S. Schienberg in the early 1950s but not by the studies of Albertini et al. with T lymphocytes mentioned above.

The papers in the fourth section by Glickman, Meuth et al., Chasin et al., Hozier and Applegate, Little et al., Gibbs and Caskey, and O'Neill et al. described the molecular analysis of mutation at the selectable markers (APRT, HPRT, TK, DHFR) in human and rodent cell lines as well as human T lymphocytes. In general, Southern blotting and DNA hybridization are used to delineate mutations on the basis of changes in restriction cleavage sites that either occur spontaneously or are induced by various DNA damaging agents; however, in several laboratories, nucleotide sequence analysis has been completed in an impressive collection of mutants. These results have permitted conclusions to be drawn regarding the relative frequencies (mutant spectrum) of base pair substitutions, large and small deletions, and rearrangement in drug selected mutants of different origin, as well as the relationship between nucleotide sequence and mutability (hot spots). They also allow a comparison of mutagenic specificity in similar nucleotide sequences that are present in pro- and eukaryotes.

The origin of "spontaneous" mutations was discussed with interest. Gibbs and Caskey described the newer techniques including ribonuclease-A cleavage, denaturing gradient gel electrophoresis and polymerase chain extension that are beginning to be used for mutation studies. Chasin et al. showed an interesting relationship between the transcribed message and the activity and amounts of the enzyme in individual DHFR mutants of CHO cells. This represents another level of sophistication being first applied to mutant analysis in mammalian cells.

The use of shuttle vectors for mutational studies in which bacterial genes are introduced to, and rescued from, mammalian cells in culture seems to be more and more successful, and information is rapidly accumulating. Tindall et al. and Ashman and Davidson reported the experimental approach and results of the *E. coli gpt* gene inserted into bacteriophage λ or a retrovirus vector. D.J. Vanderberg et al. used either an SV40 or EBV shuttle vector carrying the *E. coli lacI* gene for an elegant genetic and molecular analysis of mutants. The vector is introduced and amplified in human host cells, which are treated or not with various mutagens. The *lacI* gene has been thoroughly defined by J. Miller through analysis of thousands of mutants at the locus in *E. coli.* This information was used for molecular analysis of new mutants arising at the same gene sequence carried by the shuttle vector in human cells. Precise nucleotide changes can be inferred by means of genetic analysis without actual nucleotide sequencing. More recently, the replacement of the SV40 vector with the one (developed by W. Sugden) carrying an EBV origin of replication and the transacting EB nuclear antigen (EBNA) sequence has led to a reduced level of spontaneous mutation compared to that of the native gene in *E. coli*.

Dixon et al. analyzed mutations arising at the *sup*F gene carried by an SV40-based shuttle vector. The advantageous features of this target tRNA gene are its short total length (~ 220 bp) amenable for nucleotide sequencing and the possibility of mutagen treatment either before or after its introduction into the host monkey cells. Sarasin et al. employed SV40 to analyze spontaneous or mutagen (e.g., AAAF, MNNG) induced mutations either at two gene loci of the virus itself or in nonviral genes packaged into the shuttle virus. Obviously, each designer gene or vector construct has its own advantages and disadvantages, but all have yielded valuable and complementary information.

An analogy was made in this volume that working on a particular experimental system or a gene locus is like looking through a narrow window, like the assessment of the external morphology of an elephant by blind men. This volume allows us to compare notes, make connections, and widen our windows.

It would not be possible to enumerate all the significant facts that have been presented. Rather, I've attempted to crystallize the collective wisdom expressed in this book by contributors who are among the principal players in the field. Grouped below under three headings are some of the outstanding questions that remain to be answered and a few among the infinite number of investigative possibilities.

MODULATION OF THE MUTATIONAL PROCESS

Mutation is influenced by genetic and extrinsic factors. Among the genetic factors, DNA repair (cloning of repair genes as presented by Thompson et al.) and DNA replication (e.g., a defect in DNA ligase I may be associated with the high-mutation rates in Bloom's syndrome) certainly play important roles. Genes affecting the dNTP pools in the cell (e.g., thymidylate synthetase and CTP synthetase) would produce epistatic effects on mutation. It is to be expected that mutation varies among loci, tissues, and species, and in somatic as compared with germ cells. Comparative mutagenesis is therefore not only important but urgently needed. Development of newer techniques for rapid and accurate measurement of mutations at all types of endpoints would facilitate such quantitative studies.

Examples of extrinsic factors may include mutagens, inhibitors, and base analogs. The nature and origin of spontaneous mutations deserve further exploration.

STRUCTURE AND FUNCTION OF THE GENE

Nucleotide sequence information on an increasing number of mutant genes should allow conclusions to be drawn on mutagenic specificity and mutability at specific sites of DNA in mammalian cells, as compared to lower (and small) eukaryotes and prokaryotes. The functional implications can be studied further by deletion mapping and/or directed mutagenesis. It would be particularly interesting to find the kinds and consequences of DNA alterations in the regulatory sequences, introns, and splicing sites. Molecular analysis of the essential (and nonessential) genes opposite the deleted chromosomal regions in the mouse promises to yield significant findings relevant to man.

SOMATIC VARIATION RESULTING FROM PROCESSES OTHER THAN MUTATION

Maher presented experimental evidence that indicated the possible induction by chemical carcinogens of mitotic recombination and gene conversion at the two complementary HSV-I TK sequences incorporated in mammalian cells. This exciting initial finding needs to be confirmed, especially with flanking outside genetic markers and accompanied by sequence analysis. The possible existence of transposition in mammalian cells should be explored. A good deal of experimental evidence is available on gene amplification and epigenetic changes (e.g., DNA methylation), but these phenomena should be studied in conjunction or comparison with the mutation process.

In the next few years, breakthroughs in any of these or other areas could constitute the subject matter for another Banbury conference or even a Cold Spring Harbor symposium.

Author Index

Subject Index